Alternative Cars in the 21st Century

A New Personal Transportation Paradigm

Second Edition

Other SAE titles of interest:

2003 Alternative Fuels Technology Collection on CD-ROM
(Order No. ALTFUEL2003)

Fuel Cell Powered Vehicles:
Automotive Technology of the Future—Update
by Daniel J. Holt
(Order No. T-114)

Hydrogen and Its Future as a Transportation Fuel
Edited by Daniel J. Holt
(Order No. PT-95)

Intelligent Vehicle Technologies
by L. Vlacic, F. Harashima, and M. Parent
(Order No. R-310)

Lightweight Electric/Hybrid Vehicle Design
by Ron Hodkinson and John Fenton
(Order No. R-316)

Nonlinear and Hybrid Systems in Automotive Control
Edited by Rolf Johansson and Anders Rantzer
(Order No. R-348)

For more information or to order this book, contact SAE at
400 Commonwealth Drive, Warrendale, PA 15096-0001;
(724) 776-4970; fax (724) 776-0790; e-mail: CustomerService@sae.org;
website: http://store.sae.org

Alternative Cars in the 21st Century

A New Personal Transportation Paradigm

Second Edition

Robert Q. Riley
with contributions by Dr. Wonshik Chee

SAE *International*™
Warrendale, Pa.

For permission and licensing requests, contact:

SAE Permissions
400 Commonwealth Drive
Warrendale, PA 15096-0001 USA
Tel: 724-772-4028
Fax: 724-772-4891
E-mail: permissions@sae.org

Library of Congress Cataloging-in-Publication Data

Riley, Robert Q., 1940–
    Alternative cars in the twenty-first century : a new personal transportation paradigm / Robert Q. Riley ; with contributions by Dr. Wonshik Chee.—2nd ed.
        p. cm.
    Includes bibliographical references and index.
    ISBN 0-7680-0874-3
    1. Automobiles—Research. 2. Automobiles—Technological innovations. 3. Alternative fuel vehicles. 4. Automobile industry and trade—Technological innovations. I. Title: Alternative cars in the 21st century. II. Title.

TL158.R55   2003
333.4'7629222—dc21   2003050557

SAE
400 Commonwealth Drive
Warrendale, PA 15096-0001 USA
Tel: 877-606-7323   (inside USA and Canada)
          724-776-4970   (outside USA)
Fax: 724-776-1615
E-mail: CustomerService@sae.org

Copyright © 1994, 2004 SAE International
ISBN: 0-7680-0874-3
SAE Order No. R-227
Printed in the United States of America.

To Cindy

My daughter and single mother of four,
from whom I continue to learn
the true meaning of love and commitment

Said the Eye one day, "I see beyond
these valleys a mountain veiled with blue mist.
Is it not beautiful?
The Ear listened and after listening intently
awhile, said, "But where is any mountain?
I do not hear it."
Then the Hand spoke and said, "I am trying
in vain to feel it or touch it, and I can find
no mountain."
And the Nose said, "There is no mountain,
I cannot smell it."
Then the Eye turned the other way, and
they all began to talk together about the Eye's
strange delusion. And they said, "Something
must be the matter with the Eye."

—*Kahlil Gibran*, The Madman

Reprinted with permission from *The Madman*, by Kahlil Gibran, ©1982, Random House

CONTENTS

FOREWORD

Robert Riley's original 1994 edition of *Alternative Cars in the 21st Century: A New Personal Transportation Paradigm* arrived at a critical time and served a vital need. A new priority on electric cars was then stimulating public interest, regulatory attention, and many technological developments. Information to the public was characterized by an overload of conjecture, complexity, and alternatives. Riley brought some clarity to the field. He put things in perspective, backed up by quantitative exploration. Now it is nine years later. Many more forces and technologies have entered the ring. The stakes are higher. This new, considerably revised edition appears at an even more critical time. There are now many more complexities, alternatives, and overload from information and misinformation. Clarifying realities and providing perspective become more vital. Riley handles the task well.

Throughout the 20th century, cars have been infiltrating into the very soul of the United States. In a symbiotic relationship, cars and highways created us as we created them. Cars came to define us: where we live relative to where we work, our recreations, our mating patterns, and our self-esteem. A galactic observer, upon first looking down on the United States, might assume cars to be the dominant lifeform: they travel widely, avoid jostling each other, but congregate closely and rest regularly in giant meeting lots. Little two-legged subunits might be their slaves. Are the galactic observer's insights mistaken? In any case, use of cars grew huge because of the many benefits they provided us. When automotive technology changed, it was mostly by small steps, with an occasional kick from regulations about exhaust emissions, fuel economy (corporate average fuel economy [CAFE] requirements), and safety. The 1990s initiated a decade of faster change. Sport utility vehicles (SUVs) and pickups, insulated from CAFE rules, became increasingly popular with customers whose interest in being in the heavier vehicle when collisions occurred outweighed concerns about poor fuel efficiency (or the fate of the occupants of the lighter vehicle). Manufacturers who had continually lost money trying to market small, fuel-efficient cars now found profits in the heavier, low-mpg vehicles. In a broad sense, the people who design cars are the customers (aided of course by auto company engineers), and they have designed well, although more for individual than societal benefit.

An underlying theme of alternative cars is the combination of continuing to meet our needs for safe and convenient personal mobility while decreasing local and global pollution and our dependence on nonrenewable energy sources. An obvious and practical strategy is adopting vehicles that do their job with much less energy. This cuts consumption (and preserves reserves) of conventional fuels and opens opportunities for alternative energies (mostly renewable, with low pollutant emissions) that may initially be more expensive and less convenient. Riley's entrance into the mobility field came through small, very efficient vehicles—a better starting point for his ideas than if he had been embedded in the midst of the standard car field. I am perhaps biased in his favor because of having personally entered the serious mobility field from the standpoint of very efficient specialty vehicles that had to rely on the puny power of human muscles or photovoltaic cells.

The alternative car field has evolved from a century featuring many car developments, both technological and societal. A significant modern change element emerged on January 3, 1990. The battery-powered GM Impact was first presented to the public at a press conference with Roger Smith, GM's chairman, presiding. I will never forget one of his prophetic remarks. He noted that auto companies had some hesitancy about introducing a new technology because of their experience that regulation of the technology usually followed quickly. On Earth Day 1990, he announced that GM would actually commit to producing the car commercially. Within several months the Zero Emission Vehicle (ZEV) mandate was established by the California Air Resources Board. This required that in a few years every manufacturer sell some ZEVs (presumably battery-powered) along with their regular products. In a sense it was a flawed concept, requiring manufacturers to develop, manufacture, sell, and warrant the revolutionary vehicles, but not requiring anyone to buy them.

Nevertheless, as the next generation looks back at what initiated healthy fundamental change in mobility technologies and applications for the United States and even the rest of the world, the importance of the ZEV mandate as a catalyst, a wake-up call, will be appreciated. The mandate's details evolved, adapting to a combination of changing technological reality and the vested interest of political and economic entities. For GM, the pioneering task of turning the Impact demonstrator into the commercial EV-1 was formidable. The total vehicle systems design, with high priority on vehicle efficiency dictated by the low energy capability of batteries, represented dramatic change. For the first time, virtually every part needed to be made of unconventional material, fashioned by new production techniques. It was a daring step into the future by GM. The EV-1's commercial viability is unimportant compared to the value of its initiation of significant dedication to change throughout the industry.

Globally the major car companies began serious exploration of alternative power technologies. Small entrepreneurial groups sensed the emergence of what looked like big new opportunities. Government funding, government-industry partnerships, and government laboratory support appeared. This period proved to be an education for all, as the small entrepreneurs slowly began to appreciate the magnitude of resources required for meeting reliability standards and production economies of the auto industry, while the large entities began exploring the uncomfortable introduction of revolutionary technologies into an industry (and to customers) more accustomed to advancing by small steps. The inexorable rise of U.S. dependence on foreign oil sources (including from countries with political agendas far different from ours), and the contribution of fossil fuel emissions to global climate change, emerged as serious challenges to the status quo. A new ballgame had arrived, with the rules undefined, and with new players.

In 1994, Riley's first edition was a breath of fresh air in a cacophony of information coming through all media. His new second edition is greatly expanded to address the larger list of interrelated complexities. For example, new priority is given to hybrid vehicles and intelligent transportation systems. He collects and organizes the many products/concepts technologies into a readable publication—a formidable task in a field seething with change. The first chapter takes a broad look at the big picture, not just in the United States: the challenges and opportunities that characterize the present rapidly changing situation. These concepts deserve to be explored by all students because the present student will soon be living in the rather different future that is being fashioned. This book also should be read by all adults because our priorities and actions, or inactions, will determine what happens in the next decades. The other chapters include details and quantitative assessment. The mass of information in all of these transportation areas is huge. Having summaries nicely organized into a single book is a help to everyone involved in present and future mobility.

In summary, the 2004 book is virtually an encyclopedia in its breadth, yet it still includes perspectives as well as facts. It cannot predict what the alternative cars of the 21st century will be. That subject is too complex, too dependent on present, just-emerging, or expected future technologies, and on future policies of regulators, the whim of the public, and the geopolitics of oil reserves.

Robert Riley's background includes developing some extremely efficient three-wheeled vehicles, with considerable engineering involved in the aspects of power, structure, stability, and control. Combining this solid engineering background with a gift for organizing and communicating complex topics made the book possible. His first edition helped us get to where we are now. This new edition helps provide us the tools with which a desirable future can be created.

Dr. Paul B. MacCready

ACKNOWLEDGMENTS

I would like to extend a special thanks to several people who contributed their time and expertise to this work. Dr. Paul MacCready shared many insights on a variety of subjects, and he also contributed the Foreword for this second edition. Dr. Wonshik Chee, Assistant Professor in the Department of Mechanical Engineering at the University of Wisconsin at Milwaukee, contributed the chapter on Intelligent Transportation Systems. Anthony (Tony) E. Foale, author of *Motorcycle Chassis Design*, is responsible for the material on tilting three-wheelers included in Chapter Six. John Brooks, a brilliant engineer with a lifetime of experience in powerplant design at Macullogh, Chrysler, and Garret, offered invaluable input on this highly specialized subject. Dr. Paul VanValkenberg provided several recommendations that were incorporated into the first edition. And Don Goodsell, SAE's European Editorial Consultant, did much of the research and provided most of the photos for Chapter Nine. Without his insightful technical and literary input, this book would not be the work that it is today.

I would also like to thank the companies and individuals who provided graphics, as well as technical and supportive information on a variety of subjects. These include DaimlerChrysler Corporation, Ford Motor Company, General Motors Corporation, Honda Motor Company, Nissan Motor Co., Renault, Saab-Scania of America, Subaru of America, Fugi Heavy Industries (Subaru, Japan), Volkswagen AG, and others. Both Harley-Davidson and Bob McKee of McKee Engineering Corp. provided information on Trihawk. The market research companies J. D. Power & Associates and R. L. Polk Company contributed opinions and market demographics. Dan Sperling at the Institute of Transportation Studies at Berkeley, California, Lee Schipper, co-author of *Energy Efficiency and Human Activity: Past Trends, Future Prospects*, Al Sobey, William Garrison, and many others freely offered information and personal insights about the important subject of future personal mobility options.

And then there's the unglamorous work done by the late Debi Kaiser, who labored through many weekends and weeknights putting order to the countless mechanical details that go into a complex work such as this one.

Others have made significant, although perhaps more indirect, contributions in other ways. The uniquely talented members of the Quincy-Lynn Enterprises team each contributed a piece of their personal excellence to that company's experiments with "alternative cars," and to the author's perspective on the subject. Without the personal contributions of Ed Kay, the spirit of enterprise and determination that helped make this work a reality may not have existed in sufficient measure.

And one of the most essential contributions of all came from Beverley Riley. Her unending confidence and support kept the fires burning over the two years of nearly full-time effort necessary to complete the first edition.

INTRODUCTION

Land transportation has played a significant role in the development of modern economies. The ability to freely and inexpensively move goods and people is a fundamental link in the economic chain. In addition, transportation also contributes directly to a nation's economy. In the United States, one in seven jobs is related to the automobile industry, and one in five retail dollars is spent on automotive or automotive-related products. In Organization for Economic Cooperation and Development (OECD) countries, up to 15 percent of disposable income is spent on transportation. Growth in the transportation sector is largely co-linear with the growth in the gross domestic product (GDP). In the process of providing positive benefits, transportation also produces negative byproducts. Direct effects, both positive and negative, are linked to the sector's enormous consumption of energy, resources, goods, and labor. Consumption stimulates economies, but it also depletes natural resources, and its byproducts often pollute the environment. In the process of facilitating consumption in general, the transportation sector has emerged as an enormous economic block, as well as a major consumer of natural resources and a producer of environmental pollution on its own.

In an increasingly industrialized world, the same natural resources are necessarily spread over much greater populations. Resources are therefore more rapidly depleted, and the capacity of ecosystems to absorb industrial byproducts more quickly reaches saturation. Just as deficit spending can produce temporary affluence at the expense of long-term health, spending the world's capital of natural resources without regard to global limitations can produce short-term benefits in ways that may ultimately be unhealthy for the entire system.

Attempts to quantify the environmental impact of increased consumer populations produce widely divergent results. Results depend on estimates of future populations, projections of economic trends, and assumptions regarding future technology, as well as on assessments of ecosystem tolerance. Although quantification may be elusive, hardly anyone believes that the present rate of consumption can be sustained if the current rate of industrialization and economic growth is extrapolated to a world of double the existing population. Ecological limitations to increasing consumption undoubtedly exist at some level. Based on projections by the World3 computer simulation, Donella Meadows, co-author of *Beyond the Limits*, suggests that with environmentally friendly technology, the world can support some 8 billion inhabitants at roughly the standard of living of Western Europe in the 1990s. Dr. David Pimentel estimates the population sustainable at an equivalent of U.S. lifestyle at about 2 billion.[1] The most optimistic demographers believe that population will continue to grow until it reaches 9 to 10 billion near the middle of the 21st century. Developing countries will be responsible for the most rapid increases in consumption in the coming decades, and they are also the countries that can least afford environmentally friendly technologies.

The idea of a world economy that produces, consumes, and pollutes at 20 times the present rate stretches the imagination.* There is a story about two fellow travelers who fell out of an airplane. On the way down, one of them begins to complain about falling. Finally, the other fellow looks over and says: "If you think falling is bad, suppose we survive until we reach the ground?" Likewise, suppose the world survives on its present course, prosperous and intact and replete with new energy-consuming products? How might that affect our planet, its resources, and the quality of life for its inhabitants? Fulfilling the mobility needs of this emerging new world will be a significant factor in the overall production/consumption/pollution cycle. Transportation already consumes roughly 30 percent of the world's total primary energy. According to Ove Sviden, EKI, University of Linkoping, Sweden, if energy consumption continues to grow as it has in the past, by year 2100 total primary energy required to meets the world's needs could be 10 times greater than in 1990, and transportation will be consuming 40 percent of this much larger pool. The energy consumed to move goods and people will then have to be 20 times cleaner, just to match the present burden that transportation's energy consumption places on the environment.[2]

Light-duty vehicles are the single greatest user of transportation energy. Reducing personal transportation's energy intensity may be the single most important strategy in averting the impending collision between the positive economic benefits of worldwide mobility, the world's limited natural resources, and the declining capability of the environment to absorb the byproducts of energy production.

Many people have shared their ideas for accommodating the inevitable human population while reducing the demand on private transportation.

---

* A United Nations projection for year 2100, *Newsweek*, June 1, 1992.

One of the most popular proposals is the idea of eliminating or reducing the need to travel at all. As Don Goodsell, SAE's European Editorial Consultant points out, if one has to travel more than 20 minutes to work, either one's residence or one's place of employ is in the wrong place. The ascendancy of telecommuting—working at home and communicating via the Internet—has already had a large impact on business protocols. According to a recent study by the International Telework Association and Council (ITAC), one in five U.S. workers, or roughly 28 million people, already spend at least part of their workday at remote or mobile electronic offices. It is also possible that in many of the world's burgeoning cities, private cars may become largely outmoded by expanded use of transit systems and station cars. Still others suggest that self-sufficient cities like those envisioned by Paolo Soleri might replace the sprawling conurbations of today. Dr. Paul MacCready envisions the widespread use of electric sub-cars as relieving the demand on conventional, multipurpose cars. The University of California's William Garrison sees a future served by small neighborhood cars for local travel, with commuting workers driving one- and two-passenger commuter cars along separate roadways, perhaps elevated along freeway medians or attached along the sides by outriggers. The technology of vehicles and the makeup of the overall personal mobility system may look quite unlike our own 50 years or so hence.

Chapter One provides an overview of the factors that are driving today's search for new solutions to personal mobility. Huge sums are being invested in high-risk development efforts in order to create a cleaner, more energy-efficient future. Naturally, a substantial motivation must exist for embarking on such risky pioneering efforts. Unless there is compelling evidence to the contrary, there is little justification for pioneering new technologies and products in high-risk ventures when existing technologies are obviously working; however, the status quo appears to contain great risk when one examines the subject from a global perspective. A compelling reason for change is implied by the global limitations that are sure to affect the future health of the planet, and therefore the future health of economies. Consequently, the subject is first approached as a global overview. Chapter One puts a foundation of justification under today's quest for a better way.

Chapter Two discusses business and marketing challenge, and the impact of vehicle design on consumer perceptions and acceptance in the marketplace. Technical solutions alone may not provide the complete answer to the task of supplying clean, efficient mobility to a world of rapidly expanding demand. Moreover, most complex challenges are not resolved by one sweeping answer. Instead, solutions tend to consist of several synergistic measures that in combination produce the desired result. Alternative approaches to vehicle design, packaging, and advertising appeals could cut energy demand as effectively as alternative approaches to vehicle power system technology. In addition, ideas such as station cars could make transit systems far more user friendly and work synergistically within a broader vision of public transportation. Finally, personal transportation may evolve into a simple commodity, rather than the symbol of self-expression and lifestyles that cars represent today.

Chapter Three reviews the technical challenges of improving energy efficiency, mainly with an eye toward conventional power systems. Although alternative power systems show great promise, conventional power systems are becoming cleaner and more efficient. It is conceivable that this workhorse of the 20th century could become nearly as efficient as methanol fuel cells and virtually emissions-free in another 15 years or so. From this perspective in time, it is difficult to forecast the demise of the internal combustion engine or to predict the ultimate form of automobile power systems.

Alternative fuels are reviewed in **Chapter Four**. New motor fuels are sure to be widely adopted for future vehicles. Regardless of one's assessment of the abundance of petroleum, deposits are finite and vehicle energy supplies must ultimately come from another source. Even if economically recoverable petroleum reserves are three times greater than today's known reserves, we will probably have to move away from oil as a primary source of transportation energy by the year 2020. So alternative motor fuels or energy sources are inevitable. It is also possible that alternative fuels may ultimately be used more as fuels for electrical generation, rather than fuels consumed by individual vehicles. Electric cars might then complete the loop. **Chapter Five** reviews battery-electric and hybrid vehicles and their potential to reduce overall energy consumption and atmospheric pollution.

The subject of three-wheel cars is explored in **Chapter Six**. This platform may ultimately be adopted for special-purpose vehicles such as urban cars. Lightweight urban car designs can benefit from the three-wheel layout. Although it is generally assumed otherwise, the three-wheel layout may have marketing advantages because of its natural differentiation from conventional cars. Rollover accidents, however, may be more likely with three-wheel vehicles. More generally, significant concerns regarding vehicle safety arise along with the idea of integrating ultra-low-mass cars, of either three or four wheels, into traffic with conventional high-mass vehicles. **Chapter Seven** therefore reviews vehicle safety with an emphasis on low-mass vehicles. Although it is technically feasible to protect occupants during a small-car/large-car involvement, the consequences of the unfavorable transfer of energy to the smaller vehicle are inescapable. Absolute safety must ultimately come from an increased emphasis on crash-avoidance technologies.

Intelligent transportation systems (ITS) are reviewed in **Chapter Eight**. ITS technologies have the potential to virtually eliminate traffic accidents, significantly reduce travel times, and slash energy demand in the process. The potential impact of ITS on vehicle design and on the mobility system in general is so enormous that, conceivably, manually operating an automobile could become an outmoded practice at some point in the future. But a new product liability mindset will be necessary before carmakers are likely to venture much farther into automated control of automobiles. And a new mindset among drivers will be required before motorists are likely to relinquish control over their automobiles. In reality, however, mechanical and electronic systems can do a better job of managing and controlling vehicles. Over the next quarter-century or so, automobiles may learn to get along quite well without interference from us fallible human operators.

The last chapter, **Chapter Nine**, reviews the history of small-car development in Europe. Today, the idea of ultralight, special-purpose urban and commuter cars is often compared to microcar market failures in the past. Extremely small, energy-efficient cars have often produced a poor sales history, both in North America and in Europe. Chapter Nine therefore examines the success and (mainly) failures of low-mass cars that were produced in Europe after World War II. The approach is first to place market failures in their historical perspective and then to elaborate on the impact of product theme and marketing appeals, as reviewed in Chapter Two.

Finally, there is the question of practical and dispassionate choices about lifestyles. Too often we spend our limited personal inventory of time and resources on activities that amount to little more than practical self-maintenance. Ironically, the most meaningful qualities of life often lie outside the realm of practical utility. Automobiles are remarkable transportation tools, but they are also one of the few tools of necessity that can be enjoyed for the means as well as for the end. True, they provide a measure of mobility and freedom that is unavailable in subways and on buses, but they also provide a *sense* of freedom and mobility that cannot be weighed in practical terms. There is also the pleasure of owning and admiring an exquisite work of art, which certainly applies to a well-designed and finely made machine. A beautiful automobile gives us more than mere utility. Therefore, the ideas for energy-efficient personal mobility discussed on the following pages also embrace the prospect of inventing a new pleasure rather than destroying an old one.

REFERENCES

1. David Pimentel et al., "Natural Resources and an Optimum Human Population," *Population and the Environment: A Journal of Interdisciplinary Studies,* Vol. 15, No. 5, May 1994.

2. Ove Sviden, EKI, University of Linkoping, "Sustainable Mobility: A Systems Approach to Determining the Role of Electric Vehicles," published in OECD Document, *The Urban Electric Vehicle: Policy Options, Technology Trends, and Market Prospects,* ISBN 92-64-13752-1.

CHAPTER ONE

PERSONAL MOBILITY IN CRISIS

I do not believe that we are to be flung back into the abysmal darkness by those fiercesome discoveries which human genius has made. Let us make sure that they are our servants, not our masters.

—Winston Churchill

At the beginning of the 20th century, many of the world's densely populated metropolitan areas were facing near runaway environmental problems. But in those days it was called "sanitation," and the effects were mostly localized. In New York City, for example, the burgeoning horse population was polluting the streets at the rate of 2.5 million pounds of manure and 60,000 gallons of urine each day. When summer temperatures soared, it was common to see the carcasses of horses left in the streets, having dropped in their tracks from heat exhaustion. City workers removed, on average, 15,000 dead horses from the streets of New York each year. Rains turned streets into odious soups, and dry spells filled the air with germ-laden dust. Tetanus, dysentery, and a host of other diseases were attributed to the rampant pollution caused by the city's transportation animals. In the modern metropolis of 1900, the veneer of culture had been all but shattered by nearly unmanageable sanitation problems. But new technology in the form of the self-propelled carriage gave congested cities an environmentally friendly alternative. City streets could be cleaned up and paved over, and the human resources that went to the board and care of animals could be redirected toward more productive activities.

Today, 100 years after the automobile rescued cities from their transportation animals, we are again on the verge of being overwhelmed by the byproducts of our need to move freely about. Because of the sheer magnitude of the problem, today's solution may have to be as revolutionary as when the carriage gave up the horse on its way to becoming an automobile. Although automobiles are the most effective means of transportation ever devised, the problems of growing traffic congestion, high energy-intensity, and harmful emissions ask for new solutions. For the first time it seems clear that the number of automobiles that cities can assimilate is limited. Following half a century of pro-automobile planning, a renaissance of urban and intercity transit seems poised to emerge. Planners are turning more to trains, streetcars, buses, and even bicycles, and away from private cars. Even on a grassroots level, a consensus against unlimited growth of the automobile population is gaining momentum. Cities are rebelling, and the private automobile is already an unwelcome guest in a growing number of densely packed metropolitan areas in Europe. The United States is not far behind Europe in terms of traffic congestion and commuter burnout. Consequently, pressure for a similar rethink is building in traffic-polluted metropolitan areas like Los Angeles, Chicago, and Washington, D.C. Because of poor air quality in Los Angeles, the State of California embarked on a program designed to force a technology shift in transportation, and several other U.S. cities followed suit.

After nearly 20 years of intensive efforts to find new solutions, economically viable alternatives to petroleum-fueled private cars remain elusive, and transportation fuel consumption is increasing across the board. Moreover, the automobile is invariably implicated in political difficulties involving petroleum. Regardless of rhetoric to the contrary, the massive military assault of Operation Desert Storm was undertaken mainly to insure the price and flow of oil into the world's mechanized veins. Peacetime U.S. military expense directly related to protecting Middle East oil supplies is estimated at $10 to $20 billion per year.[1] As for the automobile itself, what began as a love affair is rapidly turning into a calculated business deal throughout the industrialized world.

Problems associated with personal transportation are likely to continue to compound as population grows, industrialization spreads, resources diminish, and the environment becomes more threatened. Global trends toward industrialization and higher personal incomes translate into more cars and greater energy consumption. When a society industrializes, the burro and the rickshaw are quickly discarded in favor of their mechanized brothers. People invariably favor private cars where the local economy can support them. Consequently, the automobile's impact on the environment and on the world's natural resources has achieved global proportions, rather than being a distinguishing attribute of a few wealthy nations. Growth trends are likely to continue to offset traditionally modest and incremental improvements in vehicle efficiency. Even the recent gains in air quality caused by emission controls could be wiped out in a decade or so because of the world's rapidly growing population of cars. Problems are unlikely to yield to token improvements in automotive technology. Instead, the machine itself is in need of a holistic renewal, a wholesale overhaul, if it is to remain the central component of our modern society's transportation system.

POPULATION GROWTH AND EXPANDING INDUSTRIALIZATION

In 1968, Paul Ehrlich's first book, *The Population Bomb*, arrived on the newsstands, and within a few months everyone was talking about a new end-of-the-world. This version would happen not by fire, but by an ever-increasing population of gluttonous human beings who would literally consume the entire planet. Ehrlich pointed to an imaginary world of the distant future in which exponential growth extends beyond the possible into a nightmare of some 60 million billion people who would stand at 100 persons per square yard on the earth's surface.[2] Of course Ehrlich's subject was not a hypothetical world of the distant future, but a more real world of ecological collapse he foresaw just around the corner. At the basis of his warning was a very real runaway growth in the world's population. Today, after having nearly doubled in numbers since *The Population Bomb* was published, the world is now inhabited by some 6 billion people. It took all of human history up to year 1804 for the world's population to reach 1 billion, then only 12 years from 1987 to 1999 to leap from 5 billion to 6 billion. Some time in the 21st century, population is expected to reach nearly twice that number before stabilizing (Fig. 1.1).

If economic growth remains strong, populations continue to grow, and environmental pollution follows along in step, greater pressure on the transportation sector is inevitable. Projecting 50 years into the future, the combination

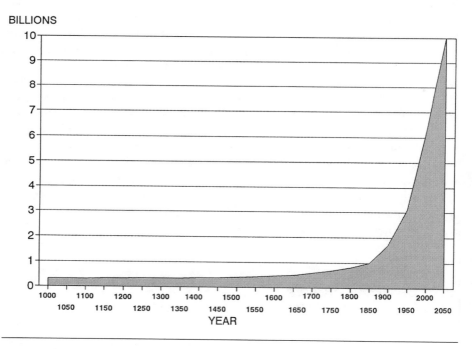

BILLIONS

FIGURE 1.1 Population Growth: 1000 A.D. to 2050 A.D.

of population growth and expanding industrialization is expected to cause an enormous increase in worldwide demand for resources, along with associated burdens on the environment. Either factor (population or industrialization) by itself contains alarming numbers for the future of transportation.

By itself, rapid population growth does not necessarily mean an equal increase in consumption. It may instead translate into a corresponding decline in living conditions. In many regions of high population growth, economic conditions have been stagnant or declining. Populations in undeveloped regions are substantially outside the loop where they neither participate in the world economy nor consume substantial resources. But contrary to popular belief, high population growth does not necessarily stifle economic growth. A key factor is industrialization. Industrialization stimulates economies, and consumption follows suit, increasing in step with the growth in gross domestic product (GDP). Normally, nations do not choose to remain undeveloped. Adopting technology generally improves a population's living conditions and helps put distance between the society and potential poverty. Technology provides a degree of control over the ruthlessness of nature, and it provides a passport into the world's economic community.

Over the past 25 years, the vast populations of Asia have been making the transition to industrialized economies. During the 1980s, developing countries maintained a steady increase in per capita income while Western economies remained relatively flat. Although Asian economies experienced an economic downturn in the 1990s, and transitional regions experienced a decline, the upward trend is expected to continue over the longer term (see Table 1.1).

Today, only 28 percent of the world's inhabitants live in developed regions. The other 72 percent reside in undeveloped and developing regions. This latter group has the potential, to varying degrees, for significantly greater increases in consumption. In a transitional or developing economy, the growth in consumption can far exceed the simple growth in population. Nations of people who were previously nonconsumers can transform into consumers almost overnight. Growth in consumption increases in step with the growth in GDP. The move toward market economies in Eastern Europe and the countries of the former U.S.S.R. will ultimately stimulate robust economic development in that relatively dormant region. Politically, it is undoubtedly in the world's best interest to encourage economic renewal in Russia. Environmentalists, however, shudder at the thought of the rapid increase in consumption that will likely accompany an economic awakening in densely populated and previously stifled regions.

TABLE 1.1 Growth in Real Gross Domestic Product (GDP)
(Annual percentage change)

Region	1982–1991	1992–2001*
World	3.3	3.4
Advanced Economies	3.1	2.8
United States	2.9	3.6
European Union	2.6	2.1
Japan	2.6	1.0
Other Advanced Economies	4.3	4.2
Developing Countries	4.3	5.5
Africa	2.2	2.8
Asia	6.9	7.4
Middle East & Europe	3.3	3.5
Western Hemisphere	1.8	3.4
Countries in Transition**	1.4	−2.4

*Includes projections for year 2000 and 2001.
**Central and Eastern Europe (excluding Belarus and Ukraine), Russia, Transcaucasus, and Central Asia
Source: International Monetary Fund (IMF), www.imf.org

The richest 20 percent of the world's inhabitants consume 86 percent of all goods and services and produce 53 percent of all carbon dioxide emissions. In contrast, the poorest 20 percent consume only 1.3 percent of goods and services and produce 3 percent of carbon dioxide emissions. If the rest of the world ate as Americans do, food production would have to be doubled. If the rest of the world drove automobiles as Americans do, all proven petroleum reserves would be exhausted within a decade. If a Vespa motor scooter were given to everyone in China, the world would be plunged into a new energy crisis overnight. Even at today's 6 billion inhabitants, many people argue that the world cannot support worldwide consumption at a level equal to that of Americans at the beginning of the 21st century.

Worldwide consumption equal to today's level in North America may never actually materialize. The superlatives do, however, provide a sense of the growing pressures on society's infrastructures and resources, as well as on the ecological stability of the planet. Although one can dispute particular forecasts, and select data may be imprecise, the overall trend and magnitude are unmistakable. Anticipation of rapidly increasing pressures on all aspects of the transportation infrastructure is already driving an increasingly intense search for a new paradigm for energy and transportation.

AN EXPANDING WORLD
WITH DIMINISHING PETROLEUM RESOURCES

A continuous supply of economical energy is fundamental to industrialized economies. Throughout the 20th century, developed countries were blessed with cheap and abundant energy supplies. Changes in the price or threats to the supply of energy send shockwaves rolling through the world's financial systems. Energy is an intrinsic part of every aspect of society, from heating, cooling, and lighting homes to providing power to manufacture goods and fuel to propel cars. Automobiles are especially large users of energy. Moreover, they are almost totally dependent on petroleum motor fuels, and fuel made from today's inexpensive petroleum has proven difficult to replace. A universally growing demand for energy in combination with finite deposits of economically recoverable petroleum is expected to ultimately drive petroleum prices upward.

Governments, oil companies, geologists, and other organizations have attempted to define the Estimated Ultimately Recoverable (EUR) deposits of oil, or the total amount of oil that will ever be taken from the earth. Experts place the earth's total original endowment of oil as high as 3 trillion barrels. As of the year 2000, almost 900 billion barrels had already been consumed, and nearly the same amount remained in the ground as proven reserves. The remaining oil has yet to be discovered, but geologists believe it's there (possibly as much as 1.3 trillion barrels of undiscovered oil).

According to the Hubbert curve—a bell-shaped curve developed by geophysicist M. King Hubbert to predict the growth and decline of a nonrenewable resource—production can be expected to begin a permanent decline when roughly half the earth's total endowment of oil has been consumed (1.2 to 1.5 trillion barrels). U.S. production has been declining since the 1970s, when roughly half of that nation's deposits had been used. Based on the Hubbert curve, escalating prices caused by a global decline in oil production could occur as early as year 2010 (at about 1.2 trillion barrels), or optimistically, as late as 2040. Several factors could push the threshold closer to year 2040. With traditional oil recovery techniques, for example, only about 40 percent of the oil in a deposit can be recovered. But new techniques developed during extraction of North Sea oil have increased recovery efficiency to more than 50 percent. Even a 10 percent increase in recovery efficiency translates into billions of barrels of additional oil on a worldwide scale. So variables in recovery technology and unknowns in the ultimately recoverable reserves make it difficult to accurately predict the growth and decline of oil production. Figure 1.2 shows production decline and shortfall curves based on peak production at year 2020.[3]

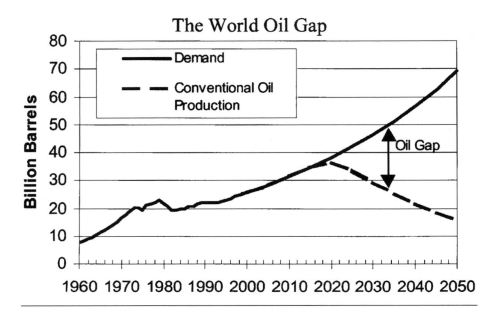

FIGURE 1.2 The World Oil Gap

Source: U.S. DOE, Office of Transportation Technologies, www.ott.doe.gov

Regardless of forecasting uncertainties, at some point in the future, production will decline and demand will outstrip supplies. With business as usual, this will likely occur during a period of unparalleled consumption and growth in demand, possibly around year 2020, as shown in Figure 1.2. If consumption were to continue to increase at today's rate past year 2020, it will then take only 21 years to consume an amount of oil equal to all of the oil consumed in the 20th century. Already, the world's oil consumption is increasing each year by an amount equal to Kuwait's entire annual production. Even at today's consumption rate, a new Prudhoe Bay discovery would extend the world's oil supply by only about six months, and a new North Sea find would equate to about three years' supply. So regardless of one's faith in future oil finds and recovery technologies, petroleum is a finite resource, and the end of the line for conventional crude oil is quickly approaching.

Motivation for a large move away from petroleum will naturally center on the economics of oil in relation to the alternatives and only indirectly on limited oil supplies. We are unlikely to wake up one Tuesday to find that the earth's oil tank has been pumped dry. Instead, increasingly difficult recovery, declining production, escalating prices, and the transfer of wealth to the holders of the largest reserves will ultimately compel a shift toward alternative technologies. The point at which petroleum becomes an economically unsound source of energy depends on the costs of alternative technologies in comparison to the economic liabilities of petroleum. Industry changeovers, however, could take 20 years or more. So we may have already arrived at the threshold for action, if economic hardship and decision making in a crisis atmosphere are to be avoided.

Estimations of demand growth vary between industrialized and developing regions. Economic conditions and energy consumption are expected to grow much more rapidly in developing countries. Developed countries will continue to move heavy industrial operations into underdeveloped and developing regions, which will stimulate economic growth and increase energy consumption in those areas. Technological progress will continue to improve energy efficiency, but mainly in developed countries where consumers can afford the technology. As a result, energy consumption is expected to grow slightly less rapidly than the world economy and the rate of industrialization; however, total world energy consumption will significantly increase through at least the first half of the 21st century as improving economic conditions in developing countries drive energy use upward.

In 1996, the world consumed an average of 71.5 million barrels of oil per day (Table 1.2).[4] As of the first half of year 2000, consumption had already reached 75.6 million barrels per day. In the United States in year 1996, petroleum supplied about 40 percent of the nation's total energy needs, which is approximately twice as much as either coal or natural gas and four times as much as nuclear, hydroelectric, and all other energy sources combined. Although the United States consumes roughly one-fourth of the world's oil, that nation's consumption growth rate is lower than in the rest of the world. As a result of much greater growth, oil consumption in Asian countries is expected to surpass that of the United States by year 2020. China, India, South Korea, and Central and South America will more than double their oil consumption by 2020.[5] World oil consumption is expected to continue to increase roughly in step with expanding populations and economies.

Market monopolies, balance of payment deficits, and the transfer of wealth to holders of the largest oil reserves are among the most troubling economic issues facing industrialized nations at the beginning of the 21st century. Transfer of wealth is one of the most direct measures of the cost of oil dependence. When the price of oil is manipulated upward through monopolistic market power, the amount of wealth transfer is equal to the quantity of oil imported times the difference between the competitive market price and the artificially inflated price. Between 1972 and 1996, for example, noncompetitive

TABLE 1.2 World Oil Consumption by Region, 1990–2020 (Million Barrels per Day)

Region	History			Projections				
	1990	1995	1996	2000	2005	2010	2015	2020
Industrialized Regions: Average annual increase = 1.0 percent								
North America	20.4	21.3	22.0	23.6	25.5	27.4	28.8	30.2
Western Europe	12.5	13.5	13.7	14.4	14.8	15.3	15.6	16.0
Industrialized Asia	6.2	7.0	7.1	6.8	7.1	7.5	7.9	8.3
Total Industrialized	39.0	41.8	42.7	44.9	47.4	50.1	52.3	54.5
EE/FSU*: Average annual increase = 0.8 percent								
Total EE/FSU	10.0	5.9	5.7	6.0	6.1	6.4	6.6	6.9
Developing Regions: Average annual increase = 3.2 percent								
Developing Asia	7.6	11.3	11.9	13.6	15.5	18.5	21.8	24.3
Middle East	3.9	4.7	4.8	5.2	6.5	7.5	8.5	9.8
Africa	2.1	2.3	2.4	2.7	3.0	3.5	4.1	4.7
Central & S. America	3.4	3.9	4.0	4.8	6.3	7.4	8.5	10.0
Total Developing	17.0	22.2	23.1	26.2	31.3	37.0	42.9	48.7
World: Average annual increase = 1.8 percent								
TOTAL WORLD	66.0	69.9	71.5	77.1	84.8	93.5	101.8	110.1

*Former Soviet Union and Eastern Europe
Source: U.S. Energy Information Administration (EIA), www.eia.doe.gov

oil pricing cost the U.S. economy alone approximately $1.4 trillion (1996 dollars) in transferred wealth.[6] This transfer of wealth resulted in a loss of potential to produce economic output. Had the $1.4 trillion remained in the U.S. economy where it could have been put to productive uses, the economy would have been more robust. On a micro level, consider the effect on a household economy when increased funds are drained away for a basic commodity with no additional benefits to household members.

Comparing regional production and demand to proven reserves provides a sense of the economic vulnerability of industrialized nations and the increasing power of oil producers. Despite the 17 percent contribution that North America now makes to the world's oil supply, that continent possesses only 5.4 percent of global reserves, yet accounts for 31 percent of total demand.

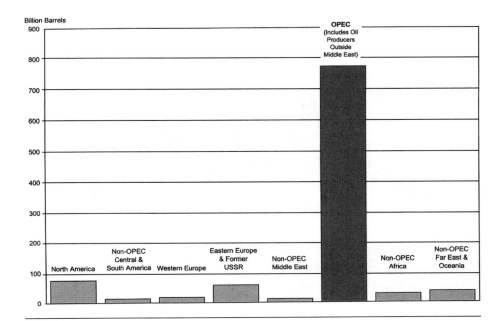

FIGURE 1.3 Worldwide Petroleum Reserves

Source: U.S. DOE, Energy Information Administration, www.eia.doe.gov

Western Europe contributes 6 percent to the world's total oil production, uses 20 percent of the world's oil output, but has less than 2 percent of the reserves. In contrast, the Organization of Petroleum Exporting Countries (OPEC) nations, which include oil producers outside the Middle East, use only 6.7 percent of the world's oil, but control 79 percent of the proven reserves (Figure 1.3). By year 2020, OPEC will be supplying more than half of the world's total oil.[7]

New oil finds cannot be relied on to head off energy and economic vulnerabilities. It is often argued that proven reserves have always been limited and that as reserves are consumed new deposits are discovered. According to the argument, new deposits will always be found because they have always been found in the past. Unfortunately, a history of discoveries does not guarantee a similar future. The United States passed the point of oil self-sufficiency in 1970. Despite the vast Alaskan fields, domestic production has been declining and dependency on imported oil has been growing. The United States is now importing more oil than it produces. Oil imports accounted for 16 percent of the U.S. trade deficit in 1998 and 25 percent in year 2000.[8] By year

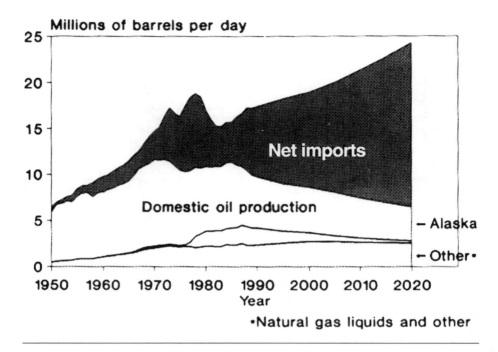

FIGURE 1.4 U.S. Oil Import Trends

Source: U.S. Office of Technology Assessment, www.ota.nap.edu

2020, as much as 70 percent of U.S. supplies may be imported, and by 2050 100 percent of U.S. oil demand will come from foreign sources (Figure 1.4).

Economic vulnerability to OPEC nations is expected to significantly increase over the next two decades as the oil deposits of industrialized nations become increasingly scarce. Manipulation of oil prices by OPEC already causes significant harm to the world's economies. Over the last 30 years, every major oil shock has been followed by a recession, and every major recession has been preceded by an oil shock. At the end of the first quarter of year 2001, the world was again teetering on the brink of recession following a period of inflated oil prices caused by OPEC's manipulation of the market.

Shifting to resources such as coal and natural gas could exert a downward pressure on the price of oil and help provide energy security over the short term, but coal-derived motor fuel would be costly to the environment. A long-term solution must center on a more fundamental change in the way in which energy is produced and consumed. The U.S. Office of Technology Assessment (OTA) recommends increased efforts to develop and integrate domestically produced alternative fuels and accelerated efforts to improve vehicle fuel efficiency. According to OTA, alternative fuels and improved vehicle fuel efficiency are two essential ingredients of any long-term resolution to the impending crisis over dwindling petroleum supplies. Although many options exist for fixed-site application, an alternative fuel or energy source for transportation presents a unique set of difficulties that are not easily resolved.

IMPACT OF TRANSPORTATION ENERGY USE

By 1999, the world was consuming energy at the rate of 402.90 exa-joules (EJ) (381.88 quads) annually (Table 1.3). Although petroleum supplies only 40 percent of the world's energy, the transportation sector is almost totally dependent on oil.

In the United States, the transportation sector consumes approximately 23.21 EJ (22 quads) annually, or about 27 percent of the total domestic energy consumption; however, transportation is responsible for 62 percent of all petroleum consumed in the United States, and personal transportation vehicles consume more than half of this oil. Motor fuel for cars, light trucks, motorcycles, and buses accounts for roughly 40 percent of the nation's total

TABLE 1.3 *World Total Energy Use by Source*

Energy Source	World Exa-Joules (Quads)	United States Exa-Joules (Quads)	U.S. Percentage of World Use
Petroleum	160.58 (152.20)	40.05 (37.96)	25
Natural Gas	91.67 (86.89)	23.52 (22.29)	26
Coal	89.44 (84.77)	22.96 (21.76)	21
Nuclear Electric	26.40 (25.25)	8.17 (7.74)	31
Hydroelectric	28.79 (27.29)	3.53 (3.35)	12
Other*	2.99 EJ (2.83)	1.066 (1.01)	13
Total**	**402.90 (381.88)**	**85.51 (81.05)**	**24**

Source: *International Energy Annual* (1999)
*Geothermal, solar, wind, wood, and waste electric power
**Categories may not equal total because of rounding.

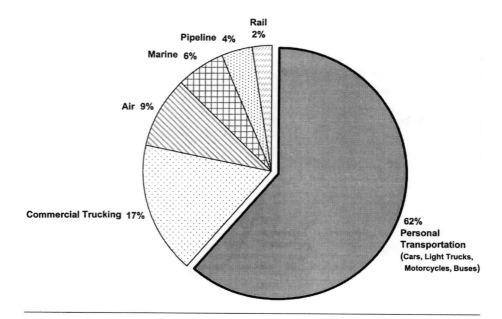

FIGURE 1.5 U.S. Transportation Energy Consumption

petroleum consumption. On a global scale, transportation consumes about one-third of the world's energy. Figure 1.5 shows U.S. transportation energy consumption by mode.

Between 1970 and 1996, world transportation energy demand increased by 110 percent, and now accounts for roughly half of the world's oil demand. Transportation energy consumption in industrialized countries is expected to grow at 1.6 percent per year through 2020. In the United States, the automobile population will grow at about twice the rate of the human population (which now averages 0.6 percent annually in the developed world). U.S. vehicle kilometers traveled is growing at about 4 percent per year. Motor fuel consumption goes up according to the increase in vehicle kilometers traveled, minus improvements in fuel economy. Taking into account estimated fuel economy gains and petroleum consumption trends in other sectors, U.S. oil consumption is expected to grow at 1.2 percent per year, reaching 24.7 million barrels per day by year 2020. Global consumption will grow from 71.5 million barrels per day in 1996 to 110.1 million barrels per day in 2020.[9]

In developing regions as a whole, transportation energy use is expected to grow at 3.9 percent annually, accounting for more than half of the worldwide growth in demand over the next two decades. In China, energy demand in the transportation sector will increase nearly 7 percent per year through the same period. The human population is increasing at approximately 2.5 percent per year in developing regions, but the automobile population is increasing much more rapidly.

The United Nations Fund for Population Activities (UNFPA) pointed out that between 1950 and 1990, the human population doubled and the world's automobile population increased by seven-fold. They estimated that between 1990 and 2010, the number of cars in the world will have doubled to about 800 million. The world's total motor vehicle population (not just cars) is expected to surpass 1.1 billion by year 2020.[10] If 1.1 billion motor vehicles were to drive by at the rate of one vehicle per second, it would take 35 years to reach the end of the line. If they lined up bumper to bumper for the driveby, the line would extend around the world 130 times.

Figure 1.6 shows worldwide automobile growth based on car populations reported by the American Automobile Manufacturers Association (AAMA).

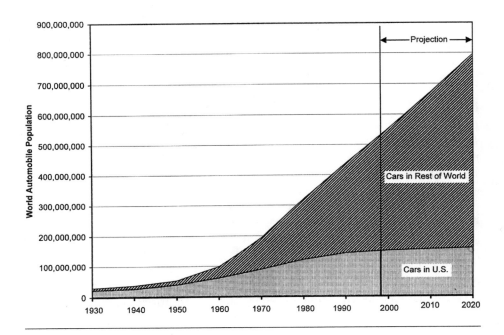

FIGURE 1.6 Passenger Cars in U.S. and World

By year 2100, transportation's share of the world's total primary energy consumption may reach 40 percent or more. According to Ove Sviden, EKI, University of Linkoping in Sweden, even if economically recoverable petroleum resources are three times larger than today's known reserves, the transportation sector must move away from petroleum motor fuel by year 2020.[11]

ENERGY DEMAND IN DEVELOPED COUNTRIES

Forecasts on a macro level depend on the sum of many long-term and difficult-to-project regional factors. One can only look at broad trends and draw general conclusions. Predictions are generally more reliable in developed economies where population growth is lower, transportation infrastructures are in place, and energy trends are mature, but change is difficult to implement. Business structures and personal habits are well entrenched and tenacious, even when external events demand better solutions. So programs designed to reduce energy intensity can end up having little effect in the face of growing populations and greater affluence. In industrialized nations at the beginning of the 1970s, an oil embargo inspired movement toward economizing on energy, and planners were optimistic that by applying known economic motivators, energy use could be significantly reduced. Technology would presumably improve energy efficiency, business would yield to economic persuasion, and people could be encouraged to modify their commuting habits. Unfortunately, efforts for change did little more than dampen demand growth.

Lee Schipper, co-author of *Energy Efficiency and Human Activity: Past Trends, Future Prospects*, surveyed energy use in eight OECD countries for the 1970–1988 period and made the following conclusions: Energy consumed for manufacturing had declined, energy consumed for household and services was about the same, and transportation energy use was up.[12] In the final analysis, Schipper found that little or no energy had been saved during the nearly two decades following the 1973–1974 energy crisis. Except for modest improvements in the United States, transportation had been the undoing of plans for lower energy appetites in industrialized nations (Table 1.4).[13]

Automobiles are the predominant mode of transportation throughout the developed world. Private cars are the most convenient mode of travel. In many cities they may be the only reliable mode. Additionally, owning a car is a sign of affluence, worldwide. It naturally follows that automobile ownership grows in step with growth in GDP. Countries with the highest GDP

TABLE 1.4 Private Cars Ownership/Use and GDP Growth Following the 1970s Energy Crisis

Country	Annual Growth in pkm%*	Car Ownership Per 1,000 Inhabitants			GDP Annual Growth %	Automobile Fuel Growth (Incl. diesel)
		1970	1987	%/yr increase		
Japan	4.3	91	247	5.9	4.2	3.49
US	1.9	420	601	2.1	2.7	0.91
West Germany	2.4	230	457	4.0	2.2	2.18
Sweden	2.1	286	438	2.5	1.8	2.68
Norway	4.6	193	392	4.2	4.0	2.89
France	3.3	240	391	2.9	2.6	3.05
U.K.	3.2	205	307	2.4	2.1	2.59
Italy (1973–1987)	3.4	190	387	3.2	3.0	3.81

Source: *Transportation Science*, February 1992

*Pkm = Passenger-Kilometers traveled

growth rate also have the highest growth rate in the population of private cars. When people can afford cars, they buy them. A saturation level of car ownership in developed countries is assumed to exist near the one-car-per-licensed-driver level. The United States, with slightly more than one car for every licensed driver, has an automobile population growth rate of 0.78 of GDP. In Japan, West Germany, and Sweden, Schipper found that the ratio of automobile growth to GDP was 1.9, 1.4, and 1.4, respectively. The high growth rate in Japan correlated with the comparatively lower ratio of cars to population and the nation's high economic growth.

A high percentage of passenger-kilometers (pkm) traveled by private car is characteristic of industrialized countries. Schipper found that in the eight OECD countries surveyed, more than 75 percent of all pkm traveled is by automobile, except for Japan, which reported 55.9 percent pkm traveled by car (Table 1.5).[14] In all countries, the absolute pkm traveled was also growing, and it continues to grow today. Next to air travel, automobiles are the most energy-intensive form of transportation.

Statistics suggest that switching to more fuel-efficient modes will improve overall energy efficiency; however, it is important to note that modal energy

TABLE 1.5 Travel by Mode in Eight OECD Countries Following the 1970s Energy Crisis

Country	Cars 1970–1987	Buses 1970–1987	Rail 1970–1987	Air 1970–1987	Total 1970–1987
Japan					
% Share pkm	42.5–55.9	14.7–9.3	41.4–31.3	1.3–3.5	100.0–100.0
Eng. Intensity (MJ/pkm)	1.73–2.09	0.48–0.55	0.20–0.20	4.29–2.63	0.95–1.37
United States					
% Share pkm	90.7–85.9	3.7–3.5	1.0–0.7	4.6–9.9	100.0–100.0
Eng. Intensity (MJ/pkm)	2.94–2.68	0.84–0.87	1.73–2.10	5.96–2.79	2.99–2.62
West Germany					
% Share pkm	77.6–83.2	10.8–8.4	10.9–7.6	0.7–0.8	100.0–100.0
Eng. Intensity (MJ/pkm)	1.91–.31	0.51–0.77	0.64–0.60	4.36–2.64	1.64–2.05
Sweden					
% Share pkm	82.4–80.0	8.2–9.1	8.5–7.9	0.9–3.0	100.0–100.0
Eng. Intensity (MJ/pkm)	1.95–2.02	0.92–1.03	0.50–0.52	6.11–3.13	1.79–1.84
Norway					
% Share pkm	74.3–81.8	15.2–7.7	7.9–5.3	2.6–5.2	100.0–100.0
Eng. Intensity (MJ/pkm)	1.72–1.69	0.68–1.29	1.00–0.80	9.19–5.97	1.70–1.83
France					
% Share pkm	79.9–81.9	7.4–6.5	12.4–10.6	0.3–1.1	100.0–100.0
Eng. Intensity (MJ/pkm)	1.49–1.53	0.52–0.70	0.46–0.35	4.53–2.13	1.30–1.36
UK					
% Share pkm	76.2–85.6	13.9–7.0	9.3–6.7	0.5–0.7	100.0–100.0
Eng. Intensity (MJ/pkm)	1.85–1.90	0.73–1.13	0.98–0.84	4.62–3.86	1.63–1.79
Italy					
% Share pkm	75.1–77.8	13.8–13.6	10.6–7.7	0.5–0.9	100.0–100.0
Eng. Intensity (MJ/pkm)	1.37–1.47	0.46–0.77	0.47–0.45	6.13–3.98	1.18–1.32

Source: Transportation Science, February 1992

intensity is significantly affected by vehicle utilization. For example, the improvement in air transport energy intensity between 1970 and 1987 was primarily the result of flights that were operated at greater load factors, rather than improved aircraft technology. (Today's aircraft are equipped with more efficient engines, as well.) Declining profits and increased competition forced airlines to become more efficient carriers. Scheduling and vehicle size were rearranged so that aircraft were flying with fewer empty seats. This approach is potentially just as effective with land transportation. Modal switching can reduce the pkm energy consumption by leaving lightly loaded automobiles at home and filling the currently unoccupied seats aboard buses and subways. Or, just as with aircraft, automobile energy intensity can be improved by operating cars at higher load factors. Energy intensity can also be improved by designing vehicles for the load factors that are typical of urban commuting.

Automobiles in industrialized nations typically operate at a load factor of 1.5 to 1.8 occupants. Doubling the load factor will reduce the energy per pkm by 50 percent. Essentially, this is the goal of carpooling. The same fuel consumption is thereby distributed over more passengers.* Carpooling also has the negative effect of reducing the driver's independence, however. Drivers may no longer come and go as they please. Instead, they are restricted to the schedules of fellow commuters. Additionally, more than 70 percent of suburban office workers use their cars to run errands on the way to and from work, and 80 percent of office workers use them at some point during the workday.[15] As a result, the reasons for the automobile's popularity—the convenience and independence it affords—are largely lost by carpooling. New personal mobility products designed especially for in-town commuting could be one of the most effective approaches to changing consumer patterns. Station cars that reduce energy intensity on their own, plus encourage greater use of transit systems, would have a compounded effect.

For transportation engineers and planners, energy efficiency is not the only consideration in modal preference. Emissions may be more easily controlled in vehicles designed for mass transit, and traffic congestion is also somewhat relieved. As much as carpooling asks commuters to modify their expectations of the independent automobile, transit systems make the break complete. Many people argue that emphasizing mass transit systems is ultimately the best way to help defuse the modern-day crisis in urban traffic.

---

*Increasing the passenger count onboard light-duty vehicles has little effect on fuel consumption.

BLIGHT OF URBAN TRAFFIC

In America they call it gridlock, and in Germany it is *Verkehrsinfarkt* (traffic infarction). Regardless of the label, there is nothing quite as effective at making automobiles unattractive as being trapped in bumper-to-bumper traffic. Waiting at the entrance to the Lincoln Tunnel or creeping along at 5 p.m. on the Ventura Freeway, one quickly begins to question whether technology might be better utilized. Romance and the automobile seem suddenly incompatible. The racehorse sleekness of shining steel melts into an oddly ridiculous sight when it becomes a burden that keeps one locked in place, rather than an instrument of mobility. There is also the practical matter of squandered resources and wasted time.

Idling along in traffic congestion is wasting the world's motor fuel and creating tons of extra pollution. The U.S. Department of Transportation (DOT) estimated that fuel wasted in traffic congestion accounted for more than 11.4^9 liters (3 billion gallons) of fuel, or 4 percent of the nation's total consumption of gasoline in 1984.[16] Cars stopped in traffic have a fuel efficiency of zero. When they begin to creep along, efficiency climbs slightly above zero, but still remains dismal. One minute at idle can consume enough fuel to propel a full-size family sedan 1 kilometer or more.

As individuals, time is one of our most scarce assets. Each of us is allocated just 24 hours per day, and there is no way to manufacture more. Commuters, however, are wasting more and more time each year sitting in traffic congestion. Traffic in most of the world's cities is moving more slowly, and trip times are increasing. In London, commuting speeds are down to 13 kilometers per hour (8 mph). The Texas Transportation Institute studied 68 U.S. urban areas and found that drivers in one-third of the cities spent half as much time each year in traffic congestion as they did on vacation.[17] In more than half the cities studied, the amount of time drivers spent stuck in traffic increased by at least 350 percent between 1982 and 1997. During this period, traffic congestion jumped 1,000 percent in Indianapolis, 900 percent in Colorado Springs, 850 percent in Albuquerque, 800 percent in Oklahoma City, 767 percent in the St. Paul/Minneapolis metropolitan area, and 56 percent in Los Angeles. Despite its comparatively modest increase, Los Angeles still rates the highest in traffic congestion of any U.S. city. L.A. drivers lost 82 hours to traffic delays in 1997 at a cost of $1,370 in wasted fuel and extra travel time per person.

On a business level, time translates into dollars, and the breakdown in personal transportation is showing up on corporate spreadsheets. Already, the average U.S. business pays approximately $1,035 annually, per employee, to compensate for lost productivity of workers stuck in traffic jams.[18] The Commission on California State Government Organization and Economy warned that growing congestion has placed California on the brink of a "transportation crisis, which will affect the economy and prosperity of the state."[19]

Increased commuting time is not proportional to the simple increase in a city's population. Urban sprawl appears to be the fundamental reason why an increase in population translates into a far greater perceived increase in traffic congestion. Table 1.6 relates population growth and increased trip times in select U.S. cities.

Sitting in traffic is sure to continue to be one of the most annoying and wasteful ways to spend an otherwise pleasant morning or afternoon. But if you must get stuck in traffic, Ken Orski, President of Urban Mobility Corporation, probably has the best attitude about this predicament. Orski had this to say about surviving traffic jams: "Get yourself a comfortable car with plenty of creature comforts—a tape deck, a telephone, a fax machine, and a glove compartment-size microwave oven (yes they really exist)—and at least suffer in comfort!" According to Orski, commuters should learn to cope with conditions because they are not likely to get much better.

The sight of private automobiles stuck in traffic, each with their single occupant, is a common occurrence throughout the developed world. The architecture of Los Angeles, with its complex network of freeways, is a striking

TABLE 1.6 Population Growth and Traffic Congestion 1982–1997

Metro Area	Increase in Population (%)	Percent Increase in Driving Time	Actual Population Growth	Perceived Population Growth	Congestion Cost per Eligible Driver
Los Angeles	24.2	56.0	2,400,000	5,544,978	$1,370
Seattle-Everett	36.1	68.9	520,000	992,230	$1,165
San Francisco-Oakland	18.5	43.1	610,000	1,419,150	$995
Washington, D.C.-MD-VA	28.3	77.4	765,000	2,088,576	$1,260
Chicago-Northwestern	12.7	87.9	900,000	6,220,291	$720
Phoenix	67.8	130.7	970,000	1,868,916	$580

Source: Texas Transportation Institute, tti.tamu.edu

example of a metropolis built around the automobile. The system has been the envy of many of the world's larger metropolitan areas. Europe in general, however, lacked the wide-open spaces typical of the United States. Their traffic systems were built in traditionally more confined and densely populated environments. Many roadways were converted from streets that were already in place before cars existed. As a result, European city-traffic systems have deteriorated more rapidly and more profoundly.

Traffic congestion is highly resistant to solutions. Traffic flow is normally expressed in terms of vehicles per hour per lane. In free-flowing traffic, doubling the number of vehicles will produce twice the flow, provided headway is proportionally reduced and speed remains unchanged. But the more tightly vehicles are packed, the slower the traffic moves. Consequently, increasing the number of vehicles will increase the flow until the effect of reduced speed becomes greater than that of additional vehicles, at which point flow begins to drop. Introducing additional vehicles then accelerates the drop in flow. At some density level, traffic will actually stop and flow becomes zero.

In high-density signalized city traffic, signal timing and sequencing come into play. Signals are sequenced so drivers do not have to stop as long as they maintain a synchronization speed. Newer computerized systems can count cars and adjust signal timing to account for traffic variations. Human controllers in Los Angeles and New York can interrupt the computer system and manually defuse stacked-up traffic. Gridlock occurs when there is no unoccupied space available for queuing. For example, when a signal turns green, there must be unoccupied space beyond the intersection for cars that had been stopped. If the space is filled with cars, traffic cannot move. This will occur at some density level in every signalized system of roadways.

With an automobile transportation system, the number of vehicles in transit varies according to the number of people and the time at which they all pour into their automobiles. A high-density population that relies on automobiles for transportation will invariably produce traffic congestion, as well as the accompanying air pollution, noise, and accidents. Regardless of improvements in traffic management, there is an absolute limit to the number of cars that can fit into cities. Theoretically, one can expand the size of streets and add more parking facilities, but at some point, there may be no downtown left. In the United States, nearly half of all urban space is dedicated to parking, servicing, and providing roadways for cars. Worldwide, the average city devotes about one-third of its landmass to the automobile.

The problems of cities will increasingly be seen in terms of transportation-related problems. Cities occupy just 2 percent of the planet's surface, yet they account for 78 percent of human-related carbon emissions. In 1900, one-tenth of the world's population lived in cities (160 million), but soon after year 2000, roughly half of the world (3.2 billion people) will live in urban areas—a 20-fold increase in roughly 100 years. If only conurbations with more than 1 million inhabitants are included, metropolitan areas will include approximately 60 percent of the urban population, or more than 1.5 billion people worldwide. According to United Nations projections, in 2025 there will be 93 metropolitan areas with populations greater than 5 million. In 1984, only 34 cities were in this category. By year 2025, Los Angeles will have 15 million, Mexico City will have 30 million, Tokyo will have 20 million, and London will have 10 million. Table 1.7 shows the worldwide growth in city populations.

CONVENTIONAL IDEAS FOR TAMING THE AUTOMOBILE

Many ideas for improving the automobile's fuel economy or reducing traffic, pollution, and space requirements are based on the idea of restricting or eliminating the car. To analysts and commuters alike, a switch to public transportation is often seen as the ultimate answer to traffic congestion. The need for restricting urban automobile densities is being acknowledged even within the automotive industry. Pehr Gyllenhammer, General Manager of Sweden's Volvo corporation, said: "Die Innenstadt (the downtown area of the city) will have to be closed to individual automobile traffic. For the car to survive as the most versatile means of transportation, it will have to be taken out of situations in which it cannot prove its advantage."[20] Fiat's Gianni Angelli stated: "Something has to be done to reorganize traffic into and around the central cities, to keep the number of cars within reason."[21]

Undoubtedly, switching from cars to mass transit can reduce traffic congestion. But when people have a choice, they overwhelmingly chose the independence of the automobile. People will not willingly give up their cars. As a result, ideas for penalizing or restricting cars in the cities and concentrating more on public transportation are becoming more popular, even with local inhabitants. Germans, who are second only to the United States in per capita car ownership and perhaps first in their love of fine machines, are becoming increasingly disenchanted with cars in the inner cities. In a 1989 poll in Munich, 93 percent of the population supported giving public transportation priority, rather than spending money on improving systems to support the automobile. Politicians predicted a different result. In a parallel poll, 62 percent of the local politicians said that people would vote in favor of cars.[22]

In Munich the situation is critical. Downtown Munich has to cope with more than 1 million cars each day, and traffic is virtually at a standstill during

TABLE 1.7 Percentage of Population Living in Urban Areas

Region	Year 1996	Year 2030
World Total	45.7	61.1
More Developed Regions	5.1	83.7
Less Developed Regions	38.2	57.3
Least Developed Countries	23.3	44.0
Europe Total	73.8	82.9
Eastern Europe	70.6	81.3
Northern Europe	83.6	88.8
Southern Europe	4.3	75.2
Western Europe	81.7	87.8
Asia Total	35.2	55.2
Eastern Asia	37.6	59.1
South-Central Asia	29.1	48.5
South-Eastern Asia	34.2	55.0
Western Asia	68.4	81.5
Latin America & Caribbean Total	73.8	83.2
Caribbean	62.2	73.7
Central America	66.5	76.1
South America	78.0	87.2
Northern America Total	76.4	84.4
Canada	76.7	83.5
United States	76.3	84.5
Oceania Total	70.1	74.5
Australia/New Zealand	84.9	88.9
Melanesia	21.2	37.7
Micronesia	42.9	60.5
Polynesia	41.9	57.6
Africa Total	35.5	54.3
Eastern Africa	23.0	42.5
Middle Africa	33.2	53.2
Northern Africa	48.6	67.2
Southern Africa	48.1	64.1
Western Africa	37.0	58.8

Source: United Nations Urban and Rural Areas 1996 Publication ST/ESA/SER.A/166, No. E.97.XIII.3 (1997)

morning and afternoon rush hours. Conditions are equally bad in Frankfurt, where 200,000 people pour into the city each day—most of them by car. Because of increasing numbers of commuters, many German cities are threatened with a total collapse of automobile traffic systems. According to a recent poll of inhabitants of the Old Federal States, 85 percent of voters favor significant restrictions on city traffic, and 53 percent would like to see the automobile completely banned from downtown areas.[23]

Similar trends have developed throughout Europe. In 1985, voters in Milan, Italy, overwhelmingly approved a plan to close the city to all automobiles during rush hour. Sienna and Bologna now prohibit all motor vehicles during certain business hours, except for taxis, buses, delivery trucks, and cars owned by local residents. Athens uses odd-and-even rules keyed to license numbers to prohibit cars from entering the central city. Other options are also being tried. In Paris, every business with more than 20 employees is required to pay a public transport levy. In Lorrach, Germany, workers who drive their cars are charged punitive parking fees. In many cities the bicycle is enjoying a renaissance. In Asia bicycles are the predominant means of local transportation. And Europeans, who never entirely abandoned the bicycle, are turning to cycling more than ever. The Swiss pharmaceutical company Ciba-Geigy gives a new bicycle to employees who give up their parking spaces. In Denmark it is common for commuters to own two bicycles, one for commuting from home to the train station, and a second at the other end for riding from the station to work.

In the United Kingdom, after decades of pro-automobile planning, the railways may ultimately win favor with commuters. Driving in downtown London is nearly impossible, and the prospect of finding a parking place is even more dismal. An acquaintance of the author's who lives near London visited the United States several years ago and actually drove an automobile into New York City—a travesty unparalleled in American driving experiences. When asked about the encounter, he replied that "compared to London it was actually quite pleasant." Intercity travel in the United Kingdom is also best by rail. On the 88-kilometer (55-mile) trip from Faversham to London, for example, a roundtrip during off-peak hours costs £7.90 ($11 in 1993), and trains run four to the hour throughout the day. For the price of a ticket, one gets unlimited use of the railways in the city for the day. A similar trip by car would cost about £17.50 ($25) for fuel and £14 ($20) for parking, in addition to the aggravation of driving in London's congested traffic.

Totally car-free cities may not be the wave of the future; however, a shift in values for those who learn to rely less on private cars may be inevitable. Cities that turn increasingly toward transit systems may become a sort of training ground for commuters—churning out graduates who no longer place the private automobile at the same high level of importance. The following quote is from an American journalist, Peter Tautfest, who lives and works in Hannover, Germany:

> I remember going out on rainy nights and driving to the movies, the theater, or a restaurant only to find that on returning I would spend up to 45 minutes circling the neighborhood—finally parking my car a half hour's walk from my apartment. I learned, and started taking taxis. After doing this the third time, I became aware of the absurdity of owning a car that I couldn't use. So when our trusty Volkswagen broke down, I took it to the junkyard instead of the mechanic and thought I'd try living without a car.

> Now I ride my bike to work, whizzing past long lines of cars and beating my car-driving colleagues. On rainy days I take the bus, comfortably reading the newspaper while my colleagues are stuck in the proverbial gridlock listening to the latest traffic update.

The transportation system within a transit-adapted culture may consist of a much different cross-section of vehicle options than today. One of the biggest challenges for U.S. city planners centers on the need for on-demand local transportation now provided by private cars. Unlike those in Europe, U.S. cities often have ill-defined edges and no discernible head. For years the author lived and worked in the San Fernando Valley, which is classified as a suburb of Los Angeles; however, the San Fernando Valley is home to more people than live in the entire state of Arizona. The "city" to residents of San Fernando Valley is just as likely to mean places like Van Nuys or Burbank, which are central business districts of areas that began as independent suburbs. Downtown Los Angeles is far away and largely irrelevant to the lifestyle of many inhabitants of the "valley." Today, the busiest stretch of L.A. freeway is the Ventura Freeway in Encino, which is located in the south-central San Fernando Valley, far from downtown Los Angeles. What happened in Los Angeles is classical of developments in many U.S. cities. In order to accommodate the existing architecture, public transportation may have to expand to include options such as station cars for local transportation needs.

CONTROLLING THE AUTOMOBILE WITH HIGHER FUEL COSTS

Ideas for controlling the negative effects of private cars are often framed in terms of controlling the costs associated with driving. A popular argument against the benefits of more fuel-efficient cars centers on the price of motor fuel. As a percentage of the total costs of owning and operating an automobile, fuel costs are relatively minor, especially in the United States. According to the argument, improving fuel economy from 13 km/L (30 mpg) to 26 km/L (60 mpg), for example, does not result in large savings in the overall operating costs. So there is no significant benefit to consumers to economize on fuel by driving less, or to purchase more fuel-efficient cars. Unfortunately, the impact of transportation costs on individual budgets is largely hidden. If the real costs of driving private cars were transferred directly to drivers, economics could become a much more significant factor.

The hidden costs of petroleum consumption include several externalities, such as the military costs of protecting oil interests in the Middle East, the costs of environmental damage, and the increased healthcare expenses related to atmospheric pollution. U.S. military expenses directly related to defending Persian Gulf oil supplies, for example, translate into an additional $9 per barrel of oil imported from that region. The cost of environmental damage is difficult to estimate but may run as high as $45 per barrel, considering the costs of emission controls, damage from acid rain, global climate change, destruction of recreation values, and the treatment of emissions-related diseases. Just considering the estimation of environmental and healthcare expenses, the "true" cost of oil may be more on the order of $70 per barrel (assuming $25 per barrel for the oil itself).

Examining some of these hidden costs, the Ecological Research Institute in Heidelberg, Germany, has calculated that in order for cars to pay for all of their related expenses, gasoline would have to sell for $12 per gallon ($3.17/L). In a similar exercise, Carlo Rubia, winner of the 1984 Nobel Prize in physics, estimated that for automobiles to pay their own way, the worldwide price of gasoline should carry a retail price of $16 per gallon ($4.22/L).[24] If these hidden costs were no longer subsidized, they would have to rise to the surface, where they would affect transportation choices. At $12 to $16 per gallon ($3.17 to $4.22/L), for example, consumer attitudes toward automobile fuel economy would quickly change.

On a more moderate level, higher fuel prices in the United States are probably inevitable. The United States has for years enjoyed artificially low motor fuel prices. Lee Schipper suggests that the electric car equates to a conventional car at $6 per gallon of fuel ($1.58/L). That is, if gasoline were $6 per gallon ($1.58/L), consumers would more likely see electric cars as much more attractive alternatives. The reduced range of EVs might then appear less important in comparison with the benefits of reduced energy appetites. To continue with artificially low fuel prices, then lament the high consumption

TABLE 1.8 Gasoline Prices for Selected Countries in Year 1999
(Per-liter price in parentheses)*

Country	Price of Gasoline USD per gallon (per liter)
Japan	$3.13 ($0.83)
France	$3.79 ($1.00)
United Kingdom	$3.97 ($1.05)
Germany	$3.36 ($0.87)
United States	$1.54 ($0.41)

Source: *Transportation Energy Data Book*, Edition 20
*Prices reported in dollars per gallon. Liter prices inserted by the author.

of motor fuel, is a policy at odds with itself. Schipper notes a connectivity between fuel costs, vehicle fuel economy, and pkm traveled in industrialized countries. Although Europeans drive only 60 to 80 percent as much as Americans, and European cars typically achieve higher fuel economy, Europeans and Americans spend roughly the same portion of their incomes on motor fuel.[25] This finding may suggest price sensitivity: When motor fuel prices increase, consumption drops. Table 1.8 compares gasoline prices in select countries.

OVERLOADING THE ENVIRONMENT WITH WASTE PRODUCTS

Greater consumption translates into more byproducts. In general, consumption's byproducts turn out to be harmful to the environment. Pollution is loosely linked to the growth of GDP and therefore goes up with improved standards of living. If living standards grow at just two-thirds the rate of the past 25 years, consumption will quadruple in 70 years. In terms of total environmental impact, reducing energy consumption may be the single most important strategy for reducing this impact.

Manufacturing automobiles produces environmental pollution on an industrial level. Once the car is in service, it then becomes a continuous source of environmental pollution throughout its service life. Petroleum motor fuels are one of the major contributors of environmental pollution throughout the world. In a typical U.S. city, motor vehicle emissions account for 30 to 50 percent of hydrocarbon (HC), 80 to 90 percent of carbon monoxide (CO),

TABLE 1.9 Percentage of Transportation's Contribution to U.S. Emissions

Year	Particulate Matter (PM10)	Sulfur Dioxide	Carbon Monoxide	Nitrogen Oxides	Volatile Organic Compounds	Lead
1980	15.3	3.8	70.5	46.9	39.6	84.1
1985	19.2	4.3	47.9	44.7	37.6	74.2
1990	24.9	6.2	76.4	48.1	41.9	24.1
1997	23.6	6.8	77.6	49.1	39.9	13.3

Source: U.S. Environmental Protection Agency, www.epa.gov

and 40 to 60 percent of nitrogen oxide (NO_x) emissions. Table 1.9 shows average vehicle emissions in the United States, nationwide.[26] In OECD countries as a whole, about 39 percent of HC emissions, 66 percent of CO emissions, and 47 percent of NO_x emissions come from motor vehicles.[27]

Additionally, several studies have shown that emissions from automobiles increase significantly at the low speeds typical of urban traffic. The U.S. Environmental Protection Agency (EPA) estimates that internal combustion engine emissions at a vehicle speed of 8 kmh (5 mph) are two to three times greater than at the Federal Urban Driving Schedule (FUDS) average speed of 31.5 kmh (19.6 mph). The California Air Resources Board estimates that at 8 kmh (5 mph), HC emissions are 23 times greater, CO emissions are 13 times greater, and NO_x emissions are about 4 times greater than at 32 kmh (20 mph).[28] Different studies indicate different magnitudes of increased harmful emissions, but universally, they show higher emissions at city-traffic speeds.

GLOBAL WARMING AND CLIMATIC CHANGE

Global warming is an environmental wildcard. Expected negative effects include flooded coastal areas caused by elevated sea levels, disappearing wildlife habitats, and a wide range of consequences caused by the impact of global warming on the planet's climatic system.

Greenhouse gases include water vapor, methane (CH_4), nitrous oxide (N_2O), ozone (O_3), and carbon dioxide (CO_2). A comparison of satellite photos taken in 1970 and 1997 shows clear evidence of changes that correlate with increased levels of carbon dioxide, methane, ozone, and other trace gases in the earth's atmosphere. The buildup of greenhouse gases since the Industrial Revolution in the 19th century has altered the earth's atmosphere and appears destined to change the global climatic system. The primary greenhouse gas from human activities is CO_2, and the largest single source of CO_2 and overall greenhouse gas emissions is the combustion of fossil fuels.

Eons ago, the earth's atmosphere contained far more carbon, and the planet was significantly hotter. Over time, biological lifeforms (along with other factors) absorbed carbon from the atmosphere and safely locked it away beneath the surface. As atmospheric carbon levels dropped, the greenhouse effect diminished and the planet cooled. But with the emergence of the Industrial Revolution, carbon-rich fossil fuels have been extracted and burned at an increasingly greater rate, releasing their pent-up carbon into the atmosphere again. Over the 10,000-year period between the last ice age and the beginning of the Industrial Revolution, the atmospheric level of carbon dioxide had varied only about 5 percent, but in the period beginning with the Industrial Revolution until present, atmospheric CO_2 has increased by 28 percent. Projecting forward to 2030, the atmospheric level of carbon dioxide could reach double the preindustrial level.

Questions about global warming include unknowns such as exactly when to expect escalating global temperatures, just how rapidly and how high temperatures will climb, and precisely what effects it will have on the world's climatic system. It appears, however, that we are already experiencing early climatic effects. Global warming may be the culprit in the surge in weather-related property losses during the 1990s. Year 1998 was the hottest ever recorded. But an accurate record of temperatures has been kept only since 1880, so recent trends, including the jump in weather-related losses, may be within the standard deviation of natural variability. Evidence, however, increasingly points to human activities as a primary contributor to recent global warming trends.

At this point, no one can predict with certainty the degree to which the industrial production of greenhouse gases will impact the global climate or whether the climatic change will be good or bad, or for whom. But laboratory tests and computer models indicate that the significant increase in atmospheric carbon dioxide (CO_2), methane, and other greenhouse gases will raise global temperatures and, for better or worse, change the world's climate. As the world warms, rainfall patterns will likely migrate toward the poles, possibly leaving many of the world's richest agricultural lands with inadequate rainfall. Weather patterns now place the best rainfall over the world's best agricultural lands. At latitudes closer to the poles, the rich topsoil necessary for high agricultural production is not as abundant, but many plants grow faster and larger in a carbon-rich atmosphere. So it is difficult to predict whether a migration of rainfall patterns toward the poles will result in a global decline in food production.[29]

Limiting the world's emissions of carbon dioxide in an attempt to ward off global warming and the impending climatic change would be the most environmentally sound approach. The task of limiting CO_2 emissions, however, is technically challenging, and mandated CO_2 reductions are likely to have negative economic impacts. If significant restrictions are placed on CO_2 emissions, the cost of energy and energy-related goods is expected to go up, and economies will suffer because of it. On the other hand, if CO_2 emissions are not restricted, climatic change will likely hasten, and the results could be disastrous for the world's economies. So either way, unpleasant consequences may be in store. In order to determine the course with the least amount of negative impacts, accurate and reliable information is essential. Unfortunately, disagreement exists over the impact of human activities on global warming and the precise effects of global warming on the planet's climatic system. As the debate goes on, however, the world may be rushing headlong and irretrievably into the most calamitous event in human history. Moreover, it may soon be too late to make a difference, regardless of future emission control mandates.

The urgent nature of the problem centers on the strong upward trend in CO_2 emissions, the long lifetime of atmospheric CO_2 (50 to 200 years), and the probability that the planet's climatic system will change suddenly, once a critical threshold is reached. In other words, climatic change will not necessarily progress incrementally according to the increase in global temperatures. Instead, the oceanic and atmospheric currents that drive global weather patterns could undergo a dramatic and irreversible change almost overnight.

Transportation is responsible for approximately 32 percent of CO_2 emissions in the United States and roughly 27 percent worldwide. The United Nations Fund for Population Activities warns that, because of rapidly increasing automobile populations, developing countries will be emitting 15.06^9 metric tons (t) of CO_2 per year by 2025. At that point, developing nations will be emitting four times as much CO_2 as the developed countries. Figure 1.7 shows annual carbon dioxide emissions from fossil fuel combustion since 1820.

Production of carbon dioxide cannot be avoided when hydrocarbon fuels are burned. Hydrocarbon fuels consist mainly of hydrogen and carbon. When the fuel is burned, hydrogen is oxidized to water and carbon is oxidized to carbon dioxide. Each liter of gasoline, for example, produces 2.34 kg of CO_2 when burned. A 75-liter (20-gallon) tank of gasoline produces 175.5 kg (386 lb.) of

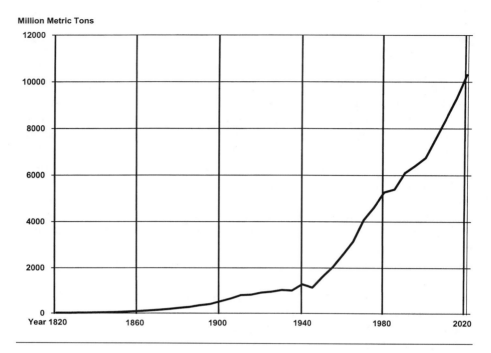

FIGURE 1.7 World CO$_2$ Emissions from Fossil Fuel Use with Projections to 2020

Source: Historical data from Oak Ridge National Laboratory, Carbon Dioxide Analysis Center, cdiac.esd.ornl.gov

CO_2. Production of CO_2 is the result of the chemistry of the fuel and has nothing to do with the design of the engine. Moreover, removing the CO_2 from exhaust gases requires nearly as much energy as is obtained from the fuel in the first place. So carbon emissions from motor vehicles running on hydrocarbon fuels are essentially proportional to the total fuel consumed. A reduction in fuel consumption, however, naturally reduces CO_2 emissions by an equal amount, and bioalcohol fuels result in a zero net gain of CO_2 because the carbon released during combustion is first absorbed from the environment by the source crops used to make the fuel.

A NEW PARADIGM FOR PERSONAL MOBILITY

Challenges facing personal mobility in the 21st century are more than just technical challenges waiting for technical solutions. On a macro level, today's sense of personal mobility problems is based on social, business, and life-style paradigms that may not necessarily project unalloyed into the future. As Orski suggested about traffic congestion, thinking in terms of a solution may be unrealistic. At least in the short term, management—or keeping an incurable disease under control through better vehicle technology—may be the most feasible approach. For long-term solutions, redefining the disease can illuminate new cures, and new cures can redefine the nature of the disease. Over time, the evolution of society and its technologies will naturally cast mobility issues in a different light.

Cutting transportation energy demand in half can be accomplished by doing the same amount of work, but doing it twice as efficiently, or by reducing the amount of work that has to be done. Today's effort to reduce automobile energy demand is mainly focused on doing an equal amount of work, but doing it more efficiently. This approach naturally follows from the present mindset about personal mobility and vehicle design. But when one considers that the purpose of an automobile is to get a 75-kg (165-lb.) person to a destination, the idea that it can only be done well using 1,400 kg (3,080 lb.) of hardware begins to appear misguided. In fact, most of an automobile's energy goes toward getting itself to the destination. So what began as a 75-kg problem has been turned into a 1,475-kg problem by the solution. A large reduction in mass, or total mass/kilometers traveled, is fundamental to any long-term solution to transportation energy demand.

Assuming that unbridled mobility is a requisite of modern societies, significantly smaller, super-efficient personal mobility products may be a viable approach. Several studies have indicated that smaller personal transportation vehicles can help relieve congestion and other traffic-related ills. Studies by Garrison and Pitstick, using GM's Lean Machine as a basis, indicated that roadway capacity is significantly improved when very small vehicles are substituted for large ones.[30] Other associated benefits include improved land and parking facilities usage, and a reduction in pollution and fuel wasted at idle. Smaller vehicles have a direct effect on lowering energy requirements over the urban driving cycle (see Chapter Three). Also, smaller vehicles are significantly less dangerous to pedestrians (see Chapter Seven).

Policies designed to encourage the development of cars that fit commuting needs might be linked to incentives that also encourage a shift in consumer perceptions. A switch to more energy-efficient vehicles, as well as new attitudes toward transportation in general, can head off the upward spiral in consumption and pollution. The idea that individuals must travel halfway around the globe each year by private car in order to lead meaningful lives may be an unfounded assumption. When considered in the abstract, modern travel habits can actually appear outlandish. For most of us, effectiveness in the world comes far more from our intellect than from our body, but transportation energy is devoted to the task of transporting our body about, even though our mental presence can produce on-demand results anywhere in the world using various electronic media. So large-scale personal mobility could ultimately be seen as unnecessarily burdensome and time-consuming, and perhaps even as an outmoded relic from a prior time.

Effectively dealing with the transportation challenges of the 21st century will require new perspectives on traditional solution options. Consider the bleak hypothesis offered by William Lee, CEO of Duke Power in North Carolina:

> Suppose that the world population stabilizes at the low end of the United Nations estimate of nine billion. Next, suppose that a majority of the world's residents reach a standard of living equal to half that enjoyed by Americans. Finally, suppose that all energy uses, from factories to sports cars, grow twice as efficient. Under these optimistic assumptions, the world will need to generate three times as much power.[31]

REFERENCES

1. Oak Ridge National Laboratory, "Estimates of 1996 U.S. Military Expenditures on Defending Oil Supplies from the Middle East: Literature Review" (Washington, DC: U.S. Department of Energy, Office of Transportation Technologies, 1997).

2. Paul Ehrlich, *The Population Bomb* (New York: Random House, 1968).

3. U.S. Department of Energy, Office of Transportation Technologies, "Future U.S. Highway Energy Use: A Fifty-Year Perspective" (Draft, February 22, 2001).

4. Energy Information Administration, "International Energy Outlook 1999," DOE/EIA-0484(99).

5. *See* note 3.

6. David L. Greene, "Transportation's Oil Dependence and Energy Security in the 21st Century," Center for Transportation Analysis, Oak Ridge National Laboratory, October 1997.

7. *See* note 4.

8. *See* note 3.

9. *See* note 4.

10. *See* note 4.

11. Ove Sviden, "Sustaining Mobility: A Systems Approach to Determining the Role of Electric Vehicles," Paper presented at the OECD/IEA Conference, Stockholm, Sweden, May 25–27, 1992, published in OECD Document, *The Urban Electric Vehicle: Policy Options, Technology Trends, and Market Prospects*, ISBN 92-64-13752-1.

12. Lee Schipper and Steve Meyers et al., *Energy Efficiency and Human Activity: Past Trends, Future Prospects* (New York: Cambridge University Press, 1992).

13. Lee Schipper et al., "Energy Use in Passenger Transport in OECD Countries: Changes Since 1970," *Transportation Science*, Vol. 19, No. 1, February 1992.

14. *Ibid.*

15. C. Kenneth Orski, "A Common Sense Look at the Problem of Traffic Congestion," *Vital Speeches of the Day*, January 1, 1990.

16. Michael Renner, "Transportation Tomorrow: Rethinking the Role of the Automobile," *The Futurist*, March-April 1989.

17. "1999 Urban Mobility Report," Texas Transportation Institute, 1999 (http://tti.tamu.edu).

18. *Bicycling Magazine*, March 1993, p. 28.

19. *See* note 16.

20. Peter Tautfest, "Clearing Up The Euro-Jam," *World Monitor*, March 1991.

21. *Ibid.*

22. *Ibid.*

23. Matthias Huthmacher, "Slowly Moving Forward," *Scala*, Aug./Sept. 1991 (Frankfurt, Germany).

24. *See* note 20.

25. Lee Schipper et al., "Fuel Prices, Automobile Fuel Economy, and Fuel Use for Land Travel: Preliminary Findings from an International Comparison," Unpublished draft #LBL 32699, for International Energy Studies, Energy Analysis Program, Lawrence Berkeley Laboratory, Berkeley, California.

26. 1980 and 1985 emissions from U.S. Environmental Protection Agency, Office of Air Quality Planning and Standards, "National Air Quality and Emissions Trends Report, 1989," Research Triangle Park, North Carolina; 1990 and 1997 emissions taken from "National Air Quality and Emissions Trends Report, 1997."

27. Asif Faiz et al., "Automotive Air Pollution: Issues and Options for Developing Countries," *The World Bank*, August 1990, WPS 492.

28. Quanlu Wang and Danilo L. Santini, "Magnitude and Value of Electric Vehicle Emissions Reductions for Six Driving Cycles in Four U.S. Cities with Varying Air Quality Problems," *University of Transportation Studies*, November 22, 1992.

29. Intergovernmental Panel on Climate Change, "IPCC Special Report, Summary for Policymakers, Emissions Scenarios, year 2000."

30. William L. Garrison and Mark E. Pitstick, "Lean Vehicles: Strategy for Introduction Emphasizing Adjustment to Parking and Road Facilities," SAE Paper No. 901485.

31. Gregg Easterbrook, "A House of Cards," *Newsweek*, June 1, 1992.

CHAPTER TWO

PERSONAL MOBILITY VEHICLES FOR THE 21ST CENTURY

Courtesy: General Motors Corp., www.gm.com

In truth, the ideas and images in men's minds are the invisible powers that constantly govern them.

—John Locke

Over the past two decades, environmental considerations have increasingly influenced the design of automobiles. Cars not only have to meet traditional manufacturing, safety, and lifestyle requirements, but they are also expected to achieve unparalleled fuel economy and produce few or zero harmful emissions. Today's challenge to simultaneously reduce harmful emissions and make a quantum leap in automobile fuel economy is normally seen as a technology problem. It will be helpful to see it in terms of a transportation systems and marketing problem as well.

Advanced automobile power systems promise to make personal transportation far more energy efficient and environmentally friendly. Production hybrid-electric vehicles already achieve fuel economy on the order of 25 kilometers per liter [km/L] (59 miles per gallon [mpg]), and vehicles achieving 35 km/L (82.5 mpg) or more are just around the corner. Practical fuel-cell vehicles are also in view. But inherently higher production costs present manufacturing and marketing hurdles that may be greater than the task of developing the technology. Vehicles designed around expensive technology that offers little in the way of increased personal benefits could squeeze corporate profits and present manufacturers with a tough sell in the marketplace. Longer amortization created by higher first costs could fundamentally change the market, perhaps forcing industry away from today's consumer market model and toward a commodity supply model of automobile manufacturing and marketing.

A New Perspective on Personal Mobility Solutions

Today's efforts focus on the idea of developing advanced-technology vehicles and then delivering them to consumers at a cost on par with conventional products. Although better technology is fundamental to any long-term solution, technology by itself may not provide the complete answer. In addition, an exclusively technology-oriented approach may lead product planners away from viable alternatives. New types of products that rely more on vehicle packaging and market positioning could play an equally important role. Significantly downsized vehicles—smaller, lighter, energy-efficient personal mobility products, both conventionally and electrically powered—could help turn the tide against escalating energy demand and open new markets in the process.

Marketing issues, however, are most often cited against the idea of producing alternative types of vehicles. Although nearly everyone understands that a switch to small, super-efficient vehicles would help the environment, industry has been slow to see marketing opportunities and create products of this category that capture the imagination of consumers. This chapter, therefore, explores vehicle packaging and design options, with an eye toward creating a different perspective on the design and market positioning of alternative personal mobility products—vehicles that might be referred to as "alternative cars."

To paraphrase the quote by John Locke presented at the beginning of the chapter, "the ideas and images in the minds of designers invisibly guide them toward particular design options." A product's character naturally emerges from the collective mindset of its designers. Consumer appeal of any alternative vehicle design depends on the ideas and images in the minds of its creators, not on the core idea of saving energy and emissions through size/mass reduction. In order to enjoy success in a consumer market, significantly different vehicle types—alternative cars—must be rendered in ways that create new appeals of their own. Energy savings and emissions reduction must be positioned as secondary benefits. Or stated differently, a consumer vehicle's environmental benefits can be an effective motivator only in terms of providing a rationale for a purchase that is, in fact, based largely on the product's emotional appeal.

Questions about the appropriate role of industry are also important. Should industry champion societal causes by developing products that run against the grain of consumer preferences or should carmakers simply make cars for profit? Should manufacturers attempt to lead consumer tastes or should they design to the tastes that already exist? Is it really in the best interests of carmakers to undermine high-end markets by promoting low-end alternative products?

Obviously, the foregoing questions contain assumed premises and implied answers, and one could easily rephrase them for the opposite conclusions. For example, should industry support the long-term economic health of its market by designing products that minimize future environmental burdens or should manufacturers simply make cars for maximum short-term profits? Should manufacturers develop products that leapfrog ahead of competitors and carve out new territories in the marketplace or should they stick with tradition? Finally, should carmakers expand their market by creating new types of personal mobility products, including high-end products for high-end markets, or should they continue to imply through design and marketing messages that car buyers need only one type of product to meet all of their transportation needs?

Invariably, the corporate mindset, the ideas and images in the minds of corporate leaders, governs whether new ideas are seen as valid and whether the company innovates and leads or holds onto tradition. Corporate philosophy, which permeates the design of its products and advertising appeals, contains implied and overt messages that invariably influence consumer perceptions. Corporations can hardly avoid leading consumer tastes and preferences. Corporations package technology in ways that define new products, and in the process stimulate new uses, new values, and even new lifestyles. At one time, the idea that consumers would give up their family sedans in favor of trucks could have been dismissed as a fantasy that was largely unsupported by the marketplace.

Consumers are becoming increasingly more sophisticated and worldly, and new attitudes offer new marketing tools as well as new challenges. In today's environment, businesses are no longer just selling products. Through their products, they are expressing company values and asking consumers to subscribe to them. Consumers are also looking for implied meanings in product designs that express their personal values. When a company connects with attractive products that address the broad concerns of today's consumers, prestige and sales are likely to soar. In the early 1990s, Global Business Network conducted a six-week-long computer forum with a distinguished group of ecologists, business planners, physicists, and community leaders. The subject was the "responsible corporation" of the 1990s and how new consumer attitudes might affect Nissan's strategic planning. The results, published as "The Nissan Report," suggested that a new relationship was emerging between business and consumers:

> A new consumer is rewriting the rules of the marketplace. The greedy "me generation" consumer of the eighties is disappearing; in her place is emerging a more community-focused and responsibility-minded consumer. This new consumer cares not only about a product and what it can do for her (or say about her), she cares about its place in society as well. . . . To these new consumers, we are not just selling cars (any more than other companies are simply selling soap or hamburgers or long-distance service), we are speaking to their values, their beliefs, their ways of being.[1]

Projects such as GM's EV-1, Chrysler's EPIC van, Ford's Electric Ranger, and the electric and hybrid vehicle offerings from Europe and Japan speak directly to this new dynamic. Today, manufacturers lose money on every electric vehicle sold, but the commercial success of these early products may ultimately be overshadowed by the broad benefit of portraying their creators as environmentally conscious leaders. In the minds of consumers, environmental and energy issues are closely associated with the automobile industry. Leadership in these areas is therefore becoming increasingly more important. Whether one chooses the label of "leading rather than following change" or that of "exploring new ideas in the marketplace," the benefits of recognizing new social and marketing dynamics and moving into them on one's own terms are undeniable.*

Pioneering vastly different vehicle types will be a risky venture, regardless of whether innovations center on vehicle technology or vehicle packaging and theme. Strong and visible support from government can play an important role. Government has already become an active partner in many of the world's cities by providing tax incentives for electric cars and alternative fuels and special freeway lanes and parking facilities. Support from government increases consumer confidence and in many ways assists and even partially underwrites pioneering efforts. By leading, industry can engage government on global interests in ways that are difficult to resist. America's Partnership for a New Generation of Vehicles (PNGV) exemplifies how an adversarial relationship (the impending increase in corporate average fuel economy [CAFE] standards before PNGV) can be transformed into a cooperative effort for achieving ambitious goals.

In addition to developing new power system technologies, expanding the array of personal mobility products to include significantly smaller and lighter energy-efficient vehicle types is a viable but little understood option. New markets can be defined, emissions can be lowered, and oil can be conserved using product personality and market positioning alone. Mercedes' Life Jet F300 tilting three-wheel vehicle is an excellent example of outside-the-box thinking in product design. Products based mainly on vehicle repackaging and new marketing appeals can simultaneously accomplish important global goals, capture the imagination of consumers, and stimulate extra corporate profits. The greater the product differentiation—the greater the departure from designs that are only variations on the conventional automobile—the more likely that consumer expectations will become unlinked from the features/benefits/costs profile of conventional automotive products.

---

* Some may argue that it is more profitable to follow and let others pay the price of pioneering. Although this is often true in a mature and stable market, it can put a company at risk during periods of change.

REDUCING THE HARDWARE OVERHEAD
OF PERSONAL MOBILITY

Automobile transportation systems worldwide have two fundamental characteristics that invite energy savings: (1) automobiles are often too large and powerful for their mission, and (2) automobiles are operated on average at about 30 percent of their payload capacity (1.5 to 1.8 occupants) throughout the industrialized world. Consistently poor load factors translate into high energy intensity. As it turns out, most of the energy used by the automobile is consumed to transport itself. The energy used to move the occupant is almost insignificant by comparison. An alien lifeform arriving on Earth might observe our automobile population and conclude that people are devices created by cars so that cars can navigate through traffic without crashing. The idea that it's the other way around—that mobility activity centers on people, not on cars—could seem grossly out of step with what is actually taking place.

The fundamental mismatch between vehicle mass/size in relation to that of its payload is often unspoken, perhaps because it is so universally accepted and difficult to quantify. Vehicle energy intensity depends on the synergy of the power system and on the operating conditions at the time the energy trail is audited. On the simplest level, when a 1,600-kg (3,500-lb.) machine transports an 80-kg (175-lb.) occupant on a local trip to the market, the available-energy pie is divided so approximately 95 percent gets the car to the market and the remaining 5 percent gets the occupant there. More specifically, about 82 percent of the energy in gasoline is wasted when it is converted into mechanical power, which just pollutes the air and gets no one to the market. Of the 18 percent left, about one-third goes to overcoming air resistance and the other two-thirds is consumed by inertia and rolling resistance, of which the occupant accounts for a small portion. In this scenario the occupant gets 0.006 of the fuel's energy, the car gets 0.174, and 0.820 is wasted. Because the automobile is responsible for 99.4 percent of the total energy consumed, on a conceptual level, minimizing the car is the most straightforward way to reduce its portion of the energy budget.

The car is being slowly minimized as curb weight is reduced. In the United States, average new car curb weight has dropped by approximately 500 kg (1,100 lb.) over the past two decades. Today, roughly 25 percent less mass is providing equal transportation to Americans who own newer cars. Extrapolated to the entire U.S. automobile fleet, reduced vehicle weight calculates out to approximately 75^9 kg (165 billion lb.) less automobile on the road, or 75^9 kg (165 billion lb.) less mass to move along on the nation's roadways. For those who like superlatives, that is roughly equal to the combined weight of nearly 1 billion adults, or about one-sixth of the world's human population. One can hardly argue with the economics of leaving that much weight sitting at the curbside. Smaller vehicles directly reduce the workload, and equally smaller engines and drivetrains work hand-in-hand to reduce the overall energy appetite.

The other major component is that of load factor. Changes in payload have little effect on fuel consumption once the automobile has been defined. Consequently, when vehicles operate more closely to full capacity, fuel consumed in relation to passenger-kilometers (pkm) traveled is significantly reduced. Unfortunately, vehicles are consistently underutilized throughout the industrialized world. As noted in Chapter One, automobiles in industrialized countries operate at roughly the same load factor, about 1.5 to 1.8 occupants. A breakdown of U.S. vehicle occupancy rates by trip category is shown in Table 2.1. The average occupancy rate of U.S. passenger cars was 1.6 persons in 1990, down from 1.9 in 1977.[2] That represents a 16 percent drop in passenger car utilization. The most recent study, the 1995 Nationwide Personal Transportation Survey (NPTS), showed average trip occupancy at 1.59, which is practically identical to the 1990 study. When vehicle occupancy drops, energy intensity goes up by a corresponding amount. The vehicle will consume roughly the same energy, regardless of whether seats are occupied or vacant. If vehicle trip profiles do not allow for adequate load factors, then new vehicle types that match travel preferences may be appropriate.

TABLE 2.1 Average Vehicle Occupancy by Trip Purpose

Trip Purpose	1977 Average %		1983 Average %		1990 Average %	
	Occup.	vkm	Occup.	vkm	Occup.	vkm
Work & Related	1.3	39.3	1.3	34.3	1.1	35.6
Shopping	2.1	11.1	1.8	13.4	1.7	11.9
Other Family & Personal	2.0	13.8	1.8	17.0	1.7	21.4
School & Church	2.0	5.2	2.1	4.1	1.7	4.5
Social & Recreational	2.4	27.3	2.1	30.0	2.1	25.8
Other	2.0	3.3	1.9	1.2	1.5	0.8
Average Occupancy/Total	1.9	100.0	1.7	100.0	1.6	100.0

Source: U.S. Office of Highway Information Management, www.fhwa.dot.gov/ohim
Note: Vehicle-kilometers (vkm) is reported in the United States as vehicle miles traveled.

One cannot simply throw out seats and affect energy intensity. The entire vehicle must be downsized accordingly. Size and mass can be reduced by the greatest amount when seating capacity is restricted to two. If seating capacity is increased to three, for example, vehicle size becomes roughly equivalent to that of a four-seater. Two-occupant and single-occupant layouts, therefore, offer the greatest opportunity for reducing vehicle size and mass. Although two-passenger and single-passenger cars are generally assumed to have marginal utility, this is not supported by vehicle use patterns.

NPTS distribution of vehicle trips by number of occupants reveals that a two-place vehicle will accommodate 87.2 percent of all trips, which accounts for 83 percent of all passenger car vkm traveled in the United States (Figure 2.1). Trip diaries maintained during the Mobility Enterprise studies at Purdue Uni-

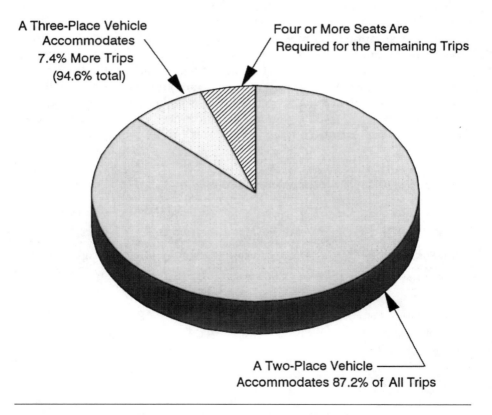

A Three-Place Vehicle
Accommodates
7.4% More Trips
(94.6% total)

Four or More Seats Are
Required for the Remaining Trips

A Two-Place Vehicle
Accommodates 87.2% of All Trips

FIGURE 2.1 Vehicle Capacity to Accommodate Trips

Source: U.S. DOT, Federal Highway Administration, www.fhwa.dot.gov

versity in the early 1980s revealed that 91.1 percent of the participants' trips were made with two or fewer occupants.[3] In Los Angeles, approximately 80 percent of all commuting trips by private car takes place with a single occupant aboard. According to GM's Al Sobey (retired), focus groups conducted in Los Angeles revealed that commuters would actually use (not *could* use, but *would* use) a single-place vehicle for 20 to 30 percent of their commuting trips.[4] When vehicle seating capacity is increased above two, benefits in terms of increased trip accommodation become increasingly marginal. For example, a three-place vehicle will accommodate only 7.4 percent additional trips, which results in a vehicle that is adequate for 94.6 percent of all trips. Expanding vehicle capacity from three to four occupants provides an additional 2.4 percent trip accommodation, which increases the total accommodation to 98 percent of trips. To accommodate the final 2 percent requires vehicles of varying capacities, up to and even greater than nine occupants.

Overweight and oversize cars and chronic underutilization of vehicles are the two most wasteful habits affecting the world's consumption of transportation energy, of which about 95 percent comes from petroleum. Using vehicles that fit driving patterns would greatly improve vehicle utilization and significantly reduce the energy intensity of personal transportation.

Figure 2.2 compares the energy intensity of a hypothetical ultra-low-mass (ULM) vehicle with that of a midsize car in various trip modes. The two-place ULM vehicle is assumed to have fuel economy of 42 km/L (100 mpg). For a vehicle in the 450-kg range (990 lbs.), this is well within the capability of conventional technology. The comparison vehicle is a 1999 Chevrolet Lumina/Monte Carlo; EPA rated at 29 mpg (12.30 km/L) highway and 20 mpg (8.5 km/L) city. Energy consumption does slightly increase as payload is added; however, its effect on fuel consumption can hardly be measured. Fuel consumption is therefore assumed to remain unchanged, regardless of the payload. Essentially, fuel consumption is determined in advance by overall system design, then occupants come along for free.

The ULM vehicle excels at reduced load factors and in city driving. As passenger count decreases, the ULM vehicle's advantage increases. In the fully laden highway mode, however, the ULM vehicle's energy advantage over the midsize car drops to about 25 percent, on a per-pkm basis, mainly because of the midsize car's greater passenger-carrying capacity. Nevertheless, its reduced fuel consumption still produces significant savings. Factored into U.S. national energy consumption, fuel savings even at the reduced highway advantage would still displace roughly 20 percent of all foreign oil imported in year 2000. The extra energy demand of the midsize car underwrites its greater mass, as well as its presumably greater use of power assist features.

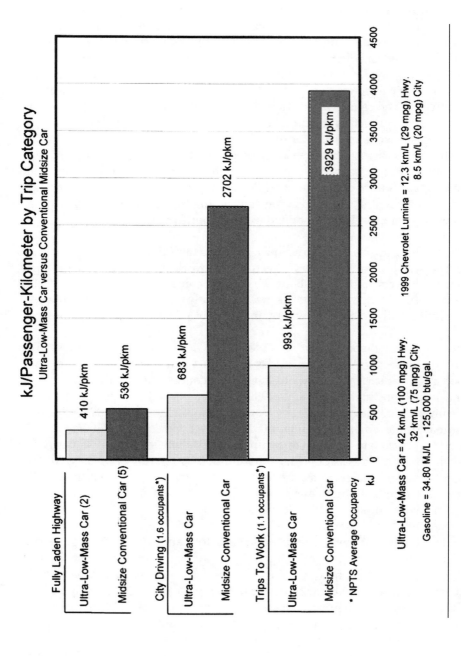

FIGURE 2.2 Ultra-Low-Mass Vehicle and Conventional Car Energy Comparison

46

This may be a good tradeoff for intercity trips with the entire family, but intercity trips are not appropriate missions for a ULM vehicle. In the suburbs or city, during single-occupant shopping trips, errands, or commutes to work, the underutilized midsize car is vastly outmatched by the smaller and lighter special-purpose car.

Cutting the hardware overhead of cars can significantly reduce the cost of mobility, in terms of traffic congestion, pollution, and natural resource depletion. On the most fundamental level, lower vehicle mass translates into reduced transportation energy intensity. The idea of improving load factors by carpooling is valid but difficult to implement. Travel statistics merely reflect the needs and wants of travelers in an essentially free-choice environment, and choices run strongly against carpooling. Attempts to modify travel habits have been largely unsuccessful. Government policies do, however, influence local model preferences where consumers are rewarded or penalized according to their choice of vehicles. In Japan, for example, K-car (midget car) sales increased from 4.1 percent market share in 1988 to 17.3 percent market share in 1991 when tax reforms and new models encouraged consumers to buy smaller cars.[5] By 1999, sales had reached 1.9 million vehicles annually.

Commercial transportation has also successfully improved load factors, especially in aviation. Policies have been unsuccessful, however, in changing the load factors of private automobiles. As noted earlier, low load factors are not unique to any particular region. Vehicle occupancy is relatively consistent throughout the industrialized world. Almost universally, travel statistics in developed nations show a trend toward lower, rather than higher load factors. One can therefore assume that automobile use patterns are fundamental, rather than idiosyncratic. The private automobile, because it is a private and personal means of transportation, naturally translates into low occupancy rates.

IMPACT OF VEHICLE SIZE ON TRAFFIC CONGESTION

Mass transit is often seen as a panacea for urban traffic ills. A major shortfall of traditional mass transit, however, centers on the need for local transportation to and from transit stations. Despite their negative tradeoffs, individual cars provide point-to-point mobility with a level of freedom, privacy, and on-demand utility that cannot be matched by the best of today's mass transit systems. Two solutions to the automobile dilemma stand out: (1) rented station cars for local mobility between destinations and transit stations, and (2) privately owned urban/commuter cars for citywide mobility. Both ideas center on significantly smaller vehicles for urban and suburban transportation as a way to reduce energy intensity, emissions, and urban traffic congestion.

Several studies indicate that smaller vehicles relieve traffic congestion. Increased parking capacity, reduced fuel consumption, and air pollution are other benefits of smaller cars. Also, smaller cars are less dangerous to pedestrians (see Chapter Seven), which is especially important in urban traffic where pedestrians and vehicles closely intermingle. Studies by Garrison and Pitstick using the GM Lean Machine as a basis indicated that roadway capacity is significantly improved when very small vehicles are substituted for large ones.[6] They were quick to point out that the actual improvement in traffic flow is greater than the raw geometric effect of smaller vehicles, primarily because of the decreased headway for drivers of smaller cars, which was documented by other researchers, including Wasielewski in an earlier report for General Motors.[7] Garrison and Pitstick also cite another study in England where cars the size of GM's Lean Machine provided capacity increases of 10 to 15 percent in mixed traffic (small and large vehicles together) and 100 percent in segregated traffic (special lanes for small vehicles). In an earlier study in the United States, McClenahan and Simkowitz also noted that in free-flowing freeway traffic shorter cars have a small effect on reducing traffic congestion.[8]

In signalized traffic, the effects are much greater. Simple theoretical models developed at the University of Pennsylvania showed that at a single signal, cars 50 percent shorter improved traffic flow by approximately 10 to 15 percent; however, McClenahan and Simkowitz discovered that benefits compounded in urban traffic because of the synergistic effects of signalized intersections in series. Using a computer model, half-length cars increased traffic flow up to 70 percent, depending on the ratio of small to large cars (Table 2.2).[9]

Land use is another important factor that normally goes unrecognized. As individuals, we would surely resist the idea of giving up half of our valuable real estate. As a society, that is exactly what automobiles do to our cities. An enormous amount of real estate is dedicated to the automobile infrastructure. The average city worldwide devotes about one-third of its landmass to the automobile. In the United States, nearly half of all urban space is dedicated to parking, servicing, and providing roadways for automobiles. Garrison and Pitstick presented ideas for increasing roadway and parking lot utilization by restriping existing facilities to take advantage of vehicles the size of GM's Lean Machine.[10] Smaller cars can free up expensive land for other productive uses. With no change in land apportionment, smaller cars can relieve congestion and increase capacity at little or no additional cost (Figure 2.3).

TABLE 2.2 *The Effects of Short Cars on Traffic Flow at Various Traffic Densities*

Number of Cars at Light	Percentage of 10-ft (3.05-m) Cars. Remaining cars are 20 ft (6.1 m)									
	0%		10%		33%		50%		100%	
	Flow (a)	Vel (b)	Flow	Vel	Flow	Vel	Flow	Vel	Flow	Vel
5	1154	13.2	—	—	—	—	—	—	—	—
10	955	6.3	—	—	—	—	—	—	—	—
15	732	3.5	745	3.4	855	4.0	1005	4.5	1240	5.5
20	745	3.6	—	—	—	—	—	—	—	—

(a) Flow converted to vehicles per hour of green.
(b) Velocity in ft/s (1 ft/s = 0.305 m/s).
Source: *Transportation Science*

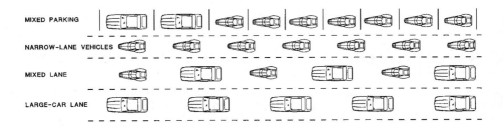

FIGURE 2.3 *The Effect of Short Cars on Free-Flowing Traffic*

IMPACT OF VEHICLE MASS ON EMISSIONS

Given equal manufacturing materials and processes, smaller, lighter vehicles require less industrial input and therefore produce fewer of the byproducts of industrial processes. Reducing mass by two-thirds (to 450 kg, for example) may not actually reduce manufacturing input by a full two-thirds. There is a strong correlation, however, between product mass and the resources required to manufacture, finish, and deliver essentially identical products.

Given equal technology, fuel-efficient ULM personal mobility vehicles would produce fewer harmful emissions. Again, all harmful exhaust emissions do not necessary decline in step with reduced vehicle mass and improved fuel economy. Carbon emissions are essentially proportional to fuel consumption, and a strong correlation exists between fuel consumption and other harmful exhaust emission, provided vehicles of equal technology are compared. Operating cycle also has a large effect on emissions. The Environmental Protection Agency (EPA) Urban Driving Cycle includes approximately 18 percent time at idle, 39.7 percent time accelerating, 22.7 percent time decelerating at close throttle, 12 percent time decelerating under power, and only 7.6 percent time at a steady-state speed.[11] Periods of acceleration and deceleration produce much higher exhaust emissions, and time at idle produces emissions with no benefits in distance traveled. Table 2.3 shows the effects of driving mode on exhaust emissions. Much urban driving time is spent in operating modes that produce extremely high emissions. Smaller cars with smaller engines directly reduce emissions during all driving modes, but most important, during acceleration and deceleration when emissions are highest and reduced vehicle mass and engine size have the most direct effect on both.

TABLE 2.3 Exhaust Emissions in Various Driving Modes (U.S. in parts per million)

Driving Mode	Hydrocarbons (HC)	Carbon Monoxide (CO)	Carbon Dioxide (CO_2)	Oxides of Nitrogen (NO_x)
Idling	1.34	16.19	68.35	0.11
Acceleration*				
0–15 mph (0–24 km/h)	536.00	2997.00	10 928.00	62.00
0–30 mph (0–48 km/h)	757.00	3773.00	19 118.00	212.00
Cruising*				
15 mph (24 km/h)	5.11	67.36	374.23	0.75
30 mph (48 km/h)	2.99	30.02	323.03	2.00
45 mph (72 km/h)	2.90	27.79	355.55	4.21
60 mph (96.5 km/h)	2.85	28.50	401.60	6.35
Deceleration*				
15–0 mph (24–0 km/h)	344.00	1902.00	5241.00	21.00
30–0 mph (48–0 km/h)	353.00	1390.00	6111.00	41.00

Source: *Automotive Air Pollution: Issues and Options for Developing Countries*, The World Bank.
*Metric speeds in parentheses inserted by the author.

ULTRA-LOW-MASS PERSONAL MOBILITY PRODUCTS

Undoubtedly, passenger cars will continue to shed weight, eventually resulting in featherweight vehicles that are far lighter than today's family sedan. New types of power systems are poised to replace combustion systems, and vehicle packaging will naturally follow along to reflect the changes in power-system architecture. In addition, the array of personal mobility products will likely include vehicle types that are vastly different from today's multipurpose automobiles. A new category of ULM alternative cars having more narrowly defined missions may become prevalent. Names like "station car," "shopper," "neighborhood car," "narrow-lane vehicle," "commuter car," and "urban car" have already become part of the vernacular of a growing wave of alternative car proponents. Finally, the personal mobility system itself may ultimately assume a much different architecture.

ULTRA-LOW-MASS MULTIPURPOSE PASSENGER CAR

Even though it was built in the early 1990s, GM's Ultralite is still a good example of the potential for mass reduction in the family sedan (Figure 2.4). In addition to extremely low mass, the Ultralite also showcases several other trends in vehicle construction, packaging, and personality that continue today. Modular packaging, wide use of plastic composites, adjustable ride height, and aerodynamic styling are all part of the formula for the car's new-millenium performance profile. At 635 kg (1,400 lb.), the internal combustion (IC) engine–powered four-place sedan achieves 42 km/L (100 mpg) fuel economy at a steady 80 km/h (50 mph). EPA driving cycles produce 19 km/L (45 mpg) city and 34 km/L (81 mpg) highway fuel economy. Despite its diminutive energy appetite, the Ultralite can still accelerate from 0 to 97 km/h (0 to 60 mph) in 7.8 seconds and reach a top speed of 217 km/h (135 mph). The effects of low mass and low aerodynamic drag combine to deliver both high fuel economy and excellent performance.

The Ultralite seats four adults in a package that is no longer than a Mazda MX-3. Gaping gull-wing doors provide wide-open access to the car's interior. The aerodynamic shape, rearward-tapering body, and small frontal area help produce the vehicle's low, 0.192 drag coefficient. Passenger payload can account for 40 percent of gross vehicle weight, which introduces the idea of adjustable ride height to account for payload variables. Ride-height adjustment also provides other benefits. The Ultralite can hunker down on the highway or adjust the front-to-rear ride-height relationship as needed to trim the car for minimum aerodynamic drag at higher speeds.

FIGURE 2.4 *GM's Ultralite.* The smooth aerodynamic shape and cab-forward layout preview a trend in vehicle packaging and styling. Notice the separation between the body and the self-contained power module at the rear.

Courtesy: General Motors Corp., www.gm.com

Engineers designed the self-contained rear power module so it can be quickly removed. A different drive package can be installed, or the existing one can be removed for replacement or servicing. Increasingly, vehicles will be built on a single platform, each with their own personality and capabilities. Individual model runs may be lower, and economies of scale will therefore come from utilizing a common platform over many different vehicles. The Ultra-lite carries the concept a step further with a power module that can be changed in the field for service, to switch fuel types, or conceivably to switch to entirely different power systems for different markets.

Even today, the Ultralite's construction is still on the cutting edge of technology. Made entirely of carbon-fiber composites, the body weighs only 191 kg (420 lb.) (Figure 2.5). The composite is a hand-laid epoxy/carbon-fiber fabric lay-up that sandwiches a core of urethane foam. This type of composite produces an extremely rigid and lightweight structure. Automotive application of the FRP/urethane foam composite was pioneered in the early 1970s at the author's former design firm, Quincy-Lynn Enterprises, Inc.* In those days, we used a low-cost polyester/fiberglass lay-up over the urethane foam core. With either skin material, using the one-off construction technique translates into a labor-intensive finishing job. The body must be entirely covered with filler and then hand finished before it can be painted, but vacuum-bag and inflatable-bladder molding techniques can produce finished surfaces, and several foam-in-place techniques can achieve the sandwich composite in production.

The Ultralite's epoxy/carbon-fiber skin, combined with an array of other high-tech materials, also leads to high materials costs. The cost of materials alone for the General Motors prototype amounted to $13,000. Economies of scale and new production techniques are bringing this type of construction closer to the production line.

Commuter Car

A commuter car is oriented to those repetitive one- and two-occupant trips typical of commuting in the cities of most of the world's industrialized nations. A vehicle in this category is assumed to be equally at home on the freeway or in the urban environment, much like today's family sedan. Curb weight might conceivably be reduced to 450 kg (1,000 lb.) or less, with power-system output in the range of 20–30 kilowatts [kW] (27–40 horsepower

---

* In 1969, Dave L. Carey built the first road vehicle using the composite. Originally, there were concerns that the skin might delaminate from the foam core because of the twisting and vibrations typical of an automotive application. By 1973, the original vehicle had more than 60,000 miles with no sign of delamination. After forming Quincy-Lynn Enterprises (Robert "Quincy" Riley and David "Lynn" Carey) in 1973, the composite system was refined and proven over a series of one-off road vehicles built by the company. Tri-Magnum, built in 1982, was the final vehicle in the series.

FIGURE 2.5 Ultralite Body under Construction. Epoxy/carbon-fiber lay-up over urethane foam results in an extremely light and strong body. With one-off construction, the entire body is covered with a thin coating of body filler in order to level out the surface.

Courtesy: General Motors Corp., www.gm.com

[hp]). An economy car theme is generally presumed. Mainly because of the presumed vehicle theme, the potential sales of a commuter car in high-end markets is normally assumed as roughly zero, which means the first product planner who mentions the idea gets fired. But a vehicle theme based on an "econobox" theme is inherently unmarketable at the outset, regardless of the vehicle's personal mobility benefits or its mechanical attributes.

Honda's EP-X concept car is an example of a more marketable commuter car theme (Figures 2.6, 2.7, and 2.8). Occupants are located centrally and in tandem to minimize air resistance and provide maximum side-impact protection. The interior is fully padded and smoothly contoured to eliminate corners and projections. A front air bag protects the driver. In Honda's words, the car offers "a totally new driving experience in an exciting new package." The appeal is appropriately directed toward the driving experience, not the family spreadsheet or the environment. In a consumer market, basic utility and conservation are essentially unmarketable attributes for a highly image-based product, such as an automobile. Instead, these attributes serve more as justifications for owning "a totally new driving experience."

FIGURE 2.6 Honda EP-X: Efficient Personal Transportation Vehicle. The EP-X concept car was designed to be highly fuel efficient and fun to drive. The 52-kW (70 hp), 1.0-liter VTEC-E (variable timing and lift electronic control—tuned for economy) engine utilizes variable lift and variable timing valves. Curb weight is 620 kg (1,300 lb.).

Courtesy: Honda Motor Company, www.honda.com

FIGURE 2.7 EP-X Interior. The smooth, flowing lines of the interior were designed for comfort and safety. Both driver and passenger are protected by air bags. Note the complete absence of edges or sharp projections.

Courtesy: Honda Motor Company, www.honda.com

FIGURE 2.8 Easy Access to Tandem Seating. Conventional doors are combined with a canopy that tilts to either side. The arrangement allows both the driver and the passenger to enter from either side of the vehicle. Exceptionally thick doors provide side impact protection.

Courtesy: Honda Motor Company, www.honda.com

Narrow-Lane Vehicle

A narrow-lane vehicle fills essentially the same niche as the commuter car but is an even greater departure from traditional design. The concept emerged from experiments at General Motors in the early 1980s with a vehicle called the Lean Machine. Vehicles like the Lean Machine are still being studied. Today, vehicles of this type are commonly known as tilting three-wheelers (TTWs). Technical information on TTWs is provided in Chapter Six.

The Lean Machine's passenger module pivots side-to-side around a longitudinal beam that extends forward from the separate power module behind. The passenger module thereby leans into turns like a motorcycle, while the power module corners flat like a conventional car. This layout results in a very narrow three-wheel vehicle with many of the handling characteristics of a motorcycle and much of the stability of a conventional car. The single-place version shown here (Figure 2.9) measures 2,616 mm (103 in.) long, and 1,194 mm (47 in.) tall. A separate power module houses the entire drivetrain

FIGURE 2.9 General Motors Lean Machine. GM's Lean Machine packs awesome performance into a 159-kg (350-lb.) package. Motorcycle-like turns produce 1.2g lateral acceleration at a 50-degree lean. Rear view shows Lean Machine's passenger module tilted slightly to the right.

Courtesy: General Motors Corp., www.gm.com

FIGURE 2.10 Vandenbrink Carver Man-Wide Vehicle. Carver is a modern version of a narrow-lane vehicle, which the company describes as a "man-wide vehicle." Maximum speed is 180 km/h (112 mph). More information on Carver is available in Chapter Six.

Courtesy: Vandenbrink bv, www.carver.nl

in a low-profile package within a tread of only 711 mm (28 in.). Wheelbase is 1,829 mm (72 in.) and curb weight is about 159 kg (350 lbs.). The 11.2-kW (15-hp) engine pushed the prototype to about 129 km/h (80 mph) and delivered fuel economy of about 51 km/L (120 mpg) at a steady 64 km/h (40 mph). Engineers at GM computer-modeled an advanced version with reduced aerodynamic drag and a 28-kW (38-hp) engine. They came up with steady-state fuel economy of more than 85 km/L (200 mpg) and 0 to 97 km/h (0 to 60 mph) acceleration in 6.8 seconds.

Vandenrink's Carver TTW is a modern version of a narrow-lane vehicle (Figure 2.10). Carver recently went into limited production in the Netherlands. Vandenbrink coined the name "man-wide vehicle" (MWV) to refer to this category of extremely narrow space-saving machines. Unlike GM's free-leaning Lean Machine, Carver uses an electronic control system called dynamic vehicle control (DVC) to control the vehicle's lean angle in turns. The operator simply steers and the DVC system does the rest.

Urban Car

Urban cars, also called "city cars," include a category of in-town cars that may or may not have the capability for freeway travel. Urban cars are specifically designed for the city environment. Because range requirements are limited, vehicles in this category are ideally suited for battery-electric power. Recharging might be available in downtown areas at specially equipped parking spaces, and fees for parking and electrical power could be paid electronically using Smart Cards. Navigational aids might also be oriented to the city environment, showing the most direct route to selected destinations and perhaps even the location of vacant parking spaces in the vicinity.

Ford's newly formed TH!NK Group is an environmentally oriented organization with the goal of providing clean and energy-efficient solutions for personal mobility. Ford's TH!NK City, shown in Figure 2.11, was introduced in

FIGURE 2.11 Ford TH!NK City. Ford's TH!NK City was introduced in Norway and is one of the first mass-produced battery-electric cars to be popularly accepted.

Courtesy: Ford Motor Co., www.ford.com

FIGURE 2.12 Mercedes Smart Car. The Smart Car is only 2.5 meters (8.2 ft.) long, 1.5 meters wide (4.92 ft.), and about 1.5 meters in height. Sales are rapidly gaining momentum in Europe.

Courtesy: MCC Smart GmbH, www.smart.com

Norway. Before Ford's acquisition, the battery-electric vehicle was made in low quantities by the Norwegian-based carmaker PIVCO.

The Smart Car (Figure 2.12) is a conventionally powered urban car with a maximum speed of 135 km/h (84 mph), which gives it freeway capability. At only 2.5 meters long (8.2 ft.), it is short enough to park nose-in to the curb in parallel parking spaces. With its rear-mounted 32.7-kW (44-hp) turbocharged three-cylinder gasoline spark-ignition (SI) engine, it will accelerate from 0 to 100 km/h (0 to 62.5 mph) in 18.9 seconds. Fuel economy is 5 liters per 100 km (47 mpg). The diesel-powered model is much more thrifty on fuel at 3.4 liters per 100 km (69 mpg).

Sub-Car

The sub-car category covers a wide range of road vehicles that may not be properly designated as cars. This category includes featherweight two-, three-, and four-wheel vehicles designed primarily for local and neighborhood use. Bicycles and mopeds establish the lower extreme, and golf car derivatives establish the upper extreme.

Nexus is a striking example of the bicycle taken to its ultimate, powered form (Figure 2.13). Inspired by the nearly standstill London traffic congestion, designers at Seymour/Powell conceived Nexus as the ultimate efficient and compact personal transportation device.

FIGURE 2.13 Nexus Human-Assist Vehicle. The 19-kg (42-lb.) Nexus incorporates a carbon-fiber frame that entirely encloses the engine and drivetrain. The down-tube serves as the fuel tank, with refueling access under the seat. A ceramic-coated, 2.2-kW (3-hp), two-stroke engine runs at very high temperatures for improved efficiency. It propels Nexus to 64 km/h (40 mph) and achieves up to 68 km/L (160 mpg).

Courtesy: Seymour/Powell, London, England

FIGURE 2.14 *Charger Electric-Assist Bicycle.* The Charger was jointly developed by GT Bicycles and AeroVironment. The bicycle's patented "Impulse System," a product of AeroVironment, senses how hard the rider is pedaling and adds electric assist according to a preselected percentage.

Courtesy: AeroVironment, Inc., www.aerovironment.com

The Charger electric bicycle (Figure 2.14) was developed as a joint venture between GT Bicycles and AeroVironment, the company responsible for GM's EV-1. Internationally, electric commuter bikes are one of the fastest-growing segments of the bicycle market. In the United States, the target market is mainly the recreational rider.

The Charger is powered by a 375-Watt electric motor. Electrical energy comes from a 24-volt lead-acid battery. Using the onboard changer, the battery can be recharged in about 4 hours. Charger's patented "Impulse Sys-

FIGURE 2.15 Ford TH!NK Neighborhood. Ford's TH!NK Neighborhood is a battery-electric golf car derivative designed for neighborhood and local shopping trips. Maximum speed is 40 km/h (25 mph).

Courtesy: Ford Motor Co., www.ford.com

tem" senses how hard the rider is pedaling and augments the rider's pedal power according to a preselected value. At 100-percent assist, electric power input equals the rider's pedaling input. So half the power comes from the rider and half comes from the electric motor. But the rider can also select 50 percent or 200 percent, for example, and change the ratio between assist power and pedaling power. At speeds above 29 km/h (18 mph), electric assist cuts out and the rider is on his or her own. Range is about 32 km (20 mi.), depending on the amount of electric assist. Ford TH!NK Neighborhood (Figure 2.15) defines the upper extreme of the sub-car category. Motor Suite (Figure 2.16) is an example of a performance sub-car.

In many countries, sub-cars are already part of the transportation system, especially in Asia where bicycles are the predominant means of personal transportation. If vehicles of this type are encouraged for neighborhood and local shopping trips, they could displace a significant amount of petroleum.

Bicycle Transportation Systems' Transglide 2000 is a bicycle transit system that, according to the developer, can increase urban mobility at a fraction of

FIGURE 2.16 Motor Suite Three-Wheel Sports Cycle. Motor Suite is an example of a performance-oriented sub-car. It was designed by Chiaki Maruyama for a contest conducted by *Car Styling* magazine, Tokyo, Japan. This three-wheel neighborhood sports vehicle would be classified as a motorcycle in the United States.

Courtesy: Car Styling, www.nos.net/carstyling/

the cost of conventional transit systems (Figure 2.17). The system consists of elevated pathways supplied with a continuously moving volume of air. Air may be heated or cooled, depending on ambient temperature, to provide cyclists with a comfortable environment. At 32 km/h (20 mph), air resistance accounts for about 90 percent of a bicycle's road load. By moving the air in the direction of travel, cycling air resistance is significantly reduced. This allows cyclists to maintain speeds on the order of 40 km/h (25 mph) with far less exertion.

Lifestyle marketing could have a large impact on the popularity of human-powered transportation in regions where the bicycle is now used mostly for recreation. Today, cycling enthusiasts pay $2,000 or more for high-tech bicycles that provide little additional benefit in actual performance, once they break the $1,000-dollar price barrier. Most of the appeal of these bicycles has to do with the pleasure of owning and enjoying fine craftsmanship and beautiful design, and being turned on to a mental image of what it means to be part of an elite group.

FIGURE 2.17 *Transglide 2000 Bicycle Transit System.* Bicycle transit system could increase urban mobility at a fraction of the costs of traditional urban transit systems. Notice that pathways are divided down the center so air can move in opposite directions.

Courtesy: Bicycle Transit Systems, www.biketrans.com

By creating an image that connects, people can become turned on to activities that may not have seemed appropriate before. Before the release of the movie *Pumping Iron*, for example, weightlifting was largely confined to a group of hard-core enthusiasts. Today, resistance training, along with a variety of other fitness activities, has become part of the mainstream culture. Moreover, today's fitness offerings include equipment styled to emulate works of art and an array of stylish apparel for every conceivable activity.

So with the right practical support and innovative design and marketing, combining fresh-air fitness with practical transportation could become the "in" thing to do. Like smoking or eating to obesity, using a ton of hardware to travel a few kilometers to work could ultimately be seen as socially inappropriate or even abusive by a fitness-minded culture. These shifts in mindset can be encouraged through product design and advertising messages. Or they can be propelled by real-world events, such as when the 1973–1974 oil crisis suddenly turned consumers toward smaller cars and gave Japanese carmakers an unexpected passport into new markets.

NEW PERSONAL MOBILITY PRODUCTS AND THE MARKETPLACE

Consensus within the industry is that the market potential of special-purpose ultra-light cars, especially in high-end markets, is extremely limited. Market studies invariably produce a thumbs-down response for the idea of a limited-utility car. But studies based on abstract product ideas and biased by negative hypothetical attributes cannot show the market potential of real products that have yet to be developed. Consumers can hardly be expected to envision the appeal of an unfamiliar product idea before it has been rendered in the real world. If motorcycles had never been invented, consider the result of a study that polled car buyers for their reaction to a new product idea for a two-wheeled bodiless car. Naturally, one would not expect much support for such a silly product idea. But today, motorcycles (two-wheeled bodiless cars) enjoy an enormous worldwide market.

A sense of vision can hardly be found in product clinics, focus groups, or consumer surveys. Moreover, to conceive and execute visionary new product ideas requires breakthrough efforts of even the most talented product planners and designers. If a group of designers who had never seen a motorcycle were asked to create a design for a two-wheeled bodiless car, it is doubtful that a product like today's motorcycle would emerge. The corporate mindset, expressed in the language of the period (a two-wheeled bodiless car, for example), tends to put boundaries around imagination. Even the best designers are compelled to work under the handicap of their own preconceived ideas, their own design paradigms based on prior experience. But breakthroughs to visionary new designs and advertising appeals are essential if alternative personal mobility products are to gain favor with consumers.

Corporate landscapes are littered with self-evident "truths" that invariably create roadblocks to imagination. The pitfalls of relying on paradigmatic assumptions stand out more clearly when one examines similar reasoning processes in a field outside of one's immediate experience. This helps avoid the trap of "knowing" too much, of being mentally entrained by one's own specialized knowledge. So it may be helpful to discuss, in nonautomotive terms, the interplay between traditional marketing rationales and visionary product design and positioning.

If, before the advent of inline skates, a market study had been done to determine what consumers wanted in their roller skates, the study would have undoubtedly shown they wanted more durable and quieter wheels, better bearings, and a more comfortable fit. But the study would not have shown that consumers wanted skates with a single row of wheels down the center. If the idea of inline skates had been proposed in a product clinic, most likely participants would have rejected it. One could have certainly argued against the wisdom of introducing a totally new type of skate, requiring a new set of skills, to a universe of consumers who wanted only a better version of an existing and familiar product. Only through a sense of vision can product planners leapfrog over this impasse of logic.

By the mid-1990s, inline skates had grown to an enormous worldwide market with nearly $500 million in annual sales in the United States alone, far exceeding previous sales of conventional skates. According to the Sporting Goods Manufacturers Association, inline skating was the fastest-growing sport of the 1990s, with nearly 30 million U.S. participants in 1997. This vibrant market was created almost entirely by visionary design and advertising appeals (Figure 2.18). Inline skates are as much a mindset and a lifestyle as they are a mechanical product. This does not imply that function is unimportant, only that actual mechanical utility is often secondary to the power of design in capturing the imagination of consumers. Design tells consumers how the product will be experienced, and it tells the world how to experience the users. Through the language of design, corporations create new lifestyles and new values. But when decision makers conceptually collapse a product's emotional meaning into its mechanical function, it becomes difficult to see the power and impact of design on a product's success. Design may then be seen as an ornament rather than the prime motivator that it often is.

Consider the success of personal watercraft. Imagine a discussion between product planners in the boating field about the success potential of personal watercraft, before their actual introduction. Such a discussion would surely have included, in some form or other, the following perfectly reasonable arguments against the idea:

- Nobody can ride one without practice.
- People can buy a "real" boat for just slightly more.
- Product clinics with boat owners show that hardly anyone would buy one.

FIGURE 2.18 Inline Skating Markets to a Lifestyle Image. Inline skates are more than just a product. They speak to a lifestyle. By wearing the "uniform," one becomes identified with the obvious athletic image of health, vigor, and youth. Because of the worldwide popularity of inline skating, officials are considering the idea of creating an Olympic category for the sport.

Courtesy: Rollerblade, Inc.; photo by Charlie Samuels, www.rollerblade.com

- Boaters are not asking for a product like this.
- Every mini-boat introduced in the past has been a marketing failure.
- Boat dealers will not carry them, so no dealer network exists.
- They really don't have much practical use.

- You can't fish from one.
- Even if they did sell, they leave very little profit margin, and they would take sales away from profitable lines.

Naturally, boat manufacturers were not especially enthusiastic about Clayton Jacobsen's idea for a personal watercraft (PWC).* But Kawasaki went ahead, sold them through their network of all-terrain vehicle (ATV) dealers, and a new industry was born. In 1998, 130,000 PWCs were sold in the United States alone, for total revenue of $868,499,000. In addition, U.S. PWC owners spend more than $300 million annually on ancillary items such as boating registration fees, launch fees, trailers, fuel, insurance, clothing, accessories, travel, and watercraft-oriented vacations.

It is important to focus on the fact that personal watercraft are not positioned as "mini-boats," although they are indeed miniature boats. Through design and positioning, these small boats define a totally different type of product and a totally new market. When Kawasaki ads implore readers to "let the good times roll," they are describing how the product will be experienced. They are selling the image, the lifestyle, and the experience, and only indirectly selling the product. True, the product must deliver, but making a product work is the easiest part of the process.

Consider PWCs from the reverse perspective. Imagine for a moment that PWCs were positioned as mini-boats costing somewhat less than full-size boats. And let's make them environmentally friendly mini-boats by installing cleaner, more fuel-efficient engines, and outfitting them with recyclable hulls and biodegradable upholstery. Instead of racy yellows, reds, and blues, they'll come trimmed in "environmental green." In our ads, we'll replace "let the good times roll" with a sales pitch on cleaning up the waterways by buying our little boats and leaving big boats at home. Assuming these hypothetical mini-boats performed just like today's PWCs, what kind of product have they become and what is their potential market? Through the power of design and positioning, PWCs, redefined as environmentally friendly mini-boats, have become essentially unmarketable products. And the hypothetical arguments against the idea of PWCs (in the discussion between watercraft product planners) have actually become true.

---

* When Jacobson was promoting the idea, the term "personal watercraft" had yet to be coined. Later, Kawasaki used the trade name "Jet Ski" to describe their standup-type watercraft, which was based on Jacobsen's design. Today, most PWCs are of the multipassenger, sit-down type, a design that was pioneered by Bombardier. "Personal watercraft" is a generic term that refers to all of the various inboard-powered watercraft of this category.

The foregoing scenarios can be extrapolated to new ideas for alternative personal mobility products. For example, the negative arguments against Jacobsen's idea by watercraft product planners are similar to arguments that might be overheard in automotive conference rooms against alternative personal mobility products. If automotive product planners were to see "environmental green and biodegradable upholstery" as primary motivators and ask designers to create a vehicle around such a theme, they will be profoundly disappointed when their new product reaches the marketplace. But if they were to "let the good times roll" with alternative vehicle designs, then the good times would likely roll in the marketplace as well.

Unfortunately, there are no guarantees, regardless of whether one chooses to stick with tradition, invest in new power-system technologies, or pioneer new product themes and marketing appeals. Winston Churchill was referring to the uncertainties of war when he said: "I much prefer to make predictions after the event has happened." A technique like that could save fortunes in misdirected development and marketing expenses, if only it would work.

Although the author could lose a fortune in spent credibility, it might be helpful to mention a success that is about ready to happen. Recumbent cycling is a fledgling market that is currently being held back by the absence of clear market positioning and unimaginative design. Today, recumbent bicycles are sold mainly on the basis of mechanical function to a loyal following of devotees. The recumbent layout, however, does offer several benefits over conventional upright bicycles, and plenty of fertile ground exists on which to build a powerful marketing message. Recumbents are faster and sleeker, they are safer in a spill, and they are more comfortable to a large segment of riders. These benefits are especially relevant to the growing number of riders in the 40-plus age group and to those with carpal tunnel syndrome and low-back problems. For these and other reasons, the recumbent layout holds the promise of making large inroads into the existing market and capturing consumers who might otherwise never have ventured into cycling at all. Today, recumbent cycling is primed for a huge burst of sales, once manufacturers connect with consumers through the language of design (Figure 2.19).

In the 1990s, vehicle personality influenced automobile energy intensity in the negative with lifestyle vehicles that pushed energy use upward. The popularity of sport utility vehicles (SUVs) illustrates how an image can influence choices, and those choices can lead to significantly higher fuel consumption. New-car average fuel economy plunged in recent years, mainly because of the power of design to influence purchasing choices. As of year 2000, light trucks accounted for roughly half of all new-car sales. Consumers are turned

FIGURE 2.19. Ground Hugger XR2 carbon-fiber recumbent bicycle designed by the author, www.rqriley.com

on to the personality of these trucks, and they buy them to drive around the city. Regardless of the rough-and-ready vehicle theme, hardly anyone buys a four-wheel-drive SUV to go trailblazing. The real reason for the popularity of SUVs has to do with the intangible messages woven into the vehicles' personality, not with the cost/value profile. This same dynamic can work in the reverse.

People buy SUVs because of the messages communicated through the language of design. They buy new VW Beetles because they strike an emotional cord. After a lackluster launch, Smart Car sales are now gaining momentum in Europe, with 101,000 units sold in year 2000, and expectations of a 20 percent increase in 2001 (Figure 2.20). If vehicles like Smart Car gain a large market share, energy demand will drop, and it will drop by a lot, and it will have little to do with actual technology.

Creative product design and innovative appeals are essential if alternative cars are to become popular with consumers. Consumers who reject the idea of such vehicles in the abstract can quickly flip to the other side when a real product offers a new bundle of appeals that were not envisioned before. Product clinics with GM's Lean Machine produced twice the level of favorable responses normally predicted by studies based on abstract product

FIGURE 2.20 German Ad for Smart Car. Smart Car ad centers on emotional appeals. It sells the "passionate experience," and the car's environmental and fuel-saving benefits are secondary. Smart Car is on the verge of achieving an almost cult-like following and status appeal with motorists in many European cities.

ideas. It would be legitimate to ask whether Lean Machine was the best that designers could do, or whether a product with more appeal might produce four, six, or ten times the expected favorable response. According to Jerry Williams, Project Engineer on the Lean Machine, the team had designs on the drawing boards that were far superior to the vehicle used in focus groups, but the project was cancelled before they could be built.

It may be useful to briefly discuss vehicle theme and packaging with an eye toward subliminal messages and product differentiation.

Vehicle Theme

Vehicle theme establishes the personality of the product. It is the broad product attribute that attracts a particular segment of buyers. The theme is strongly identified with the purchaser's self-image. One segment of consumers may identify with the "independent man-of-the-earth" image and some individuals might carry that out by wearing cowboy boots and driving a four-wheel-drive pickup. A different self-image might lead one to the showroom with the Corvettes on display. The basic attraction to vehicle types has to do with the compatibility, and even magnetism, between one's ideal self-image and the vehicle theme. That the vehicle is the best choice in terms of utility is often rationalized so purchases can be made in accordance with self-image. Meanings inherent in a product's design go far beyond its mechanical function.

Peter Dormer, in his fine book *The Meanings of Modern Design,* discusses the camera as an object of value that transcends its mere function. Dormer describes the camera as having a precision craftsmanship of a quality that can come only from the automated processes of modern production methods—a technology of creating products that goes far beyond the flawed labors of human hands to create a pristine beauty that is uniquely its own. The feel of the camera in the hand, the sound of the shutter measuring time in thousandths of a second, the immaculately detailed case, and the crystalline purity of the lens are all imbued with the same flawless precision and add to the pleasure of owning and handling the product. According to Dormer, the technical capabilities of the tools of photography far exceed utility to become an end in itself. The actual photographs become almost a secondary benefit. Photography's accouterments embody a theme that gives the purchaser a pleasurable feeling whenever he or she partakes in it. Cars speak to consumers in the same subliminal language.

Alternative vehicle types must have an intrinsic appeal of their own. Conservation and sacrifice are unmarketable themes. New personal mobility products cannot be positioned as cheap imitations of "real" cars. Designs should exemplify an upscale chic, perhaps like casual designer clothes that everyone knows are not really casual at all. The modern road bicycle is certainly environmentally friendly and energy efficient, but it speaks to the owner's appreciation of quality and fine workmanship in its appeals. A feature-by-feature tour of a carbon-fiber high-performance bicycle will likely awaken new eyes of appreciation for beautiful, simple design and attention to detail.

Alternative ultra-low-mass vehicles should also have a defining aesthetic appeal. Designs might convey a sense of solid, high-tech competency that would elevate the products' perceived value. The product should not step down to accommodate scarcity and conservation. Instead, it must move up to conquer waste and pollution through superior design. A theme of compact precession and technical transcendence thereby replaces the often-envisioned theme of economy and basic utility. These messages must be obvious when consumers experience the design. The theme and its execution imbue the vehicle with qualities that consumers want to own. New personal mobility products must be fun to own, touch, care for, and operate. Aesthetic appeal and pride of ownership are independent of vehicle size and energy consumption.

The vehicle theme establishes a connection with the consumer. It conveys through the language of form and texture an image of what the vehicle is and what it does, and it tells the consumer how the product will be experienced. The vehicle theme lets the consumer know whether to expect an exciting driving experience or just an uneventful trip to work. It also subtly suggests a price category and thereby makes a statement about its affordability and the prestige of ownership. Creating an enticing vehicle theme—one that positions the alternative car apart from the conventional automobile and imbues it with a distinctive image of its own—is perhaps the greatest single challenge to the designer. A motor home is more than just a large automobile with beds, and a motorcycle is more than a two-wheeled bodiless automobile. These vehicles make an emphatic statement of their own. That statement, carried by the appearance, feel, performance, and sound of the vehicle, establishes the perspective from which the product is seen. It breathes a personality into the machine.

The personality of an alternative "personal transporter," for example, should define a category that removes it from direct comparison with the conventional automobile. The defining attributes of an automobile already exist in the minds of consumers, and the label will transfer those attributes to the new product. An expectation of price will then have been established by comparison to the price/features profile of the standard automobile. New vehicle types therefore should stand apart and create a new category of products with a price/features profile of their own. In theme, they should not compete with existing cars, but instead become a new addition to the household, a new object of ownership and a new experience on the roadways. New vehicle types should have new descriptive names. Personal watercraft, for example, are indeed miniature boats, but to label them mini-boats would detract from their appeal, and in the process, create a size/price comparison with full-size boats.

As for executing the theme, a variety of pertinent vehicle attributes provide a natural opportunity to speak to consumers in the language of design.

Occupant Zone

The occupant zone is often considered little more than a space for the vehicle's human cargo. In reality, however, it sets the ambience of the driving experience, and it defines the quality of the connection between human and

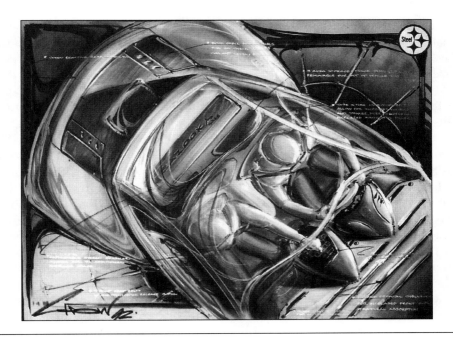

FIGURE 2.21 High-Tech Personality Centers on the Occupant. The occupant zone is more than just a space for human cargo. It creates the connection between driver and machine. Through this connection, the driver "becomes" the machine and thereby transcends human limitations of power, speed, and agility.

Courtesy: Jimmy Chow, the Art Center College of Design, Pasadena, California

machine (Figure 2.21). The design of the interior determines whether the human is merely riding in a mechanical object or whether he or she becomes an integral part of the machine. Good ergonomic design leads to a fluid type of interface wherein the driver becomes unaware of the separation between human biology and the mechanical nature of the vehicle. The machine becomes an extension of the operator's body. In effect, the driver assumes the personality of the machine as part of his or her own.

Seating Layout

Seating layout and vehicle capacity can help convey the image of an entirely new type of personal mobility product, but capacity and layout also deal with practical matters. In practical terms, seating capacity has the greatest impact on overall vehicle size. Adult humans come in discrete packages of approximately the same size, and not much can be done to change it. So the number of humans a vehicle must accommodate establishes much of its overall size and shape.

A three-place vehicle might be designed around a 2+1 theme, where behind-the-seat storage space would also serve as an emergency third seat. Providing for three occupants will accommodate 7.4 percent more vehicle trips, in comparison to a two-place design; however, several challenges are introduced with a three-place seating arrangement. The obvious difficulty is that of providing adequate occupant space without unduly increasing the size of the vehicle. If the third occupant were placed in a conventional position behind the two front occupants, the passenger compartment might easily double in size. Facing the third occupant rearward can improve space utilization, but it still may not entirely resolve the problem. Room for a fourth seat is likely to be a natural byproduct.

Another important consideration of a three-place layout is the increase in payload mass over a two-place layout. With a 450-kg (990-lb.) vehicle, for example, a payload on the order of 240 kg (three 80-kg occupants) could detrimentally affect ride height and even change vehicle center of gravity (cg). Large load variations tend to create difficulties in controlling ride, handling, and performance characteristics throughout the vehicle's load range. Adjustable ride height is often considered the only alternative to an otherwise overly stiff suspension; however, compromises can produce acceptable results. Quincy-Lynn's Urbacar at 295 kg (650 lb.) had an acceptably stable ride height and a relatively smooth ride, with either one or two occupants aboard. Urbacar's payload and vehicle mass relationships are roughly equivalent to the hypothetical 450-kg (990-lb.), three-place vehicle described previously.

A two-place vehicle simplifies the design task and still results in a seating capacity that will accommodate approximately 90 percent of all vehicle trips (in the United States). Occupants may be located side-by-side (Figure 2.22) or in tandem. Several interesting concept cars used tandem seating. Honda's EP-X and Mercedes's F300 are two examples. The primary advantage of tandem seating is that it reduces frontal area by about 40 percent. The disadvantage comes from the greater longitudinal displacement of the cg between one- and two-passenger loads. A longitudinally stable cg is an important consideration for a three-wheel design, but it is not particularly important with a four-wheel platform.

The "1+1" concept is also an option. The vehicle might be designed essentially as a one-place machine with a minimal second seat. GK Industrial

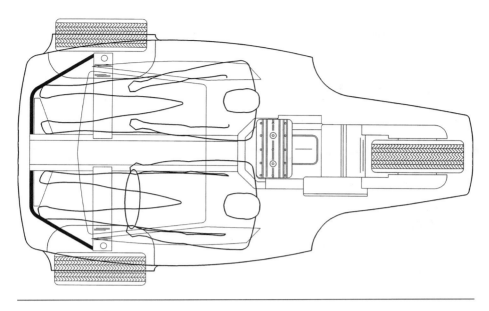

FIGURE 2.22 Side-By-Side Seating. Sitting two abreast limits longitudinal cg displacement between one- and two-up loads.

Design Associates in Japan produced a 1+1 microcar in the mid-1970s. Several mockups were built, and a final prototype was designed under the code name of GK-0. From the Japanese perspective, space is a valuable resource. This idea ultimately leads to the concept of "small, therefore good," which is a theme that runs throughout the range of Japanese products. The GK-0 utilized a slightly staggered seating arrangement that produced a sense of side-by-side seating without the crowded feeling. The resulting vehicle package was exceptionally space efficient (Figure 2.23).

Rumble seats and longitudinal bench seats that allow occupants to squeeze together and make room for an extra passenger are other options, but safety issues will undoubtedly arise with any design that uses highly unconventional seating. Occupant crash protection must be equally provided for all potential passengers. Providing crash protection for an occupant in a minimal jump seat or a rumble seat is likely to present difficult engineering challenges. In this regard, an abbreviated tandem arrangement offers interesting possibilities. The rear passenger in a 1+1 tandem layout could be protected by an energy-adsorbing device built into the back of the driver's seat (Figure 2.24).

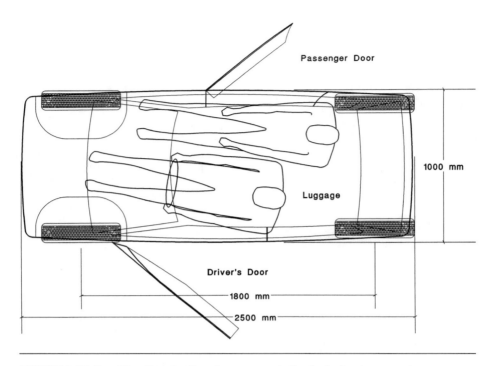

FIGURE 2.23 One-Plus-One Seating Arrangement. By displacing the second occupant slightly toward the rear, a relatively narrow vehicle provides adequate shoulder room.

Seat construction and even the concept of movable seats are features that might be reevaluated. Quincy-Lynn's Tri-Magnum, Trimuter, and Centurion were designed with seats formed into the body in a fixed position. Upholstery was attached directly to the body over high-density foam. This gave the interior an attractive molded appearance. Seats were form-fitting and comfortable, even though they lacked a traditional suspension system. When the seats are fixed, however, steering, brake, and throttle controls must be mounted to a movable carrier in order to provide adjustment.

Ingress and Egress

Ingress and egress present interesting opportunities for distinctive design. The gull-wing door and the clamshell canopy are attractive and practical alternatives to conventional doors. When a vehicle opens like a jet fighter, an otherwise bland design can become exciting.

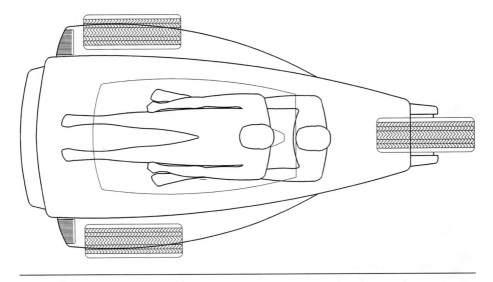

FIGURE 2.24 A Condensed Tandem Arrangement. Space is efficiently utilized in a one-plus-one layout by allowing the rear passenger's legs to extend along the outside of the driver.

On a practical level, a very small and low-to-the-ground vehicle presents difficulties for the conventional side-entry door. Gull-wing doors provide greatly improved access, and they usually can be fully opened even in confined parking spaces. Also, lightweight doors that end several inches above the bottom of the body can be removable to create a convertible effect. Removable gull-wing doors were featured on the three-wheel VW Scooter. But lacking a practical justification, gull-wing doors can stand on the basis of aesthetic appeal alone. The Mercedes 300SL, the Bricklin, and the Delorean are excellent examples of exotic vehicles that were made more intriguing by the appeal of gull-wing doors. In GM's Ultralite, they are both practical and visually striking (Figure 2.25).

When it comes to providing easy access to a very low vehicle, the clamshell canopy is one of the most practical and attractive, albeit highly unconventional, approaches. Bob McKee used the clamshell on his beautifully executed prototype electric car, the Sundancer. Both GM and Ford have used it on several concept cars. With this arrangement, the entire top of the vehicle hinges out of the way to expose a gaping access to the interior. One simply steps over the side of the body and slides into place (Figure 2.26).

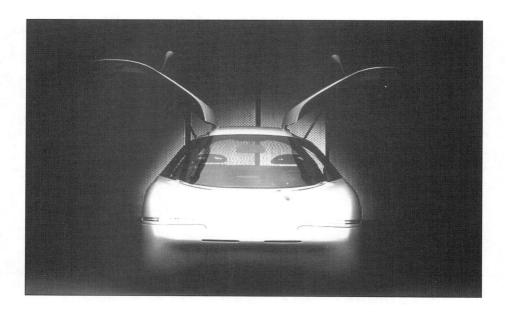

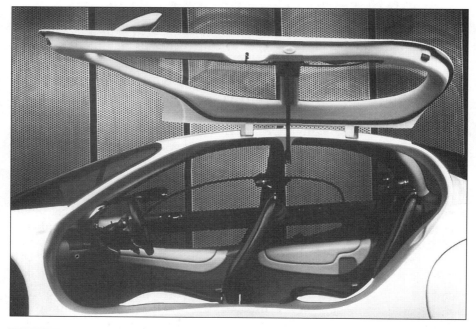

FIGURE 2.25 Ultralite's Gull-Wing Door. GM's concept car Ultralite is fitted with large gull-wing doors that provide wide, unobstructed entry into the car's interior.

Courtesy: General Motors Corp., www.gm.com

FIGURE 2.26 *Quincy-Lynn Enterprises' Tri-Magnum.* Quincy-Lynn's Tri-Magnum used the clamshell canopy for both practical and styling purposes. Occupants are seated side-by-side. www.rqriley.com

Perhaps the most important decision with the clamshell design is determining the most appropriate cut-line: the point at which the canopy separates from the body. Extending the canopy deeply into the body sides has the benefit of reducing the step-over height, but a canopy cut too deeply into the body can compromise structural integrity, create a sealing problem around the perimeter, and result in a heavier canopy. Deep canopy sides can also extend into the free access space and actually interfere with entry. Ideally, the canopy should separate at the beltline or further down along an appropriate character line. The perceived difficulty in stepping over a high body side is largely overrated in the author's opinion. Tri-Magnum's canopy was cut at the beltline, which resulted in a high 711-mm (28-in.) step-over. The high sill gave the car a strong race car flavor that enhanced the design's sporty feel (Figure 2.26).

Gull-wing and clamshell systems also have disadvantages. With both systems the weight of the door must be counterbalanced for easy opening. In addition, the space needed to house retractable windows is usually quite limited. Windows must often be removable, hinged, or sliding—all less than

FIGURE 2.27 Three-Place Sport and Recreational Vehicle. Doors pivot upward and forward in this sport and recreation vehicle (SRV) concept by industrial designer Antanas Kirov Yanev. Notice that occupants are arranged with the driver in the center and the two passengers to the sides and behind. Yanev's fuel-cell-powered SRV is equipped with adjustable ride height for increased ground clearance while off the road.

Courtesy: Antanas Kirov Yanev, industrial designer

ideal solutions. Another problem centers on safety during a rollover. A door that opens upward will not open when the vehicle is resting on its roof.

Gull-wing and clamshell doors are not the only options. Subaru's EM-1 and Jo-Car dispensed with doors entirely, and in their place installed hinged side-beams designed to keep occupants in and other vehicles out. Antanas Kirov

Yanev, a Bulgarian industrial designer, designed the vehicle shown in Figure 2.27 so doors swing up and forward.

Modular Design

Modular design is a concept that has been permeating the universe of consumer and industrial products for many years. Today, products are increasingly designed around a basic reconfigurable platform to save costs and broaden product lines. Styling often emphasizes, rather than disguises, a product's modularity. In this way, the designer uses the language of form to speak to the consumer about the product's capabilities. Modularity also suggests a type of efficient engineering that can imply superior design. Concept drawings of the Esoro E301 electric urban car developed by the Swiss firm Esoro-AG in Zurich shows a basic platform that can be reconfigured to produce as many as five different vehicles, three of which are shown in Figure 2.28.

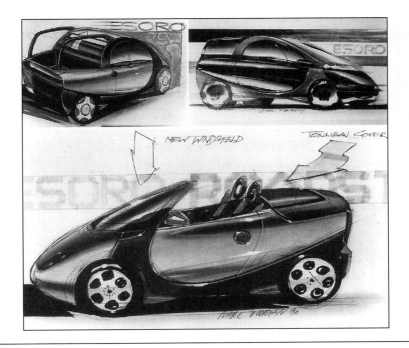

FIGURE 2.28 *Esoro E301 Concept Drawings.* The Esoro E301 was developed by Esoro AG, Switzerland. The Esoro E301 utilizes a basically two-seater platform that can be configured as a pickup or van, or as a coupe, hardtop, or even a four-seat passenger car.

Courtesy: Esoro AG, www.esoro.ch/esoro/

FIGURE 2.29 The Totally Modular Vehicle. An illustration created by student Jimmy Chow while he was a student at the Art Center College of Design shows an idea for a totally modular automobile. Front suspension, occupant compartment, and rear power pod are self-contained modules with controls that interface electronically, rather than mechanically. The vehicle can be reconfigured for different markets and operating environments by combining a variety of different modules.

Courtesy: Jimmy Chow, the Art Center College of Design, www.artcenter.edu

The modular aspect of GM's Ultralite (see Figure 2.4) is emphasized by the strong cutline around the rear. Alternative personal mobility products might incorporate reconfigurable body styles and alternative drive packages to let users select power systems according to local conditions and individual needs. A totally modular vehicle might consist of a central passenger module, a front suspension/steering module, and a rear power module. Modules could then be interchanged to customize the vehicle for different markets or different consumer preferences. Figure 2.29 shows a totally modular vehicle designed by Jimmy Chow while he was studying at the Art Center College of Design in Pasadena, California. Today's drive-by-wire technologies bring concepts like this one within the realm of practical feasibility.

Flexible/Plastic Body Panels

Conversion to plastic body components has been under way for many years and is gaining momentum. Major incentives include the following:

- Corrosion and damage resistance
- Weight reduction on the order of 30 to 50 percent
- A 30 to 50 percent reduction in annual tooling costs
- Low-cost facelifts
- Increased styling freedom
- Conducive to just-in-time manufacturing at or near assembly plants

Plastic components account for approximately 15 percent of the mass of today's automobile, and the applications are growing each year. Plastics are especially compatible with the concept of ULM alternative cars. Outer body panels, particularly at the front and rear, and even the fenders, might be flexible urethane or a similar compound that will resist damage from minor contacts with other vehicles. Honeycomb structures and rigid foams can create light, energy-adsorbing substructures for appropriate energy management at key locations.

The use of molded foam/vinyl laminates and integral skin foams is also being expanded to include even the seating areas (Figure 2.30). It is possible to replace conventional fabricated upholstery with thermoformed panels made of laminates or integral skin foams. New materials and processes can simultaneously reduce manufacturing costs and make products more aesthetically appealing.

Electronic Automobile

Once microprocessor technology is introduced, a new world of opportunities become available to the designer. Electronics open the way for new approaches to systems management, driver information, collision avoidance, and active controls that were not possible before (Figure 2.31). In addition, significant psychological benefits are also intrinsic to the nature of modern electronics. Microprocessor technology offers a precise and visually impressive interface with the operator that is uniquely its own. Vehicles now come to life in ways that go far beyond the task of providing information and control. Subtle qualities of sight and sound create a sense of vehicle competency, adroitness, and even luxury. Audible communication, head-up displays, and voice activation and identification are tools for adding perceived value to the product and more fully executing the vehicle theme. Active controls such as steer-by-wire and brake-by-wire replace mechanical connections with electronic terminals making it possible to more fully implement the concept of modular design. Ultimately, vehicle electronics may supplant occupant-protection systems as collision avoidance and advanced vehicle control systems virtually eliminate automobile crashes.

FIGURE 2.30 *Interior of Jeep Willys Concept Vehicle.* Extensive use of thermoformed panels and integral skin foams reduces manufacturing costs while enhancing the high-tech look of the interior of DaimlerChrysler's Jeep Willys concept vehicle.

Courtesy: DaimlerChrysler Corp., www.daimlerchrysler.com

"Active Steering" has been around for some time in various forms. In the early 1990s, Saab developed the steer-by-wire system shown in Figure 2.32. A joystick replaces the conventional steering wheel, and steering is accomplished by an electrohydraulic steering servo. Mercedes F300 Life Jet uses a similar system (Figure 2.34). Other versions utilize a conventional steering yoke or wheel connected to an electromechanical power steering unit. Steer-by-wire could improve vehicle safety by eliminating the steering column.

Intelligent vehicle highway systems, advanced vehicle control systems, navigation, and communication electronics can work together to create attitudes and perceptions that are likely to eliminate any remaining perceptual association between vehicle size and vehicle quality. Like fine watches and personal

SAFETY & CONVENIENCE	POWERTRAIN	ENTERTAINMENT& COMMUNICATION	DRIVER INFORMATION	BODY CONTROL
Climate Control	Dynamic Mounts	Noise Reduction	Head-up Display	Multiplex Wiring
Cruise Control	Electronic Camshaft	Cellular Telephone	Navigation/Maps	Inter-Module Network
Keyless Entry	Ignition Timing	CB Radio	Digital Gauges	Body System Diag.
Light Reminder	Spark Distribution	Digital Audiotape	Diagnostics Display	Smart Power Drivers
Memory Seat	Fuel Delivery	Fax	Service Reminders	Dynamic Ride Control
Sensory Wiper	Turbo Control	Television	Clock	Active Suspension
Auto Door Lock	Emission Monitoring		Trip Computer	Load Leveling
Headlight Dimming	Transmission Shift		Intelligent Highways	Anti-Theft
Traction Control	On-Board Diagnostics		Collision Avoidance	Electronic Steering
Antiskid Braking	Operational Adaptation		Drowse/DWI Alert	Electronic Throttle
Load-Sensing Braking	Energy Recovery			Electronic Braking
Window Control				Electronic Muffler
Air Bag Restraints				
Advanced Vehicle				
Control Systems				
Voice Recognition				

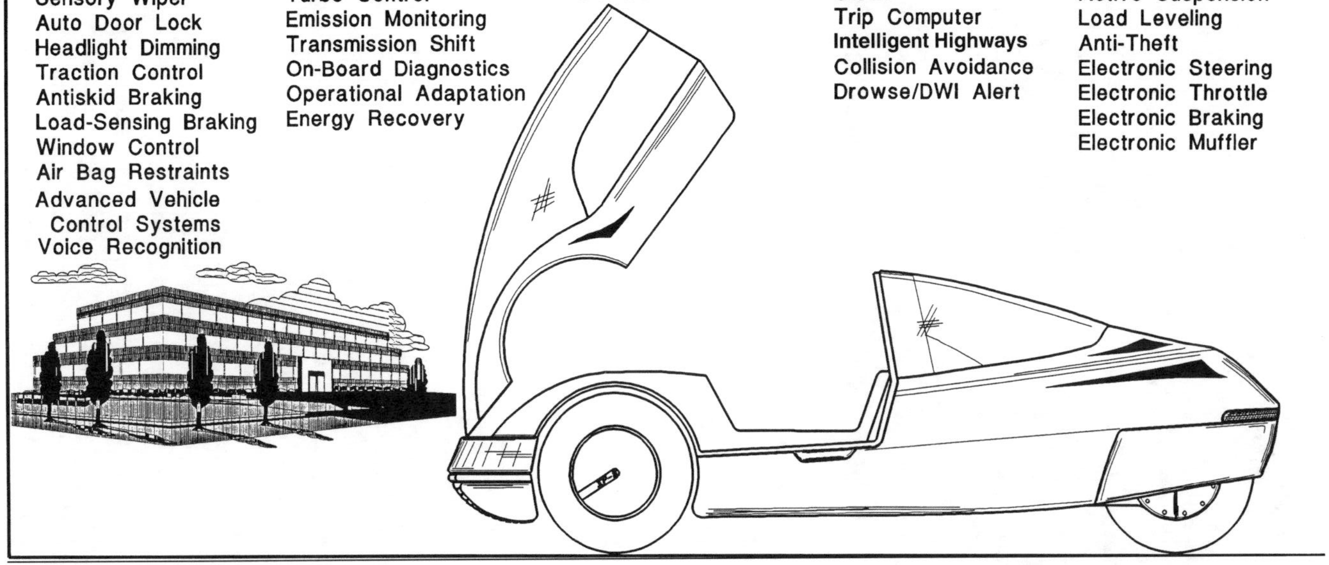

FIGURE 2.31 The Electronic Automobile

FIGURE 2.32 *Active Steering from Saab.* Electrohydraulic joystick steering system relies on system redundancy and self-testing and monitoring to provide extreme reliability.

Courtesy: Saab Cars, USA, Inc., www.saabusa.com

computers, compact vehicle packaging can imply superior engineering and high product quality.

Three-Wheel Platform as a Marketing Tool

Although three-wheelers are inherently less costly and lighter, the layout is normally seen as a detriment in the marketplace. In the author's experience, however, the three-wheel platform can actually become an asset with the types of alternative personal mobility products and market positioning under discussion here. From a marketing perspective, the three-wheel platform can provide a natural distinction between new types of mobility products and the larger vehicles that consumers identify as automobiles. This differentiation gives designers much greater freedom to make bold design statements. Designs that may be too unconventional for an automotive product can become refreshingly appealing when consumers experience them as belonging to an entirely new category of products. Advertising messages can guide consumers to a vision of where the new vehicle fits into their personal mobility options. The platform helps convey the message that this is not just another version of an automobile.

FIGURE 2.33 XR3 Tilting Three-Wheel Vehicle Concept. The author's XR3 tilting three-wheel vehicle concept is an example of the extent to which product differentiation is possible with the three-wheel platform. Marker sketches like this one by artist Bill Ortis provide early-stage visualizations of a prospective vehicle's personality.

The ultimate demise of England's Reliant Robin is a case in point. Despite regulatory distinctions in the United Kingdom, the Robin was positioned as an automobile having one less wheel, and it was marketed on the basis of its tax- and cost-saving benefits. In this way, it was similar to the environmentally friendly personal watercraft that became redefined as a mini-boat earlier in the chapter. In contrast, the XR3 design study (Figure 2.33), the Mercedes' F300 Life Jet concept vehicle (Figure 2.34), and the Tempest (Figure 2.35) are positioned at the opposite end of the design spectrum. Mercedes's F300 is essentially an undefined type of personal mobility product. If Mercedes chooses to produce a vehicle like the F300, the company would be free to guide perceptions with advertising messages that speak to lifestyle qualities that consumers want to own. The Tempest illustrates a styling theme that is essentially a concept for a type of hybrid three-wheel motorcycle, but such extreme departures are unnecessary in order to provide a sense of differentiation. Transit Innovations' Project 32 Slalom (Figure 2.36), for example, appears quite automotive-like while still standing apart from the conventional automobile.

FIGURE 2.34 Mercedes F300 Life Jet. Mercedes introduced the F300 concept vehicle at the 57th Frankfurt International Motor Show. This high-technology three-wheeler combines the cornering dynamics and feel of a motorcycle with the safety and comfort of a car. The body and front wheels automatically lean into turns when cornering. Notice the aircraft-style steering wheel, which allows full-lock turns with only a 90-degree turn of the wheel.

Courtesy: Mercedes-Benz USA, www.mercedes-benz.com

FIGURE 2.35 Tempest Tilting Three-Wheel Vehicle Concept. The Tempest tilting three-wheel vehicle concept is such a radical departure from automotive design that it actually becomes more like a hybrid motorcycle, along the lines of the XR3. Tempest was designed by Paul Morrow, an automobile designer who lives in the UK.

Courtesy: Paul Morrow, UK automotive designer

FIGURE 2.36 Transit Innovations' Project 32 Slalom. Transit Innovations' Larry Edwards wanted a design for Slalom that would closely follow traditional automotive design. In contrast to the F300 and Tempest, the nonautomotive distinction of the three-wheel platform has been subdued with Slalom. For more information, visit the author's website at www.rqriley.com.

Courtesy: Transit Innovations

Potential Market for Alternative Cars

The best that can be done regarding market potential is to speak in terms of estimates and general market demographics. Estimating market potential for new products is especially difficult because both buyers and sellers lack experience. There is no sales history on which to base an analysis. Consumer surveys often produce inaccurate results because consumers have no product experience to guide them, and the waters get extremely murky when the product is hypothetical rather than actual. Market potential cannot be determined in the abstract. Much depends on consumer response, not to the idea, but to the actual product.

Demographic factors and trends, as well as consumer preferences and attitudes regarding conventional cars, are probably inapplicable when radically different transportation products are discussed. Undoubtedly, such information can be helpful, but traditional market analyses are likely to be off the mark in estimating a product's success potential. Sub-cars and nonautomotive-type products, for example, are so far removed from conventional products that automobile demographics and automotive-based projections of market potential may be useful only in the most general way. Still, it will be helpful to review a few demographics that might be expected to have an impact on market potential.

Vehicle Ownership and Use Trends

Households are becoming smaller and people are driving more (Table 2.4). This trend is consistent throughout the industrialized world. In the United States from 1969 to 1990, persons per household decreased by 20 percent and vehicle miles traveled increased by approximately 22 percent. Households have fewer members, they have more cars available to them, and they are making more trips and driving alone more often.[12] In 1990, average vehicle occupancy was 1.6 persons during all trips, and 1.1 persons during trips to work. The NPTS report also reveals that in 1990, 88.1 percent of person-miles traveled (pmt) in the United States was aboard private vehicles and 67.1 percent of all vehicle-miles traveled (vmt) occurred with one occupant aboard. Approximately 83 percent of the nation's driving was done with only two occupants aboard.

ULM Vehicle and Multivehicle Households

Sales of limited-utility alternative mobility products will undoubtedly come from that segment of households with multiple vehicles. In single-vehicle

TABLE 2.4 U.S. Household Travel Rates in Miles (Kilometers inserted in parentheses by the author)

	1969*	1977	1983	1990
Annual vehicle trips	1,396	1,443	1,486	1,702
Annual vehicle-miles traveled	12,414	12,035	11,739	15,096
	(19,974)	(19,364)	(18,888)	(24,289)
Annual person trips	2,322	2,808	2,628	2,707
Annual person-miles traveled	22,465	24,919	22,802	24,816
	(36,146)	(40,947)	(36,688)	(39,929)
Persons per household	3.2	2.8	2.7	2.6

Source: USDOT, Federal Highway Administration, www.fhwa.dot.gov
*In the 1969 survey, only auto and van trips were collected as private vehicle trips. In 1977, 1983, and 1990 surveys, the definition of private vehicle was expanded to include pickups and other light trucks, recreational vehicles, motorcycles, and mopeds.

households, it is presumed that a single car has to serve the broadest possible utility. Therefore, a multipassenger, conventional car is the car of necessity. Today, multicar households are common, and the type of car purchased often depends on preference, rather than utility. Many consumers have different models for different uses and different occasions. When two or more cars are available, criteria other than maximum utility may apply to the second, third, and fourth vehicle. Statistics indicate that nearly 60 percent of U.S. households already have at least two vehicles, and about one-third of that group have three or more. In Europe, about 20 to 30 percent of households have two or more vehicles. A limited-utility vehicle would therefore target this market. Table 2.5 shows the acceleration in multivehicle ownership in the United States.

TABLE 2.5 Households by Number of Vehicles

Vehicles Available	Percent in 1969	Percent in 1977	Percent in 1983	Percent in 1990
Zero	20.6	15.7	13.5	9.2
One	48.8	34.6	33.8	32.8
Two	26.6	34.4	33.5	38.4
Three +	4.6	15.3	19.2	19.6

Source: U.S. Department of Transportation, Federal Highway Administration, www.fhwa.dot.gov

The potential market is likely to be some percentage of total multivehicle households. NPTS figures combine all light-duty vehicles including vans and pickup trucks. There are fewer multiautomobile households, and consumers are probably less likely to exchange a personal van or pickup for an ultralight alternative car. Today, light trucks account for about half of total new-car sales in the United States. Sixty percent of U.S. households have two or more cars, according to 1996 estimates by Urban Decision Systems. Household income also has a large effect on the degree to which households may be considered as potential new-car purchasers. R. L. Polk & Co. lists 47.9 percent of all multicar households as having an annual income of $50,000 per year or more. The potential market is therefore reduced to about 10.72 million, or about 11.5 percent of the total 92.5 million U.S. households.

The potential market for an alternative personal mobility product is not necessarily limited to new-car purchasers. Commenting on the potential U.S. market for micro-mini cars, John Hemphill of J.D. Power and Associates estimated that 25 percent of buyers would be diverted from the used-car market.[13] This assumption is based on the idea that a micro-mini car would be less expensive than a standard car. Many second cars are purchased secondhand, primarily because of lower used-car prices. The degree to which used-car buyers are attracted would then depend on the household's typical budget for a used car, as compared to the price of an alternative car. Research has shown that price is extremely important, and consumers will often choose a less expensive model even if long-term operating costs are higher.

Other variables include the degree to which household members are willing to swap vehicles to accommodate those days on which an alternative car does not fill a member's driving needs. For example, vehicle image is important to nearly everyone, and in at least some of these households, the second car may be a teenager's hot rod or jacked-up truck. Consequently, the idea of swapping vehicles could conflict with someone's self-image.

Effect of Reduced Seating Capacity

Limited seating capacity is generally believed to have a strong, negative impact on a vehicle's potential market. Although two-seater vehicles will accommodate approximately 90 percent of vehicle trips, consumers prefer four-passenger cars by a large margin. Studies by J.D. Power and Associates place the ratio at about 5 to 1 in favor of four seats. Other vehicle attributes, however, can combine to outweigh this preference, and certain attributes may be intrinsic to the vehicle theme—the quality that attracts consumers in the first place. The "Z" car provides a real-world example of how consumer preferences may not apply in specific cases. Nissan's "Z" car was offered in both two-passenger and four-passenger versions, yet two-passenger sales far surpassed four-passenger sales. With the "Z" car, seating capacity is closely associated with vehicle theme.

When considered in the abstract, it must be assumed that reduced seating capacity will limit the universe of potential purchasers. The most effective tool for minimizing this adverse effect would be a strong vehicle identity (theme) that separates an alternative personal mobility product from comparison with the standard car. If consumers perceive the new product as essentially another version of a standard car, reduced seating capacity is likely to have a strong negative impact on the potential market. If it is perceived as an entirely new type of transportation product, the alternative car's seating capacity could become unlinked from traditional expectations. Seating capacity might then be perceived as one of the defining attributes of the vehicle type, just as it is with sports cars.

Market Impact of Battery-Electric Power

Before the introduction of battery-electric cars, carmakers commissioned studies in an attempt to forecast sales. Two-car households are normally seen as the consumer base for battery-electric cars. When defining the initial market, however, additional criteria were normally applied because of the electric car's limited range and the need for a recharging site. Researchers at the Institute of Transportation Studies, University of California at Davis, applied the following criteria to define potential purchasers:[14]

1. The family must own their own home.
2. The home must have a garage or secure carport for overnight charging.
3. There must be two or more vehicles in the household.
4. There must be at least one person in the household who drives less than 70 miles (113 km) roundtrip to work.

Based on these criteria, the potential market included 28 percent of U.S. households. Vehicle occupancy was not considered. This segment was then reduced by the $50,000 annual income criterion in order to filter out potential new-car purchasers. The remaining segment then totaled 13 percent of U.S. households. When consumer preferences having to do with the electric car's reduced range, limited luggage space, reduced power, and higher first costs were applied, market penetration dropped to approximately 1 percent.

This finding was also supported by an unpublished 1991 study by Ford Motor Company on the potential market for electric cars, which indicated a 1 percent initial market penetration as well. For their study, Ford defined the electric car as being 100 percent emissions free, having a top speed of 75 mph (121 km/h), with 50 percent less space, reduced range, and costing $3,000 more than a conventional car.

In the United States, technical constraint studies (the degree to which trip lengths limit electric cars as a transportation option) show that battery-electric cars can serve a high percentage of consumer driving needs; however, the market limitations of electric cars are linked more to consumer perceptions than to vehicle technical capabilities. Consequently, market estimations ultimately favored stated preference studies in which participants make hypothetical choices from sets of product attributes. Although many consumers do not know the range of their present cars, consumer reaction is strongly negative to limitations in vehicle range. Disutilities of $10,000 to $15,000 have been reported by some researchers as resulting from limited range and long recharge times.[15]

Seating capacity is also closely associated with perceived limitations, rather than actual limitations. One significant difference, however, is the fact that limited range can leave the occupants stranded and without a car at all. Limited seating capacity might be an inconvenience at times, but limited range represents the potential for total failure of the automobile. Consequently, the range limitation of battery-electric vehicles is a powerful disincentive in the marketplace.

Although successes have occurred in Europe, the North American experience with battery-electric vehicles has been disappointing. Table 2.6 shows North American battery-electric car sales through the end of year 2000.

It is important to understand that the sales figures in Table 2.6 are for battery-electric cars that are positioned as direct competitors to conventional cars. On

**TABLE 2.6 Battery-Electric Light-Duty Passenger Vehicles
(Major OEM Sales/Leasing 1996–2000)**

Chrysler EPIC	Ford Ranger EV	GM EV1	GM S-10 Electric	Honda EV Plus	Nissan Altra EV	Toyota RAV4-EV	Total
206	1,259	853	500	300	110	789	4,017

Source: Electric Vehicle Association of the Americas (EVAA), www.evaa.org

the basis of cost versus utility, and in a world of large multipurpose automobiles and cheap motor fuel, conventional IC engine–powered cars have a large advantage in the marketplace. A shift in vehicle personality and positioning could help take electric vehicles out of direct comparison with conventional vehicles.

Regardless of their environmental concerns, consumers are naturally self-serving when it comes to spending $20,000 to $30,000 on a product that must fill emotional needs, as well as practical transportation needs. To paraphrase Dr. Paul MacCready's remark at the 1996 World Car Conference: "We all think it's a good idea for our neighbors to switch to environmentally friendly alternatives, as long as we can keep on driving our gas guzzlers."

SUBLIMINAL MESSAGES AND NEW CONSUMER VALUES

Much has been written about new roles for industry, the responsible corporation, and new consumer values. The message a product projects goes beyond simple utility to an abundant world of implied meanings. In this new age, the corporation has new responsibilities, and its products have new meanings that were not even considered just 20 or 30 years ago. Today's consumer is aware of the environmental effects of technology and wants to know about the integrity of the institutions that manage the technology. A product is no longer just a product, but it speaks about the values of the people who created it. When consumers purchase a product, they also subscribe to the product's implied values.

An automobile expresses company values in the language of its design. Its products tell consumers whether the company values quality and durability or whether they think in terms of throwing things away and making more. Presumably, a well-made product comes from a solid company. Abundant safety features imply a concern for the welfare of consumers. A company that is concerned about the welfare of the human species will likely build products that are kind to the environment. Stereotyping is a powerful byproduct of prevailing attitudes. Consumers will increasingly ask: "Does this product represent values to which I want to subscribe, or am I being sold on something that does not match my belief system?"

Likewise, there are also subliminal meanings in consumer attitudes. Consumers naturally want to protect the environment, but they do not want to sacrifice. People are more affluent and they want to be healthier, happier, and more free. Self-discipline is not the wave of the future, and consumers are unlikely to become environmental purists and sacrifice their own lifestyles for the planet. To paraphrase Dormer's characterization, regardless of the emerging green consumer, no one will make a lot of money selling hair shirts. People are concerned about consumption and pollution, but they want manufacturers to resolve problems in ways that merge with lifestyles.

The automobile market is especially sensitive to the self-indulgent factor. A popular theory of the 1980s suggested that educating consumers about the costs of energy and the benefits of conservation would ultimately convince them to accept more inconvenience and expect less from their cars; however, econobox cars, ride sharing, and public transportation were largely unsuccessful even immediately following the 1970s energy crisis. Buying habits and studies of psychological motivations show that conservation and sacrifice are not marketable attributes. Green market appeals must carry a different message. Instead, consumers must be motivated by the inherent desirability of the product. Conservation must add in some way to the product's attractiveness. A green orientation must be an upscale attribute—a chic product theme that in no way speaks to limited utility or reduced aesthetic appeal. Astute manufacturers will listen to consumers' needs and concerns and then package products in ways that appeal to consumers' fantasies. Consumers must be able to satisfy their moral and social concerns in the process of filling their more basic ego needs.

A NEW PARADIGM FOR PERSONAL TRANSPORTATION

The author visited a class at the Art Center College of Design in Pasadena, California, where students were working on their ideas for the "car of the future." One of the renderings, a beautiful work-in-progress of an urban car design, stood out from the others. When asked about his design, the student released an avalanche of information about an unseen world to come. He talked for several minutes about future lifestyles and the new generation's awareness of the environment and the planet's limited resources. He described how the desire for fun transportation would merge with environmental concerns and ultimately guide the design of vehicles. Instinctively, this young creator was setting the context for his design. He was creating a perspective from which to experience his vehicle concept—a set of conditions and attitudes, or a paradigm within which his design would be appropriate and attractive. This ability to see through different eyes can be a powerful tool for reshaping perceptions and opening new horizons.

The paradigm, once accepted, determines how ideas and designs will be perceived and how they will fit into the world. It determines which ideas are valid, and it sets guidelines for interpreting data. A new world view can lead creative work in new directions, and it can reveal products and markets that may not have been apparent before. There is a story about two market researchers who were sent, independently, to one of the world's less developed countries by one of the world's largest manufacturers of shoes. Their instructions were to report back on the potential market for their product. When the first researcher arrived and observed the conditions, he immediately faxed the following message to corporate headquarters: "No market here. Nobody wears shoes." Within a few hours, headquarters received word from the second researcher. His message read: "Great market here. Nobody has any shoes."

This humorous story about diametrically opposed conclusions from identical observations illustrates the degree to which perspective can influence interpretation. By changing focus, interpretations can be turned upside down and inside out. Of course, neither researcher had adequate information, but determining which of their conclusions would ultimately prove correct is unimportant to this discussion. The focus here is on individual perceptions and how a change in perspective can reveal different and even seemingly contradictory facets of the same phenomenon.

A paradigm shift, which Joel Barker described in "Future Edge" as a switch to a different framework of rules, can sharpen the judgment of the entrenched practitioner, but more important, it can give birth to electrifying new visions. Einstein's theories introduced a new set of rules that opened new frontiers in human understanding. Before having seen the other side of the mountain, Einstein's brilliant revelation described its treasures in glorifying detail. Singularities, quarks, mesons, and the connectivity of time and space are real phenomena that were first visited by the mind's eye before they were seen in the physical world. The idea that they existed at all flowed from the new scientific paradigm. The objects and relationships that were ultimately revealed through Einstein's paradigm shift would have remained forever invisible through the filters of Newtonian physics.

By adopting new filters, new opportunities and new product ideas can begin to take shape. Through the filters of the existing paradigm, significantly downsized vehicles are generally envisioned as unsafe, motor-scooter cars that have little appeal in the marketplace. But new designs for alternative personal mobility products—products that have large market and profit potential for their creators and provide safe transportation to their purchasers—can emerge through new ways of seeing.

Today, attitudes tend to cast extremely small and lightweight vehicles as low-cost, utilitarian versions of full-size cars. Once this premise is accepted, vehicle attributes consistent with the vision naturally emerge and an outline of market potential, profitability, and even vehicle safety then follows suit according to the core idea. These details can quickly change when the hypothetical vehicle is seen from a different perspective: as an essentially new and separate vehicle type with its own personality, its unique place in the transportation system, and its individual potential in the marketplace. By emphasizing safety features, visually impressive driver information systems, advanced vehicle control and crash-avoidance systems, and distinctive and attractive vehicle layouts and styling, alternative cars can be positioned as superior transportation products.

An outmoded perspective of the recent past failed to see trucks as widely appealing alternatives to conventional passenger cars. But today, half of new-car sales are trucks of a different nature. A low-end alternative car in a high-end market is likely to produce disappointing sales. When the vehicle theme and the market are correctly matched, however, both sales and corporate profits are likely to soar.

CARS AS A SUBSYSTEM
OF THE TOTAL TRANSPORTATION SYSTEM

The foregoing discussion focuses on consumer vehicles and explores the appeals necessary to attract a large segment of consumers. But like the story of the researchers who were sent to evaluate the market for shoes in a less-developed country, a different perspective can turn the problem upside down and inside out.

Consider that we now have a somewhat haphazard personal transportation system wherein private cars, which are designed mainly for their emotional appeal to lay consumers, are responsible for 91 percent of total pkm traveled. Contrast this perspective with the idea that personal mobility is a fundamental commodity in a modern society. Imagine the problems that would likely arise with a commercial transportation system consisting of vehicles chosen mainly on the basis of emotional appeal. In reality, the concept of personal transportation as a consumer item may become outmoded. Assuming that transportation is an essential commodity, it may be more appropriate to see long-term solutions in terms of an integrated systems approach. When the paradigm is redefined in this way, it suggests that our present system of consumer-oriented private cars may be out of step with the realities of the 21st century.

The assumption presented earlier in the chapter, based on the consumer market model, is that purchasers must be motivated through traditional

FIGURE 2.37 Station Cars. These prototype electric cars by Horlacher AG in Switzerland are examples of the types of vehicles that would be available at station car depots. Cars will be clean, charged up, and ready for local trips by those using public transit systems.

Courtesy: Horlacher AG, Switzerland

appeals in order to realize the energy and environmental benefits of alternative personal mobility products. But this assumption evaporates when personal transportation is seen as a commodity and the network of individual vehicles is seen as part of the total transportation system. With traditional mass transit systems, vehicles are selected on the basis of cost effectiveness, not on the basis of style, image, and emotional appeal. The same shift in values occurs with respect to cars when they are redefined as a subsystem within a citywide transportation system. In the United States, this concept is being developed and demonstrated by the National Station Car Association.[16]

A network of station cars would consist mainly of a fleet of small, energy-efficient vehicles that are, in effect, an extension of the mass transit system. Larger vehicles are also envisioned for those who need them. Each local mobility system, comprising station cars of various sizes, would be individually tailored to the particular transportation needs of the community (Figure 2.37). Station cars would be available for transport to the mass transit station, and at the other end, for transport to various local destinations. The fleet, comprised mostly of battery-electric vehicles, would be cleaned, serviced, and charged at the station renting them. Initially, rental stations would be located at mass transit stations. But ultimately, stations would be set up at university campuses, convention centers, airports, and residential and commercial complexes. Station car users would have the option of selecting from various sizes of vehicles, depending on the needs of the particular trip. In this way, the system provides the ultimate flexibility in vehicle capabilities, while allowing for the use of the smallest, lightest, most energy-efficient vehicle possible. Underutilization of large multipurpose cars is therefore minimized.

Although station cars are significantly more energy efficient and environmentally friendly, they also make mass transit more convenient and usable. With station cars, commuters no longer have to worry about getting to the transit station, then from the destination station to work or to a local shopping complex. So mass transit systems become more efficient as well. By encouraging the use of mass transit, the energy and environmental benefits of station cars are compounded.

For many of us, the idea of renting a car only when needed may initially present an emotional roadblock. But when the economics are considered, particularly within multivehicle households as a replacement for the second or third car, the idea makes even better sense. Assuming 10,000 miles per year and an average speed of 35 mph, operating time for a car calculates to 286 hours per year, or 3.26 percent of the total hours available for use. During the other 8,474 hours, the car is simply parked and taking up space. But one pays for it all the same, whether it's in use or not. Compared to other household appliances, the car is used less and costs far more. The National Station Car Association's figures on the relative costs per hour of use are shown in Table 2.7.

TABLE 2.7 Cost of Various Household Appliances

Item	Annual Hours in Use	Percent of Total Hours	Cost per Hour
Car	286	3.26	$14.50
Refrigerator	8,760	100	$0.03
Clothes Dryer	156	1.8	$0.58
Electric Heating	2,333	26.6	$0.33
House (used 24 hr./day)*	8,760	100	$0.25
House (used 14 hr./day)**	5,110	58.3	$0.43

Source: National Station Car Association, www.stncar.com
*Assumes the house is in use even when no one is home.
**Assumes the house is not in use when no one is home.

The cost of instant mobility turns out to be many times greater, on an hourly use basis, than any other common utility. Even if the costs to the environment and balance of payments deficits (due to imported energy) are disregarded, and only the direct costs of owning and operating an automobile are considered, station cars turn out to be far less costly than privately owned multipurpose vehicles.

In its most conservative form, a network of station cars could replace a significant portion of a society's second or third family vehicles. But in its most ambitious form, station cars might ultimately supplant private cars altogether, in which case transit systems would be more widely deployed and far more extensively used. Although the energy and emissions benefits of a comprehensive system are obvious, such a scenario could also have far-reaching impacts on carmakers and society as a whole.

Vehicle design and manufacturing would naturally shift from today's consumer-market model to a commodity-supply model. Carmakers would no longer have to battle it out in the marketplace in an effort to attract the largest number of consumers. Dollars now spent on advertising and cosmetic model changes could be redirected into producing better-quality vehicles. Platforms and components could become more standardized, and tooling could be amortized over a greater number of units. The focus in vehicle design would naturally shift from cosmetics and styling to building greater functionality and cost effectiveness into products. Emerging intelligent transportation systems that now stretch the limits of consumer affordability would become more affordable when vehicles are working and earning revenue throughout a greater portion of each day. Simply doubling a vehicle's hours in use, from 286 hours to 572 hours per year, for example, cuts operating expenses nearly in half.*

With comprehensive transit systems in place, large multipurpose cars might ultimately disappear from communities and urban areas. Shopping might then be done largely on the Internet and goods delivered to community distribution centers. From there, packages would be routed to individual dwellings on automated conveyors or on small electric runabouts. Consider the impact of this scenario on the design of residential communities. Conventional roadways could entirely disappear from communities, leaving walkways and cycling lanes through parklike expanses between homes. On closer examination, a personal mobility system that is initially envisioned only as

---

* Although energy consumption goes up when the car spends more hours in use, the cost of energy represents only a small portion of a vehicle's total operating expenses.

an adjunct to mass transit contains the seeds of change that could ultimately transform lifestyles and societies.

LARGE-SCALE PERSONAL MOBILITY AS AN OUTMODED CONCEPT

Although the discussion is about automobiles, it may be interesting to go further out on the limb of speculation and consider the idea that unbridled personal mobility could ultimately become outmoded. In the immortal words of Yogi Berra: "The problem is that everybody wants to be somewhere else."* It is certainly possible that the time could come when there is no real point in being somewhere else. Today, people communicate and work together, even when they are located on opposite sides of the globe. Although it is still in its infancy, the Internet has already evolved into a worldwide educational, informational, business, and social tool that could have hardly been imagined at the outset. Undoubtedly, this low-cost, worldwide networking medium will have far-reaching impacts on society over the next decade or two, not the least of which will be to significantly reduce the need for personal mobility.**

Regardless of actual need, it seems reasonable to expect that people may want the freedom of mobility, just for the experience of going somewhere and being there. Consider, however, that it may not be necessary to actually go somewhere in order to have the experience. One need only project current developments in computer technology 20 years or so into the future to envision a time when virtual experiences may substitute for actual experiences in many areas. In fact, virtual reality could open up many opportunities for experiences that may otherwise have been impractical or unavailable to the average person. Already, computer operators can reach into thin air while watching the duplicate motions of a virtual hand grasp an object and move it about on a monitor. With computer point-cloud modeling, one can actually feel the resistance of the electronic "clay" as it is pulled, carved, and shaped in three-dimensional electronic space. Once it becomes possible to

---

* Yogi Berra may not have actually said that, but if someone had asked, that's probably what he would have said about today's transportation problems.

** The author continued to send computer disks across town via courier for at least two years after it was possible to simply send the electronic files as an e-mail attachment.

realistically deliver the appropriate stimuli to all of the senses, our vacation experiences might conceivably be downloaded from the Internet.*

Many readers will undoubtedly feel that virtual experiences can never be fulfilling, simply because they are artificial. But consider that radio, television, CD players, video players, and movies are all artificial reproductions. The music from a CD player is only an artificial reproduction of the real thing, and there are no actual actors on a movie screen or inside a television set. Yet these reproductions are entertaining and fulfilling, even though particular media play to only one or two of the senses. If all of the senses were satisfied, it would be questionable whether one could tell the difference between a virtual experience and an actual experience, except for the absence of waiting in lines, fighting crowds, and struggling for a better view. In reality, many virtual experiences may be far better and more fulfilling than similar experiences in the real world. Even now, at sporting events, television audiences usually have the best seat in the house.

Regardless of whether the foregoing speculations are realistic, nontransportation technologies are sure to reduce the importance of travel in our everyday work and play. Looking far enough into the future, personal mobility of the magnitude common in today's world could become an outmoded practice. The time may come when there is little benefit in using precious time and energy just to transport our physical presence to various locations around the globe.

REFERENCES

1. "The Nissan Report," edited by Steve Barnett, 1992.

2. "1990 Nationwide Personal Transportation Study," Early Results, Office of Highway Information Management.

---

* Ideas that seem unlikely at first often have a way of becoming ordinary later on. About 1980, a friend, who was a bit of a computer nerd, encouraged me to computerize my design firm, Quincy-Lynn Enterprises. But at that time, it was difficult to see what one would actually do with a PC in the product design field. Today, everything is designed on 3D computer-aided design (CAD) systems, and a designer can hardly make a living without the aid of modern computers. According to experts in the field, computer technology is about at the stage of the automobile when Henry Ford first introduced the Model A.

3. F. T. Sparrow, "Automotive Transportation Productivity: Feasibility and Safety Concepts of the Urban Automobile," Final Report to The Lilly Endowment, Inc., December 1984.

4. This report was not publicly distributed. This figure was obtained through conversation with one of the study's facilitators.

5. "1992 The Motor Industry of Japan," published by Japan Automobile Manufacturers Association, Inc., Washington, D.C.

6. William L. Garrison and Mark E. Pitstick, "Lean Vehicles: Strategy for Introduction Emphasizing Adjustment to Parking and Road Facilities," SAE Paper No. 901485.

7. Paul F. Wasielewski, "The Effect of Car Size on Headway in Freely Flowing Freeway Traffic." Report No. GMR-3366, General Motors Research Laboratories, 1980.

8. John W. McClenahan and Howard J. Simkowitz, "The Effect of Short Cars on Flow and Speed in Downtown Traffic: A Simulation Model and Some Results," *Transportation Science*, Vol. 3, No. 2, May 1969.

9. *Ibid.*

10. *See* note 6.

11. Gino Sovran and Mark S. Bohn, "Formulae for the Tractive-Energy Requirements of Vehicles Driving the EPA Schedules," SAE Paper No. 810184.

12. *See* note 2.

13. John Hemphill, "The Market Potential for Micro-Mini Cars in the United States," speech delivered before the Mini and Microautomobile Forum: Overview and Potential Problems, Transportation Research Board, National Academy of Sciences, Circular No. 264 (September 1983).

14. Thomas Turrentine et al., "Market Potential of Electric and Natural Gas Vehicles: Final Report for One Year," Institute of Transportation Studies, University of California, Davis (UCD-ITS-RR-92-8).

15. Thomas Turrentine et al., "Household Decision Behavior and Demand for Limited Range Vehicles: Results of PIREG, a Diary-Based, Interview Game for Evaluation of the Electric Vehicle Market," unpublished paper submitted for presentation at 1993 TRB Annual Meeting, July 1992.

16. See National Station Car Association's website at www.stncar.com.

CHAPTER THREE

THE TECHNOLOGY
OF FUEL ECONOMY

Courtesy: DaimlerChrysler, www.daimlerchrysler.com

*[T]o myself I seem to have been like a boy playing on the seashore, and
diverting myself in now and then finding a smoother pebble or a prettier
shell than ordinary, whilst the great ocean of truth lay all undiscovered
before me.*

—Sir Isaac Newton

Automobile power systems are on the verge of a leap in technology that could rival the transition from the horse to the self-propelled carriage at the beginning of the 20th century. But despite today's new high-tech power systems, if it were possible to look back from a vantage point in the 22nd century, one would probably see that the great ocean of truth had been all undiscovered before us. Fuel economy as a measure of vehicle efficiency could become as outmoded as expressing the cost of mobility in terms of bales of hay. Today's attempts to more precisely control the brute-force events inside a combustion engine may seem as hopelessly inadequate as breeding stronger workhorses for greater agricultural production. But the demise of the combustion engine may not come as quickly as the current enthusiasm over fuel cells seems to foretell, and fuel cells are not necessarily the ultimate power source for transportation. On this side of change, it is difficult to envision the curves in a road that has yet to be traveled. Like the futurists of the early 1900s, attempts to render the future from our perspective in time will invariably be flavored by today's paradigm of technical feasibility (Figure 3.1). So it

"Things the Motorist Wants to Know"- 1911 vision of 1961 car

FIGURE 3.1 1911 Vision of 1961 Car. The artist who made this drawing was obviously limited by the ideas and images prevalent from his perspective in time.

Courtesy: Access Magazine, University of California Transportation Center, www.its.berkeley.edu

is best to submit to it and begin where we are—to explore automobile power systems in familiar terms and concepts.

Today, the popular measure of good environmental power-system design is one that produces few or zero harmful emissions and achieves the highest possible fuel economy. But these are just two of the many attributes that must be woven into the fabric of a marketable consumer vehicle. An experimental car built by British Ford several years ago would cruise at 24 km/h (15 mph) with fuel economy on the order of 892 km/L (2,100 mpg). The difference between experimental, super-fuel-efficient cars and those on the showroom floor depends largely on the degree to which fuel economy is emphasized over other vehicle attributes. With an experimental vehicle, designers have free rein to maximize fuel economy without the normal constraints imposed by manufacturing and marketing realities. When it comes to a production design, hundreds of additional criteria become relevant and a different product emerges. Attributes such as reliability, manufacturing costs, creature comforts, and vehicle performance—attributes that may not be essential in an experimental vehicle—become crucial in a production design. Ultralight composites give way to more affordable materials, a sleek aerodynamic body sacrifices to styling and room considerations, power-hungry accessories are installed, and high-tech experimental drivetrains forfeit to more buildable, more affordable, and more proven hardware. In effect, the design gives up one excellence in favor of others.

The cooperative U.S. government/industry effort under the Partnership for a New Generation of Vehicles (PNGV) is designed to triple the fuel economy of conventional-size passenger cars through improvements in technology. The primary program metric is achieving production-ready prototypes capable of 35 km/L (83 mpg) by 2004. Although the program's target fuel economy goal has already been achieved, the task of keeping the cost on par with that of a conventional family sedan still presents difficult challenges.

This chapter explores the various factors affecting fuel economy, with an eye mainly toward conventional power systems. But conventional power systems appear to be limited in terms of potential fuel economy. If conventional internal combustion (IC) engine power systems were optimized to the practical limits of technology, it may be possible to double efficiency from today's 15 percent or so (overall vehicle efficiency) to perhaps 30 percent before we run out of ideas. At 30 percent efficiency, the average 10-km/L (25-mpg) sedan would then achieve 20 km/L (50 mpg). That appears to be about the best we can hope for with conventional technology, which will not solve energy and emissions problems over the long term. Just to hold our own against growth in demand, the world's automobile fleet will have to be in the

35- to 45-km/L (83- to 106-mpg) range in about 25 years. PNGV calls for a 35-km/L (83-mpg) production-ready prototype by 2004, which puts today's efforts into perspective.

So to make the necessary gains, we may have to make the leap to advanced power systems, and we have to build lightweight vehicles so there is less total mass to transport around the world's cities. Depending on the ultimate successes with fuel-cell development, conventional power plants may continue for some time as the predominant prime mover for hybrid-electric vehicles (hybrids are treated separately in Chapter Five).

Achieving triple the present fuel economy with a production design is also possible with conventional power systems, provided vehicles are significantly smaller and lighter. The idea of defining a new vehicle category with a different mass/performance profile was explored in the last chapter. Using conventional technology, it is possible to design personal mobility products with fuel economy on the order of 40–50 km/L (94–118 mpg) and performance equal to today's compact cars. With advanced technology power systems, ultralight mobility products could be significantly more thrifty on fuel.

Minimum fuel consumption per passenger-kilometers traveled occurs when vehicle capabilities most closely match mission requirements. When vehicle capabilities significantly exceed mission requirements, transportation energy intensity is generally greater, regardless of the basic technology. Energy intensity drops when smaller, limited-capacity vehicles for one- and two-occupant local trips are substituted for large multipurpose cars. Although downsizing is not normally considered a technology, significant downsizing requires an increased emphasis on compact packaging, cost-effective and lightweight materials, more efficient processing, and smaller and lighter powertrains. In addition, improved strategies for managing the forces of collisions or avoiding them altogether are also essential. The natural evolution of automotive safety design could offset any inherent imbalance in safety between low-mass and high-mass vehicles in as little as 10 to 15 years.

INGREDIENTS OF FUEL ECONOMY

In the most fundamental sense, fuel economy is the product of load times system efficiency. Consequently, there are two ways to reduce vehicle fuel consumption: (1) increase the efficiency of the powertrain in order to deliver more work from the fuel consumed, or (2) reduce the required work (load). Work required to propel the vehicle is defined as "road load." Road load is normally divided into the three categories of inertia weight, aerodynamic

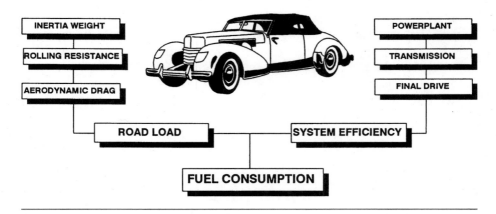

FIGURE 3.2 Ingredients of Fuel Economy

drag, and rolling resistance. Together, they determine the energy needed to propel the vehicle throughout the course of the trip (Figure 3.2).

The vehicle's total energy supply comes from the fuel it consumes; however, only a portion of the fuel is actually transformed into useful work. The percentage of the fuel's latent energy that is delivered to the drive wheels as useful work is determined by "system efficiency." The primary subsystems that determine overall system efficiency are the engine, transmission, and final drive. Each of these subsystems is affected to varying degrees by the values of thermal efficiency, friction, and the viscous drag of lubricants, as well as by the operating variables over the course of the driving schedule.

Modifying the details of a vehicle's design affects the balance of energy flow throughout the system. Individual changes have complex effects, which are often difficult to analyze. An engine with greater thermal efficiency may actually consume more fuel if its operating characteristics are not well matched to the vehicle's demand for power (operating schedule). Altering the engine's brake specific fuel consumption (bsfc) curve, or changing the transmission shift schedule to take advantage of engine operating characteristics, can significantly improve fuel economy, even though engine thermal efficiency remains unchanged. When matched by complementary changes in the powertrain, reducing vehicle mass is a simple and relatively low-tech modification that can have a significant effect on fuel economy. Given the high cost of advanced power systems, unconventional technology should not be overemphasized in relation to the effects of vehicle downsizing and systems engineering on energy requirements. Mass effects alone are responsible for roughly 82 percent of the net energy consumed over the urban driving cycle. Eliminating load is the most direct route to greater fuel economy. In addition, most (but not all) harmful emissions tend to be proportional to the total fuel/energy consumed, assuming that vehicles of equal technology are compared. In the future, reducing vehicle energy requirements may be the single most important strategy in further reducing transportation's harmful byproducts.

Fuel economy is normally expressed as distance traveled per unit of fuel consumed (km/L, L/100 km, or mpg). Except under controlled conditions, however, distance/fuel-consumed provides only an approximate measure of system thermal efficiency. In actual practice, fuel consumption is affected by a variety of interrelated factors, including ambient temperature, trip duration, driving cycle, and driver technique. Total fuel consumed throughout a trip is factored into the total distance traveled to obtain a fuel economy figure. Even though km/L or mpg is a highly variable figure, it nevertheless serves as a convenient and easily understood measure.

Urban Driving Cycle Patterns' Effect on Fuel Economy

Over the city driving cycle, fuel economy plummets. Early experiments with traffic density and its effect on fuel economy showed that light traffic conditions with an average speed of 29 km/h (18 mph) resulted in fuel economy that was only 60 percent of the steady-state 29 km/h value. When increased traffic density reduced average speed to 19 km/h (12 mph), fuel economy fell to 44 percent of the steady-state 29-km/h value.[1] Degraded urban fuel economy comes primarily from light engine loading, high inertia loads, and frequent periods of braking and time spent at idle.

During city driving, conditions such as acceleration, engine loading, and time spent braking or at idle are continually changing across a wide range. These variations result in equally wide swings in fuel consumption. Studies by Gino Sovran at General Motors Research Laboratories in the early 1980s revealed the breakdown shown in Table 3.1 in times spent under various driving modes.[2]

Vehicle mass reduction first affects fuel consumption by reducing steady-state road load, but over the urban driving cycle, the vehicle and its rotating components reside at a steady-state speed only 7.6 percent of the time. The effects of inertia are much more significant in city driving. Inertia loads are directly proportional to the mass of the vehicle and its rotating components. Nearly 40 percent of urban-cycle driving time is spent under acceleration.

TABLE 3.1 Time Spent in Various Operating Modes

Operating Mode	Urban		Highway	
	Seconds	Percent Time	Seconds	Percent Time
Acceleration	544	39.7	252.8	33.0
Cruise	104.5	7.6	292.7	38.3
Powered Deceleration	164	12.0	159.2	20.8
Braking	311	22.7	57.3	7.5
Idle	248.5	18.0	3	0.4

Source: SAE Paper No. 810184, www.sae.org

Time spent at idle and during closed-throttle deceleration accounts for 41 percent of the total operating time over the urban cycle. Fuel consumption at idle and during closed-throttle deceleration is directly proportional to engine size. Engine size is determined primarily by vehicle acceleration requirements, which in turn is proportional to vehicle mass.

Figure 3.3 shows the relationship between the total energy (instead of total time) required for inertia, aerodynamic drag, and rolling resistance over urban and highway operating cycles. Inertia loads are less significant during highway driving; however, inertia consumes 60 percent of the total vehicle energy over the Environmental Protection Agency (EPA) urban driving cycle.[3] Inertia loads and rolling resistance together account for 82 percent of the total energy needed to propel the vehicle over the urban cycle. A reduction in vehicle mass, therefore, has a direct effect on 82 percent of the vehicle's total energy requirements in the urban environment. According to tests at Oak Ridge National Laboratory, the effects of mass account for 75 percent of a vehicle's total energy use over combined urban and highway driving schedules.

Idle shut-off systems have been known for some time to reduce fuel consumption in heavy stop-and-start traffic. Today's advanced hybrid vehicles almost universally employ idle shut-off systems. At a predetermined threshold the engine stops, and then it is automatically restarted on demand by pressing the accelerator pedal. Conventional 12-volt starting systems using flywheel pinions are noisy and cannot be relied on to repeatedly and rapidly restart the engine. But the current effort to switch to a 42-volt electrical system makes possible quiet, reliable engine restarting using the integrated flywheel starter/alternator, which opens the way for universal adoption of idle

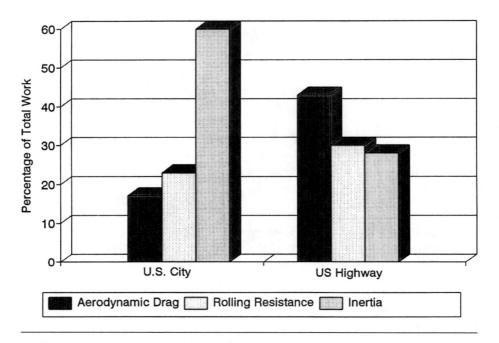

FIGURE 3.3 Driving Cycle Load Comparison

Source: Fundamentals of Vehicle Dynamics, www.sae.org

shut-off. Such systems can have a large impact on the fuel economy of conventional vehicles.

Cold-running characteristics are another important consideration for cars that operate primarily in the urban environment. Most urban trips are short, and consequently the vehicle may never become fully warmed up. This causes poor fuel economy and increased exhaust emissions. Modern IC automobile engines run relatively clean once they are fully warmed up. With today's engines, most hydrocarbon and carbon monoxide emissions occur during the warm-up period while the charge is enriched and before catalyst light-off. Tests at GM several years ago demonstrated that on a cold winter day (–12°C), a four-mile trip that began with a cold startup resulted in an average fuel economy that was only 72 percent of a similar, cold-startup 24-kilometer trip. Engineers discovered that even after 24 kilometers (15 mi.),

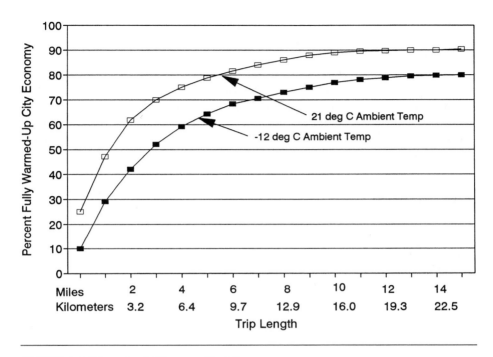

FIGURE 3.4 Urban Cycle Warm-up Fuel Economy

the vehicle was not fully warmed up and fuel economy was still approximately 10 percent below optimum.[4]

Data for the curves in Figure 3.4 were obtained from vehicles that were not equipped with charge-air heaters. Such preheaters use exhaust heat to reduce engine warm-up time; however, charge-air heaters have no effect on bearings, seals, lubricants, transmission, final drive, and tires. Warm-up energy can come only from burning fuel. The amount of energy required to reach a fully warmed-up state depends primarily on the heated mass. Reducing the mass of the powertrain and lowering the quantity of lubricants, as well as utilizing waste heat to augment the warm-up process, can significantly reduce fuel consumption during short-duration trips. Preheaters can also significantly reduce catalyst light-off time.

Another option for reducing warm-up and catalyst light-off time is latent heat storage. With smaller power plants, latent heat storage devices become more effective tools for rapidly increasing engine temperature after a cold startup. One type of heat battery looks like a large tin can tucked inside the engine compartment. A typical device might measure 170 mm in diameter by 370 mm long, and weigh approximately 10 kg. A heat battery of this size can release 50–100 kW of heat during the first 10 seconds. Heat capacity is in the order of 600 W/h when cooled from 80°C to 50°C. The storage medium is a water/salt mixture, which is insulated by a conventional vacuum barrier. Heat is delivered to the engine coolant to induce rapid warm-up. Reheating is done with waste heat returned by the coolant, and stored heat can be used several days later.

Accessory Loads

Accessories account for a significant portion of an automobile's total energy requirements (Figure 3.5). The power consumed by accessories can equal or exceed the power required to actually propel the vehicle at cruise. Loads imposed by the air conditioner, power steering, alternator, and cooling system can total as much as 3.5–7.5 kW in a full-size sedan, and accessory loads are increasing with the addition of modern electronics. A switch to the 42-volt electrical system will provide the necessary power and improve the efficiency of electrical subsystems.

Part of the technology of fuel economy includes a close scrutiny of energy-robbing convenience items. Eighty percent of new cars sold in the United States are equipped with air conditioning. Today, air conditioning is perceived as a necessity, and consumers are not likely to give it up in order to save fuel. The air-conditioning system, however, is the largest single energy parasite aboard the automobile. Power required to cool the interior of a full-size automobile can equal the power required for all other accessories combined. Thermal load on the air-conditioning system depends on the thermal characteristics of the cabin. A cabin with high surface-to-volume ratio (large surface area in relation to interior volume) is more difficult to cool. An appropriate heat barrier between the passenger compartment and the engine room and exhaust system is crucial.

Contrary to popular belief, experiments have shown that exterior color makes little difference in vehicle heat gain. Lighter interiors do, however, help reduce the demand on space conditioning. Glazing is undoubtedly the largest culprit of all. A solar heater consists of a flat box (high surface-to-volume ratio) with a darkened interior covered on one side with glass. Air conditioning a modern automobile with its large expanse of glazing is somewhat like air conditioning a solar heater. Solar control glazing can reduce heat gain by approximately 30 percent, but it cannot completely solve the problem.

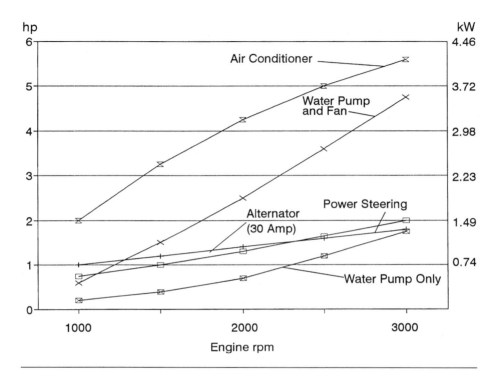

FIGURE 3.5 Accessory Power Requirements

Road Load

Road load defines the power required to move the vehicle. It has a direct effect on fuel consumption. The greater the road load, the greater the appetite for fuel. A road load graph typically shows steady-state demand over a range of driving speeds. In actual practice, however, road load is much less consistent than the image portrayed by the smooth lines of a road load curve. Instead, it swings widely according to the variations in vehicle payload and speed, weather conditions, and the grade and condition of the road surface. The engine converts fuel into work to meet the continuously changing demands of the road load. Reducing the road load places less demand on the engine and thereby reduces fuel consumption.*

---

* Fuel consumption will be reduced if the engine is appropriately downsized.

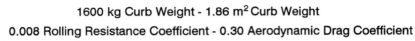

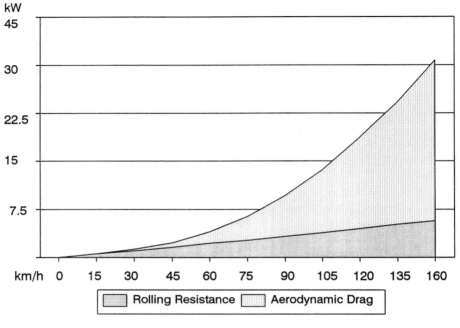

FIGURE 3.6 Steady-State Road Load of Standard-Size Car

During steady-state cruise, road load consists of rolling resistance and aero-dynamic drag. Rolling resistance is directly proportional to the weight and speed of the vehicle. Consequently, if vehicle weight or speed doubles, rolling resistance also doubles. The power required to overcome aerody-namic drag is directly proportional to the frontal area of the vehicle but increases to the cube of vehicle speed. At any given speed, if vehicle frontal area is doubled, twice the power will be required to overcome aerodynamic drag. As speed increases, aerodynamic drag increases to the cube. The two following road load graphs compare the steady-state aerodynamic drag and rolling resistance of a hypothetical large car (Figure 3.6) to those of a hypo-thetical ultra-low-mass (ULM) vehicle of about the same size and weight as Quincy-Lynn's Trimuter (Figure 3.7). Both graphs use the same criteria, and neither graph takes into account the drivetrain losses or the power required for parasite systems.

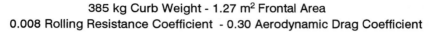

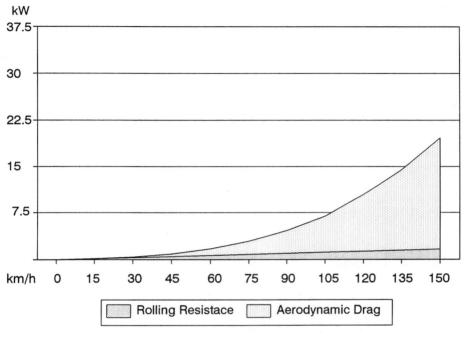

FIGURE 3.7 Steady-State Road Load of ULM Vehicle

Developing a Road Load Graph

A road load graph can be developed for an existing vehicle by performing a coastdown or a draw-bar test. A coastdown test is performed by allowing the vehicle to coast to a stop from a predetermined speed. If the inertia weight of the vehicle is known, road load can be calculated from the deceleration curve.[5] A draw-bar test is performed by towing the test vehicle with a rigid bar equipped to measure the pulling force. Readings are taken in velocity increments throughout a range of relevant speeds. These points, which indicate velocity and force values, are used to develop a curve that represents the road load in horsepower or kW.

More often a road load graph is developed for a hypothetical vehicle as an engineering aid. Such a graph provides the basis for selecting the correct engine, configuring the drivetrain, and predicting performance and fuel economy. In order to develop an engineering road load graph, the velocity/ force values must be calculated on the basis of the target vehicle weight, the projected rolling resistance, and the estimated aerodynamic drag.

Rolling Resistance

Rolling resistance is the force it takes to make the vehicle roll. It is a linear function that is directly proportional to the vehicle weight and the velocity at which it is moving. The rolling resistance coefficient can vary widely depending on the chassis design, roadway surface conditions, and most important, the design and inflation pressure of the tires. Although pneumatic tires are essential for good handling and ride, they are responsible for approximately 85 percent of a vehicle's rolling resistance. The remaining 15 percent comes from bearings, seals, gear friction, and the viscous drag of lubricants.

Tires: Balance Between Ride, Handling, and Drag

A steel wheel running on a steel rail provides the least rolling resistance of any wheel design. Steel wheels, however, have no inherent damping and very poor adhesion. Unlike a pneumatic tire, a steel wheel does not flatten across the area of contact with the surface (contact patch). The effective radius remains unchanged throughout the circumference, and as a result, rolling resistance is minimal. By deflecting on contact with an irregularity, a pneumatic tire acts as a sort of air spring to smooth out the ride. This same flexibility also allows the tire to flatten at the contact patch, which places a greater portion of the tire in contact with the road (Figure 3.8). The tire's pneumatic resilience effect is one of the sources of its excellent adhesion, ride, and handling characteristics. Pneumatic resilience is also the primary source of its high rolling resistance. Much also depends on sidewall flexibility and materials compounding.

At 90 km/h (55 mph), each tire on the vehicle is forced though a complete radial distortion at the rate of nearly 800 times per minute. Such flexing requires a significant input of energy, which can only come from fuel consumed by the engine. The energy required for tire distortion is normally expressed as the coefficient of rolling resistance. A full-size passenger car equipped with soft-riding bias-ply tires might have a rolling resistance coefficient as high as 0.015. A vehicle equipped with more efficient state-of-the-art radial tires could exhibit a rolling resistance coefficient as low as 0.008. Rolling resistance can be significantly reduced by increasing the inflation

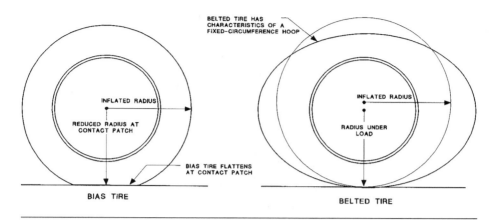

FIGURE 3.8 Tire Distortion Characteristics

pressure, but normally at the expense of ride comfort. Improved tire design offers the greatest potential for reducing rolling resistance.

Tire manufacturers have been hard at work for many years to develop low rolling-resistance tires while still maintaining superior adhesion and ride qualities. During the past two decades, tire rolling resistance has fallen by approximately one-third. Today, new tires have been developed that have even less rolling resistance. Goodyear's most recent contribution is the low rolling resistance for GM's electric car, EV-1 (Figure 3.9). The tire has a rolling resistance coefficient of 0.0048, which is about 55 percent less than a conventional highway tire. Tires for EV-1 are inflated to 448 kPa (65 psig), or about twice the inflation pressure of a conventional tire.

Other Components of Rolling Resistance

Advanced lubricants can also contribute to efficiency gains. For many years, engineers have been aware of the drag on transmission and final drive components caused by the high viscosity of lubricants. High viscosity helps prevent gear pressure from squeezing lubricant away and allowing the teeth to make metal-to-metal contact; however, switching to lower-viscosity lubricants has produced mixed results. When the lubricating film can be maintained at a lower viscosity, drag is reduced. If lubricant is squeezed away, friction increases and more than offsets any reduction in viscous drag. In this respect, the lower gear pressures of a relatively low-power, low-mass vehicle present new opportunities for reducing drag with low-viscosity lubricants.

FIGURE 3.9 Low Rolling-Resistance Tire from Goodyear

Courtesy: Goodyear Tire & Rubber Company, www.goodyear.com

AUTOMOBILE AERODYNAMICS

Nearly every aspect of vehicle design, including creature comforts, vision, systems cooling, interior ventilation, noise, and styling, are affected by, or have an effect on, the aerodynamics of the design. Aerodynamic drag alone is a major factor in the vehicle's appetite for energy, and it becomes increasingly so at higher speeds (Figure 3.10). Aerodynamic drag is a product of the vehicle's frontal area, its drag coefficient, and the cube of its speed. The cubic increase of drag in relation to speed makes aerodynamics a major consideration when designing high-speed highway vehicles. At the lower speeds typical of urban traffic, aerodynamic drag is not nearly as significant.

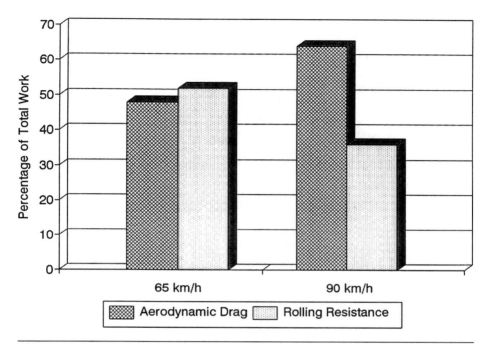

FIGURE 3.10 *The Effects of Speed on the Relationship between Aerodynamic Drag and Rolling Resistance*

Features that determine the drag profile of the vehicle include the following:

Appropriate curvatures for maintaining laminar flow

Shape of the nose	Slope of the hood	Windshield angle
Shape of pillars	Rain gutters	Gap profiles
Ground clearance	Body inclination	Backlight angle
Rear body shape	Wheel house openings	Wheels
Cooling system	Internal airflow	Underbody flow
Spoilers & dams	Mirrors	Headlights
Antenna	Recess of glazing	Wipers
Door handles	Effect of air leaks	

Drag is not the only aerodynamic effect the designer must consider. Passenger compartment flow and climate control, mud spray and particle deposits, systems cooling, exhaust dispersion, and noise are all affected by a vehicle's aerodynamic characteristics.

Working with the Relative Wind

The velocity and direction in which the vehicle and air encounter each other are referred to as the *relative wind*. Automobile aerodynamics is concerned with the effects of the relative wind as it flows around and through the vehicle. Relative wind has several effects on the vehicle in addition to the drag it imposes. It can increase or decrease the pressure at the contact patch through positive or negative lifting forces, it can stabilize or destabilize the vehicle with yaw inputs, and it can cool and ventilate or destroy cooling and ventilation flow as a result of pressure differentials around the body.

At highway speeds, the forces of induced lift can reach a magnitude of 1300 N (292 lb.) or more on a full-size sedan. These lifting forces can have a significant effect on the vehicle's stability. Crosswinds also impart destabilizing forces that cause the vehicle to yaw. High lift and yaw inputs, combined with reduced adhesion on a rain-slicked road, can cause loss of control under extreme conditions. Low-mass vehicles imply greater attention to aerodynamic effects. At freeway speeds, a lightweight car could also develop high lifting forces. As a percentage of the vehicle's gross weight, lifting forces could be proportionally higher than in vehicles of greater mass. The destabilizing effects of inertial forces in combination with crosswinds and side gusts from passing vehicles may therefore become more significant as vehicle mass decreases.

Streamlining and Aerodynamic Drag

The techniques for reducing aerodynamic drag are collectively referred to as *streamlining*. Air resistance has been studied for more than 200 years. Early on it became clear that the shape of the body has a large effect on the magnitude of the drag it produces. As recently as the wind tunnel experiments by the Wright brothers in 1902, however, experimenters were unsure about which shapes actually produced the lowest drag. It was soon discovered that the ideal aerodynamic shape was the teardrop, with the large end facing into the relative wind. The trailing end of the teardrop should taper at a shallow angle of 15 degrees or less. Figure 3.11 compares the actual drag on 30.5-meter (100-ft.) lengths of rod of various profiles when exposed to a 160-km/h (100-mph) wind. Each length of rod presents an identical frontal area to the relative wind. Shape (profile) is the only difference.

R = Aerodynamic Drag on 30.5 m Profile at 160 km/h

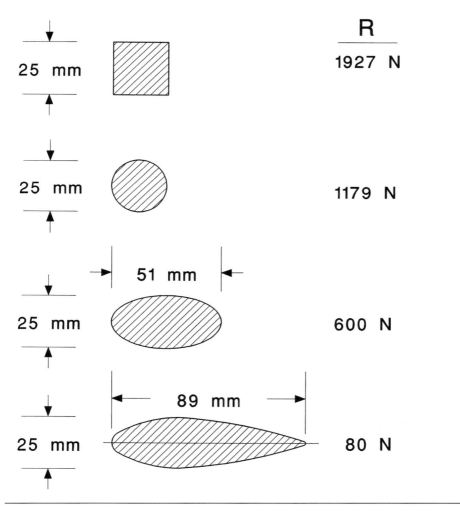

FIGURE 3.11 *The Effect of Shape on Aerodynamic Drag*

The objective of streamlining is to shape the body in a way that will smooth out the flow of air and thereby reduce the drag coefficient. The coefficient of drag (Cd) can therefore be used to compare the ability of various shapes to slip though the air with minimal resistance. Cars have become increasingly more slippery over the years (Figure 3.12). A modern, aerostyled car has a Cd

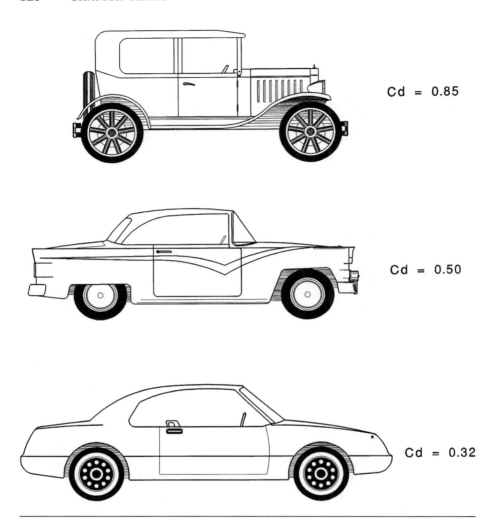

Cd = 0.85

Cd = 0.50

Cd = 0.32

FIGURE 3.12 Declining Aerodynamic Drag of Automobiles

on the order of 0.25 to 0.35. A 1930 Model A Ford had a Cd of about 0.85, and the Cd of a 1955 Ford was on the order of 0.50. A 1977 Porsche 924 had a Cd of 0.36. GM's Precept PNGV concept car has a drag coefficient of only 0.16.

Sources of aerodynamic drag are normally divided into the general categories shown in Table 3.2.

TABLE 3.2 Aerodynamic Drag by Source

Source of Drag	Percentage of Total Drag
Profile Drag	55%
Parasitic Drag	17%
Internal Flow	12%
Skin Friction	9%
Induced Drag (lift)	7%

Profile Drag

Profile drag is a function of shape. The ideal aerodynamic shape is the tear-drop, with the large end toward the relative wind. Unfortunately, the teardrop does not accommodate passengers and cargo well, nor is it very attractive. A boxlike shape better utilizes space but has poor aerodynamic characteristics. Modifications to the basic shape of the box can result in large improvements in flow characteristics.

Profile drag is minimized when laminar flow continues as far rearward as possible. *Laminar flow* refers to a condition in which the air stream follows smoothly along the body without breaking away into turbulence; however, laminar flow is difficult to maintain. As long as the air stream is accelerating across a smooth, convex surface, it has a good chance of remaining laminar. Experiments indicate that separation is most effectively delayed when the body surface's radii of curvatures vary continuously according to an equation of the second order. This implies that all curves of the body surface must be of the third order. Cubic equations applied to test model body surfaces demonstrated very low drag.[6] As the flow moves downstream, the boundary layer (a thin layer of slow-moving air next to the surface) begins to thicken, form rollers and eddies, and finally break away into vortices. At this point, separation is complete and drag sharply increases.

Any sudden break in a smoothly contoured surface can result in flow separa-tion. Wheel wells have always been problem areas for engineers because they destroy laminar flow along body sides. Recessed window glazing is another design attribute known to destroy laminar flow across the sides of the greenhouse (Figure 3.13). Once separation occurs, it may be difficult to reestablish a laminar flow. Protuberances such as mirrors, handles, bug deflectors, and antennae produce drag of their own, but they can also induce separation.

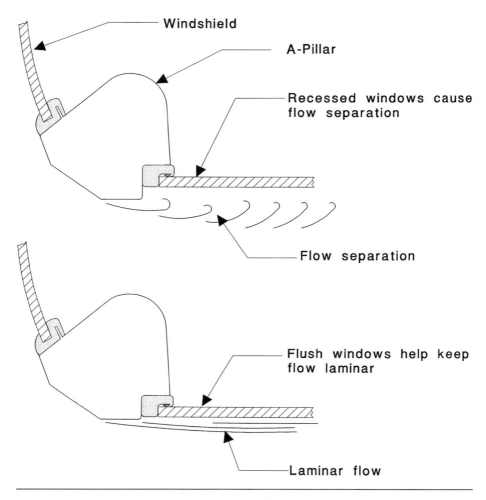

Windshield

A-Pillar

Recessed windows cause flow separation

Flow separation

Flush windows help keep flow laminar

Laminar flow

FIGURE 3.13 Flush Windows Significantly Reduce Drag

Regardless of how well the body encourages the flow to remain laminar, separation must ultimately occur at the rear of the vehicle. Separation creates a standing low-pressure area behind the vehicle. The breakaway cross-section at the rear has an effect similar to that of a flat plate of equal size presented to the air stream. Consequently, the body should taper toward the rear, similar in effect to the reverse teardrop, to form an aft section that is as small as possible (Figure 3.14). Another technique is to direct some of the relative wind into the low-pressure zone behind the car in order to reduce the magnitude of the pressure differential. Engineers used this technique with the

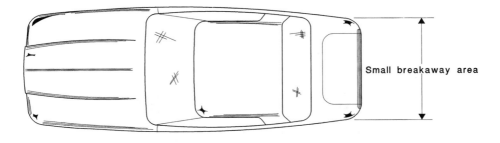

Small breakaway area

FIGURE 3.14 *Aft-Tapering Body Reduces Separation Area*

Precept concept car by taking in cooling air just behind and above the rear wheels and then directing it into the low-pressure area at the rear. This makes the airflow behave as though the body has an aft taper, when in fact it does not.

Induced Drag

An automobile is relatively flat along the bottom and curved across the top. That is also the basic shape of an airplane wing. As a result, an automobile is normally subjected to significant lifting forces at higher speeds (Figure 3.15). As mentioned earlier, a standard sedan can generate lift on the order of 1300 N (292 lb.) at highway speeds. Drag is a necessary component of lift. A design that minimizes lift will also reduce the resulting drag, provided that flow separation or turbulence is not induced.

Parasitic Drag

Parasitic drag refers to the drag induced by components that project into the air stream. This includes items such as ornaments, bug deflectors, wipers, antenna, rearview mirrors, license plates, door handles, air scoops, rain gutters, bezels and trim, and luggage racks. Exposed mechanical components underneath the vehicle also contribute to parasitic drag. The magnitude of parasitic drag can be much greater than the simple aerodynamic drag of the item. For example, a rearview mirror is typically located on the side of the body just aft of the windshield. In this location, it may be directly in line with high-speed air spilling from the sides of the windshield. This high-speed air stream can induce drag that is more than 1.5 times greater than the drag of the same mirror in free air. Furthermore, if the mirror induces

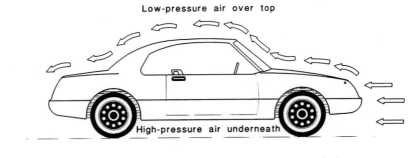

Lift results from the increase in airstream velocity
over the upper profile of car and airfoil.

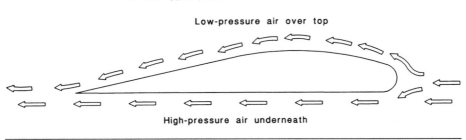

FIGURE 3.15 Lift Is Developed across Top of Both Shapes

flow separation, the combined effects become even greater. Exposed wipers cause similar effects. Exposed wipers induce drag of their own, plus they encourage flow separation across the windshield.

Enclosing the underbody with a flush pan can reduce drag by as much as 17 percent. Flush windows will reduce drag by 5 to 10 percent, depending on the overall body design. Raising the headlights of a Corvette can increase drag by approximately 3 percent. The drag coefficient of a Porsche 924 is more than 5 percent greater with the headlights raised.[7] Parasitic drag becomes much more significant on an aerodynamically clean shape. Extra effort taken to establish laminar flow may be to no avail if mirrors, door handles, or trim moldings destroy laminar flow and increase drag.

Skin Friction

The effect of viscosity produces a thin layer of relatively slow-moving air next to the surface of the body. This is called the *boundary layer,* and the friction it produces is called *skin friction.* At high speeds, a very polished surface will reduce skin friction by presenting a slippery surface to the air. At the relatively low speeds of an automobile, however, the polish of the surface has little effect on drag. The primary culprits of automotive skin friction are raised fasteners, exposed hinges, trim moldings, and the configuration of gaps.

Internal Flow

Drag caused by air flowing through the passenger compartment and cooling system can amount to as much as 12 percent of the total drag of the vehicle. A properly designed cooling and ventilation system can keep drag to a minimum and reduce energy demands by working in unison with the pressure differentials on the body surface. For example, air that is drawn into the passenger compartment from the base of the windshield will help maintain laminar flow across the windshield by relieving the high-pressure area. Allowing air to exit into the low-pressure area behind the backlight can also reduce drag. But airflow through the cabin normally has minimal effect compared to that of the cooling system.

Flow through the radiator and engine compartment produces the greatest drag from internal flow. Appropriate ducting can make a large difference (Figure 3.16). In his book, *Designing Tomorrow's Cars*, Walter Korff suggests that the principles of efficient ducting developed for the aircraft industry can be used to design more aerodynamically efficient automobiles. These principles require that the air outlet be as carefully controlled as the air inlet. Also, ducting must be sealed, or loss of energy will occur. Korff suggested that the size of the cooling air intake is much too large on most automobiles. A fast-moving vehicle with the radiator located behind the entrance of the duct at a distance equal to its own height needs an opening only 17 percent as tall as the radiator. If the vehicle spends much of the time in slow-moving urban traffic, the duct size should be increased by approximately 50 percent, or to approximately 25 percent of the height of the radiator. The closer the radiator is moved toward the intake, the taller the opening must be. Also, cooling system flow should be controlled by restricting the outlet flow, not by blocking the inlet.

Frontal Area

Reducing vehicle size is one of the most straightforward ways to reduce air drag. The frontal area—the silhouette the vehicle presents to the relative wind—has a direct effect on the drag it produces. Reduce the frontal area by half and aerodynamic drag also drops by half. This technique is limited

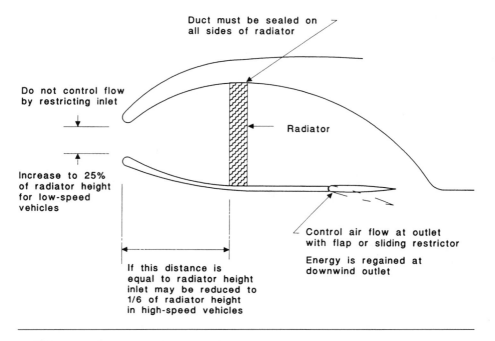

Duct must be sealed on
all sides of radiator

Do not control flow
by restricting inlet

Radiator

Increase to 25%
of radiator height
for low-speed
vehicles

If this distance is
equal to radiator height
inlet may be reduced to
1/6 of radiator height
in high-speed vehicles

Control air flow at outlet
with flap or sliding restrictor

Energy is regained at
downwind outlet

FIGURE 3.16 Minimize Air Inlet

because vehicles must have room for occupants. Frontal area can be reduced only by reducing height and width. Reducing length has very little effect on aerodynamic drag.

Minimum vehicle height is established by the sum of the ground clearance, the seat height, the seated height of the 98th percentile male, and the required headroom. These values are based on the mechanical requirements of the chassis, along with interior dimensions established through prior research into human space requirements and comfort levels. The cabin height of a typical sedan will be on the order of 1,150–1,200 mm (45–47 in.). A sports car might be limited to slightly more than 1,000 mm (40 in.) inside. Placing the occupant in a semireclining position can reduce the cabin height to approximately 900–950 mm (36–37 in.) (Figure 3.17).

Shoulder width establishes the minimum cabin width. The cabin may slant inward at the top and bottom because of the inward displacement of the occupants' head and hips; however, a close-fitting roofline will create a crowded feeling and can be hazardous in a collision. Crowding across the

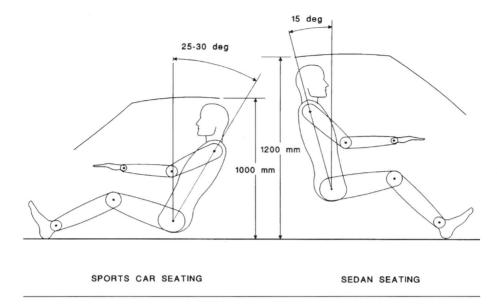

FIGURE 3.17 *Seating Position Affects Vehicle Height*

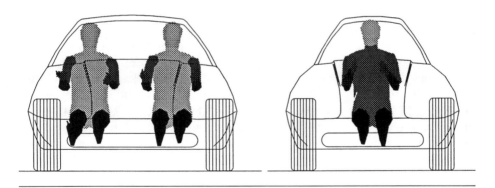

FIGURE 3.18 *Tandem and Side-by-Side Seating*

shoulders also tends to be uncomfortable. A simple method of reducing frontal area of a two-place vehicle is to place the occupants in tandem. By sitting one behind the other, vehicle frontal area can be reduced by as much as 40 percent (Figure 3.18). The Mercedes F300 is an excellent example of the tandem seating arrangement.

FUEL-EFFICIENT POWERTRAIN

The powertrain consists of the engine, transmission, and final drive. Ideally, the powertrain will be light, small, and simple to manufacture; it will have high specific power; and it will consume as little fuel as possible and produce few harmful emissions. The quality of a vehicle's performance, its fuel efficiency, and the amount of harmful emissions it produces is determined by hundreds of interrelated variables. A vehicle's performance, fuel economy, and emissions level flow from the interaction of its various subsystems as the powertrain meets the demands of the operating schedule. Advancements in powertrain design normally depend on technologies that increase the control of events and relationships, and thereby more precisely match powertrain performance to operating conditions.

Prime Mover

In Newton's time the prime movers were the beasts of burden and the water wheel. Throughout the 20th century, the prime mover of personal mobility vehicles has been the IC engine. Today's IC engine, however, bears little resemblance to its early predecessors. The design of engines has reached a level of refinement that would have been unimaginable to early pioneers. The experience of the Wright brothers is a case in point. In 1903 the Wright brothers wrote to several engine manufacturers with specifications for an engine to power their soon-to-be-built flying machine. To the brothers' dismay, not a single manufacturer could meet their requirement for an engine that would develop 7.4 kW (10 hp), yet weigh no more than 68 kg (150 lb.). The Wrights ultimately built their own engine: a 91-kg (200-lb.) machine that developed 11.7 kW (15.75 hp) at 1,200 rpm. Today, after a century of refinements, a high-performance automobile engine can develop power on the order of 1 kW/kg (1 hp/1.5 lb), and a two-stroke engine can develop nearly twice the power per weight.

The IC engine is often maligned because of its harmful exhaust emissions and poor thermal efficiency, but only in the last 25 years, since the widespread use of electronic engine management, has it been possible to gain control of events inside the engine. It is conceivable that, over time, the reciprocating engine could approach the efficiency of today's fuel cells and run virtually free of emissions.

Losses of Converting Fuel to Mechanical Power

Fuel, regardless of its form, contains a discrete amount of energy (specific heating value). Only a portion of this energy is converted to useful work by the engine. The difference between the heat-energy work-equivalent of the fuel and the mechanical work available at the output shaft represents the thermal efficiency of the engine.

The thermodynamically ideal combustion cycle for the IC reciprocating engine is the so-called constant-volume process, which is based on adiabatic and frictionless conditions and the behavior of ideal gases. Unfortunately, real engines do not operate according to theoretical ideals. Instead, they operate with incomplete combustion and with losses to flow, heat dissipation and discharge, and mechanical friction and parasitic loads. Of the energy released during combustion, much is either discarded through the cooling and exhaust systems (waste heat), used to move the fuel/air charge through the engine (pumping losses), or sacrificed to internal friction and parasitic loads (e.g., oil pump, water pump, fan, alternator). A primary objective of advanced engine technologies is to put some of this lost energy to work.

Figure 3.19 compares the fuel that is converted into useful work by a typical Otto engine against the various losses that occur. Brake horsepower is the power available to run the accessories and propel the vehicle. As the chart indicates, the largest losses are caused by rejected heat. When fuel is converted into heat energy and then into mechanical power, energy losses tend to be inversely proportional to the combustion temperature (i.e., the higher the temperature, the lower the loss). By raising the operating temperature of the engine, it is possible to recover some of the rejected heat and put it to

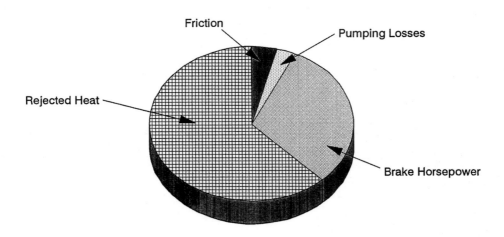

FIGURE 3.19 Distribution of Fuel Energy at Wide-Open Throttle

work; however, there are practical limitations to the ability of materials to operate at extremely high temperatures. Ultimately, low heat-rejection technologies may be perfected, and engines of greater fuel efficiency may become available. It is theoretically possible to develop adiabatic diesels (low heat rejection) that operate at twice the fuel efficiency of today's designs.

Power and Fuel Efficiency Characteristics of the Reciprocating IC Engine

Figure 3.20 shows power and fuel consumption curves that are typical of a spark-ignition gasoline engine. The values expressed are those of the engine

TYPICAL FOUR-CYCLE SI POWER CURVE

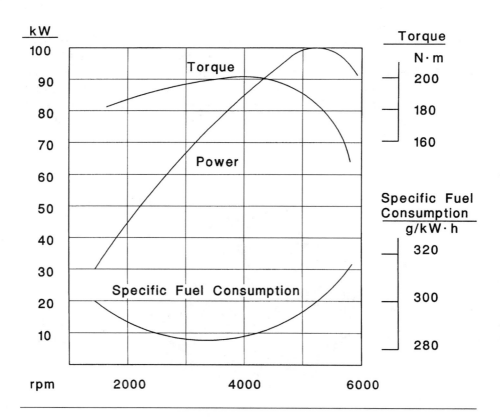

FIGURE 3.20 Power and Fuel Consumption at WOT

operating at wide-open throttle (WOT). They show the characteristics of the engine under its most efficient operating conditions. In actual practice, the engine will rarely operate according to the conditions represented in the chart. Instead, fuel economy will be degraded by part-load operation.

The power band of the engine extends from slightly above idle to the rotations per minute (rpm) at which power begins to drop. Depending on the design, a spark-ignition engine idles at 800–1,000 rpm and develops peak power at 3,500–6,000 rpm. Power, torque, and fuel consumption curves are typically established at WOT.

An important characteristic is that the engine can produce a given power across a wide rpm range by regulating the throttle. For example, in the curve above 30 kW (40 hp) is developed at approximately 1,500 rpm at WOT. Output can also be limited to 30 kW at any speed above 1,500 rpm by appropriately throttling the engine. As far as the vehicle is concerned, engine speed is unimportant as long as output matches demand; however, the rpm at which the engine delivers power has a significant effect on fuel consumption.

Challenge and Opportunity of Part-Load Fuel Consumption

A typical four-stroke, spark-ignition gasoline engine operates most efficiently near WOT. That is, the engine consumes the least fuel per unit of output when it is developing close to maximum power across its rpm band. With diesel and two-cycle gasoline engines, peak fuel efficiency occurs at slightly lower loads. A diesel engine operates most efficiently at approximately 60 to 70 percent load, and a two-cycle engine develops peak efficiency at approximately 50 percent load.

Over the course of a trip, engine efficiency is continually changing according to load. Unfortunately, it changes in a way that is essentially opposite of what is needed for good fuel economy. When engine load is reduced at cruising speeds, efficiency plummets. When engine load increases, such as during brief periods of acceleration, efficiency increases. When the engine is operating under a light load, it may consume two or three times the fuel for a given output as it does when loaded into its region of maximum efficiency (minimum bsfc). Most automobiles generally operate with the engine loaded to only a fraction of its capability.

Part-load fuel consumption characteristics are responsible for a popular myth, namely that reducing vehicle weight does not do much to improve fuel economy. There is some truth at the basis of the idea because if weight is removed from a vehicle without a corresponding reduction in engine size, fuel consumption will not necessarily decrease. This is because the engine is more lightly loaded in the lighter vehicle and therefore is operating in a region of reduced fuel efficiency. Increasing engine size and leaving vehicle weight unchanged produces the same result. As a general rule, doubling engine size in a given vehicle will result in approximately 50 percent greater fuel consumption per distance traveled.

Reducing vehicle weight has a direct effect on fuel consumption if the engine is downsized to maintain an equivalent power-to-weight ratio. Moreover, fuel economy improves when the engine is downsized for a lower power-to-weight ratio, mainly because of increased engine loading at cruise. For maximum fuel economy, one would minimize vehicle weight to keep rolling resistance and inertia loads at a minimum, reduce vehicle size for minimum aerodynamic drag, limit the power-to-weight ratio, and operate at cruise at the lowest practical engine rpm.

Poor part-load fuel economy normally results from excessive installed power, which translates into light engine loading at cruising speeds. Under lightly loaded conditions, most of the energy (fuel) is expended to keep the engine working, rather than to provide motive power. This results primarily from the fact that pumping and friction losses are speed dependent and function independently of power output. Figure 3.21 shows friction and pumping losses of an engine at WOT. At high-output levels, friction and pumping losses are relatively insignificant compared to the energy lost to rejected heat; however, if a given rpm is maintained and the engine is throttled to reduce output, the portion indicating friction and pumping losses will expand into the brake horsepower region virtually in step with the degree to which the engine is throttled. Carried to the extreme, an engine revved under no load will exhibit friction and pumping losses that entirely replace the brake horsepower band. Figure 3.21 shows how output affects the various losses.

The relationship between engine load and fuel efficiency is most clearly illustrated by a fuel consumption map. Such a map indicates fuel consumption in islands, which represent rpm per load regions of operation (Figure 3.22). Cruising power should be taken at the lowest possible rpm, where the engine will be loaded into its region of greatest fuel efficiency. Peak power demands are then met by transmission downshifting, which lets the engine accelerate into its region of greatest output.

Part-load fuel consumption characteristics offer both challenges and opportunities. At urban speeds, the typical automobile operates at very light loads,

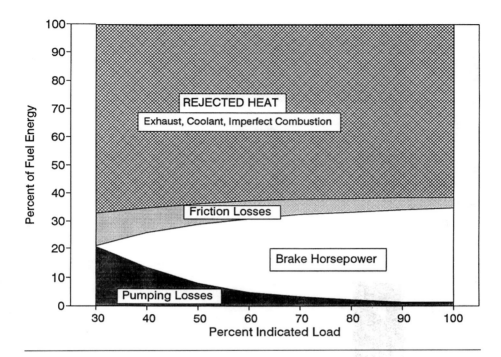

FIGURE 3.21 Distribution of Fuel Energy

wherein losses to rejected heat, friction, and pumping loads can account for 90 percent or more of the fuel consumed. Operating at greater engine loading is one of the most straightforward methods of increasing vehicle fuel economy; however, a vehicle's ability to respond to peak demands suffers when installed power is limited in order to load the engine into its region of minimum bsfc at cruise. But small, low-inertia engines that rapidly respond to peak demands, supercharging systems that enable smaller engines to provide greater peak power, and continuously variable transmissions can help offset part-load inefficiencies.

Honda's Insight hybrid electric vehicle employs a relatively new strategy for overcoming the shortfall in peak demand power with reduced installed power. Engineers at Honda fitted an underrated engine with an electric motor: essentially a pancake motor that fits between the IC engine and the transmission. The motor provides extra torque during periods of peak demand, and thereby makes up for the shortfall in torque of the underrated IC engine. By reducing installed IC engine power significantly below the

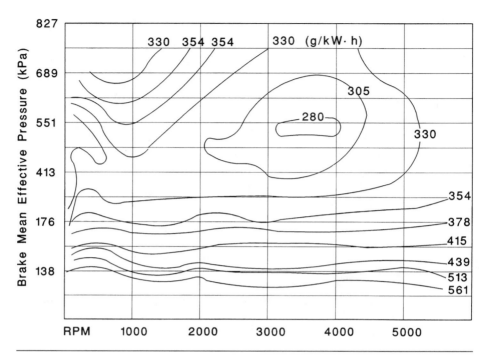

FIGURE 3.22 Fuel Consumption Map

level that would typically be required for adequate acceleration, and relying on electric assist for peak demands, engineers were able to achieve more than double the fuel economy of an equivalent vehicle relying on IC engine power alone.

The foregoing discusses part-load fuel economy from a purely applications perspective, but improved engine design can also have a large impact on part-load fuel consumption characteristics. Today, much opportunity exists within the areas of rejected heat and part-load friction and pumping losses. The energy lost to rejected heat can be reduced by higher operating temperatures and by using exhaust recovery systems, such as turbocharging. Engine pumping losses can be reduced with better valving systems. Improved valving, more efficient supercharging, better fuel delivery, variable compression, "homogeneous charge compression ignition" combustion, and even full expansion concepts offer several ways to put more of the fuel's energy to work in the reciprocating engine.

Rotary Valves

Poppet valves were adopted by early engine designers because they provided a quick and simple solution to valving needs. Poppet valves are easy to lubricate, they seal well, and they operate with relatively little load on their linear bearing surfaces. In addition, the sealing surface of the valve lifts away from, rather than slides across, its stationary mating surface. Consequently, there is no need to lubricate surfaces that are directly exposed to the hot gases of combustion, but poppet valves also have inherent disadvantages. Their characteristically high surface area and the feature of opening into the combustion chamber tend to place limitations on combustion chamber and piston design. Hot spots can develop on the exhaust valve and cause preignition. In addition, the reciprocating action of poppet valves often limits engine rpm because of their tendency to "float" at high speeds. These characteristics have been an ongoing problem for engineers since the poppet valves' inception. Many rotary-valve systems have been developed in an attempt to solve the shortfalls of the poppet valve.

Rotary valves have been around almost since the inception of the Otto engine. The design is actually a derivative of the steam engine's sliding valve. The fact that the valve rotates, rather than reciprocates, makes it inherently appealing. Because of its rotary action, the valve can operate at much greater speeds. A rotary valve also runs cooler because the same surface is not continuously exposed to hot combustion gases. This characteristic reduces the tendency for preignition, and it allows the engine to operate at higher compression ratios on lower octane fuel. The rotary valve exhibits better breathing characteristics, and therefore less pumping losses, even though the aperture opening does not provide the average aperture size (over cycle time) of a poppet valve of equal physical aperture. Rotary-valve engines tend to be more fuel efficient, they are quieter, and they offer a lower engine profile. Reduced valvetrain friction is also claimed by proponents, although comparisons are often made with older poppet valvetrain layouts. As early as 1946, Marcus Hunter, author of *Rotary Valve Engines,* calculated that rotary valves actually develop greater friction loads than poppet valves.[8]

Unfortunately, rotary-valve designs have been plagued with significant problems of their own. Typical problems include thermal distortion, excessive wear, imperfect sealing, and excessively high oil consumption because of the need to lubricate surfaces that are exposed to combustion. This, combined with the fact that poppet valves are proven and reliable, has tended to discourage research of rotary valves by major car companies. The field has therefore been left primarily to independent experimenters and to companies in related industries.

The most notable early rotary-valve designs are probably those produced by Rowland Cross. Cross began developing rotary-valve engines in 1922. These early experiments produced impressive results that even Cross did not fully appreciate at the time. Speaking in London at a meeting of the Institution of Mechanical Engineers in 1957, Cross said: "Apparently I did not realize at the time that I was setting standards only met by the finest poppet engines then in existence, and which were highly developed in comparison with my crude and very badly made rotary-valve engine." In the 1970s, Cross Manufacturing, in conjunction with Esso Research Co., developed a unique four-cylinder radial engine that used a single rotating valve at each cylinder to control both the intake and exhaust ports. The compact 1.6-liter engine is reported to have been extremely fuel efficient and would operate on low-octane gasoline.

Work over the past 15 years at GV Engine Research PTY Ltd., in Cronulla, Australia, has produced a design that developers claim eliminates many of the valve's difficulties. The key component is a specially designed sleeve that controls lubrication and provides a positive seal (Figures 3.23 and 3.24). Most of the work at GV Research has been with small displacement, four-cycle engines. Part-load specific fuel consumption of engines equipped with the new valve is lower at all points of the operating range. Most important, the rotary valve has the greatest impact on fuel efficiency at low rpm and light loads typical of the urban environment.[9] Improved part-load fuel consumption is characteristic of rotary valves (Table 3.3).

George Coates, an Irish inventor and mechanical engineer, took a different approach by entirely eliminating oil from the valve's environment. Unlike the GV cylindrical valve design, Coates uses a spherical valve made of Nitral-loy that rides on a floating carbon-ceramic pressure-activated seal. The valve is unlubricated, except for the minimal lubrication provided by the fuel (Figures 3.25 and 3.26). Higher compression ratios and lower valve temperatures

TABLE 3.3 *Rotary Valve Efficiency Improvements (Based on 500 cm³ SI Engine)*

	Percentage of Full-Load Operation			
	15%	25%	40%	60%
RPM	Percent Reduction in Brake Specific Fuel Consumption			
1,500	30%	11.8%	8.0%	9.3%
2,500	21%	12.3%	11.5%	8.0%
3,500	18%	13.7%	8.6%	5.2%

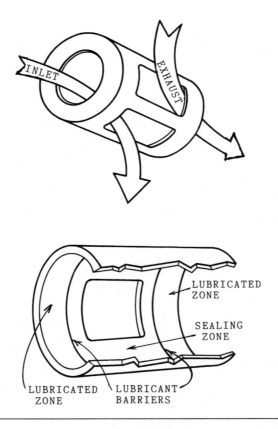

FIGURE 3.23 Rotary-Valve Gas Path and Seal Details. Top drawing shows gas path through rotor. Bottom drawing shows floating seal that eliminates oil consumption problems typical of previous designs.

Courtesy: GV Engine Research, Cronulla, Australia

are claimed. By eliminating the hot, poppet exhaust valve, both preignition and NO_x emissions are reduced. Each cylinder is equipped with two valves: one for intake and one for exhaust. Variable phasing is therefore possible. Coates claims an overall 12 percent savings in fuel and a 20 percent increase in output.

A perfected rotary valve might be especially applicable to the valved, direct-injection, two-stroke engine. A power stroke at each revolution, along with a characteristically short duration for exchange of gases, triples the speed at which gases must be cycled through the combustion chamber. The rotary

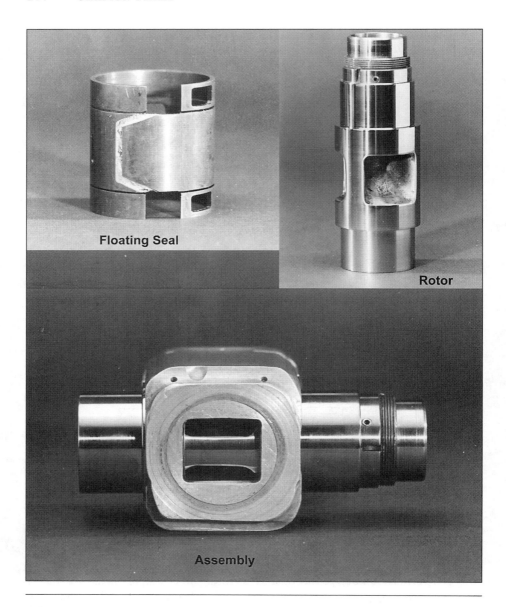

Floating Seal

Rotor

Assembly

FIGURE 3.24 Rotary-Valve Assembly. Low-profile rotary-valve assembly eliminates complicated valvetrain and allows extra freedom in combustion chamber design.

Courtesy: GV Engine Research, Cronulla, Australia

FIGURE 3.25 Coates Retrofit Rotary-Valve Cylinder Head

Courtesy: Coates Enterprises, Ltd.

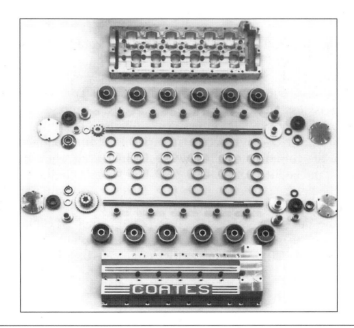

FIGURE 3.26 Coates Rotary-Valve Head Assembly

Courtesy: Coates Enterprises, Ltd.

valve's improved breathing and greater speed capability could resolve some of the difficulties of the valved, two-stroke engine. Rotary valves might also reduce manufacturing costs. One manufacturer of lightweight industrial engines has implemented a program with one of their models to eliminate the camshaft, pushrods, and rocker arms and replace them with a simple, self-contained, low-profile rotary-valve cylinder head. Lower costs are especially important in automotive two-stroke applications where conventional valving may increase efficiency but also increase costs and thereby diminish the cost advantage of the two-stroke engine. Although the rotary valve has a history of designs that have fallen short of creating a practical replacement for the poppet valve, that does not preclude the rotary's ultimate success.

Variable Compression Ratio

The compression ratio of an engine is typically fixed at a specific value. Of necessity, compression ratio is a compromise value in order to account for operational extremes. As a result, the engine is often operating at less than peak efficiency throughout the rest of its operating band. The ability to vary the compression ratio according to operating conditions can significantly improve engine performance and efficiency.

Higher compression ratios let the engine run with leaner mixtures, and they improve the engine's thermal efficiency. Detonation and overheating at high loads, especially in the lower speed ranges, normally establishes the upper compression limit, but a compression ratio that is appropriate at engine high loading is less than ideal for lighter loads. Compression ratio also affects the production of NO_x emissions, and the necessary compression ratio for minimum NO_x production also varies according to operating conditions. Ideally, the compression ratio would be increased when the engine is lightly loaded and reduced at lower speeds and greater engine loads. A continuously variable compression ratio that changes according to operating conditions could hold NO_x emissions in check and increase fuel efficiency throughout the operating range. More important, fuel efficiency would be especially enhanced in the lower load ranges typical of city driving. Variable compression ratio (VCR) technologies give engineers the ability to selectively adjust compression according to operating conditions.

Variable compression ratio can be achieved in several ways. VCR engines have used variable-height pistons, vertically sliding cylinder heads, adjustable-length connecting rods, and secondary combustion chambers with adjustable volumes. A VCR engine developed by Volkswagen utilizes a secondary

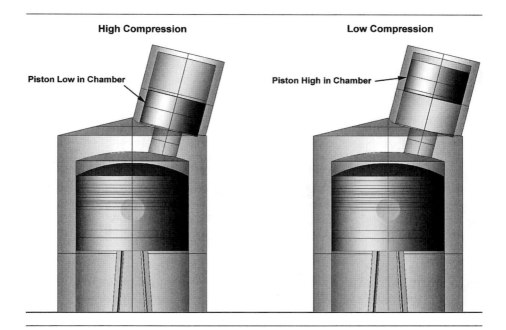

FIGURE 3.27 Secondary Chamber VCR Scheme

chamber at each cylinder, with a small piston that is adjusted to vary the volume of the chamber and thereby vary compression (Figure 3.27). Piston depth is controlled by an electric motor and worm gear arrangement.

Another VCR scheme runs along the lines of the design developed by Gregory Larsen (U.S. Patent No. 5,025,757). Larsen varies compression ratio by pivoting the cylinder block in relation to the crankcase, as shown in Figure 3.28. An advantage of the Larsen design is that it directly varies the size of the combustion chamber without resorting to secondary chambers, variable geometry of internal moving components, or variations in valve timing. Saab's prototype Saab variable compression (SVC) engine works similarly by tilting the entire upper section of the engine. Saab's "monohead" is pivoted at the crankcase using a hydraulic actuator, which varies compression ratio between 8:1 and 14:1. Like Larsen's design, a flexible bellows seals the stationary crankcase to the movable upper section of the engine.

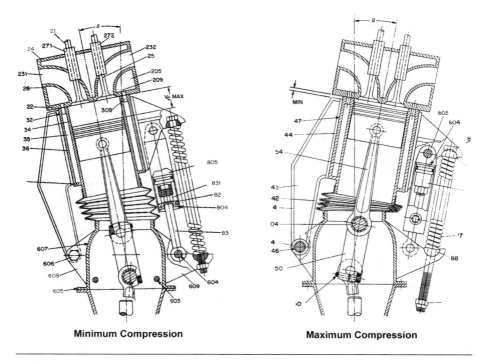

Minimum Compression   Maximum Compression

FIGURE 3.28 Larsen Variable Compression Ratio Scheme. Variable compression is achieved by pivoting the cylinder block in relation to the crankcase. Drawings are from U.S. Patent No. 5,025,757.

Variable Valve Actuation

Variable valve actuation (VVA) is another emerging technology that is already producing significant improvements in engine performance. Historically, the point at which valves open and close, the *dwell time,* and the overlap between intake and exhaust valves (phasing) are fixed to specific crankshaft rotational points (crank angles). As a result, the timing of these events is compromised throughout much of the operating band because ideal valve timing varies according to engine speed and load. Control over valve lift allows the engine to operate more efficiently over the entire speed/load range. The torque curve, as well as emissions and fuel consumption characteristics, become more fluid and controllable attributes because they are no longer fixed by the geometry of the engine.

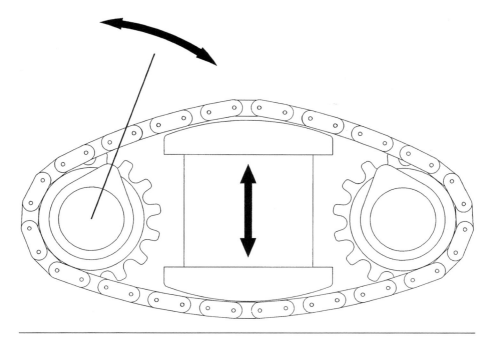

FIGURE 3.29 Porsche Variable Cam Phasing

Early versions of VVA systems enhance specific areas of engine performance through a limited number of valve timing and phasing options. The more sophisticated the system, the greater the benefits and the higher the costs. A simple two-position system of control over intake valve lift is relatively inexpensive and can still provide improved idle and a flatter torque curve. Such systems are incorporated into engines of production automobiles from Honda, Nissan, Mercedes-Benz, Alpha Romeo, and Porsche. Figure 3.29 shows the operating principle of the variable cam phasing (VCP) system used in the Porsche 968. It controls the angular relationship between intake and exhaust cams by moving the chain-tensioning device.

Continuous and nonincremental control over valve timing and phasing, controlling intake and exhaust valves independently of each other and independently of the crank angle, opens new possibilities in engine design. VVA provides engineers with control of events that was not possible before. VVA can even control compression ratio and change the effective displacement of the engine. Advanced VVA systems are expected to result in more power from smaller engines, greater low-rpm torque, a broader torque band, greatly

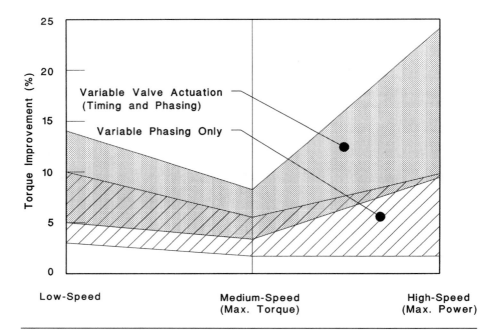

FIGURE 3.30 Potential Torque Improvement with Variable Valve Actuation

improved part-load fuel efficiency, and lower emissions. Figure 3.30 shows the torque improvements possible with an optimized VVA system.[10]

Several versions of a VVA system were built and tested at Clemson University. Professor Alvon C. Elrod, along with the help of engineering students, built several prototypes of a variable timing camshaft system that produced a 20 percent improvement in fuel efficiency in laboratory tests. The system can be retrofitted to existing engines or incorporated into an existing line of engines without redesigning the engine or the production line. The device takes the place of the existing camshaft, and on engines with dual overhead cams, it provides control over timing as well as phasing (Figures 3.31 and 3.32).[11] Similar split camlobe systems have shown up on production engines, including past-master Honda with their VTEC engines.

Development began with the premise that poor part-load fuel economy is caused largely by the throttling process of the spark-ignition engine. Part-load throttling losses occur when charge airflow is restricted (throttled) to control engine output. Restricting the intake creates a vacuum, which places an additional pumping load on the engine. At WOT, manifold vacuum is

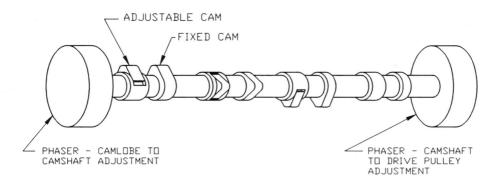

INTAKE CAMSHAFT

FIGURE 3.31 *The Clemson Variable Timing and Phasing Camshaft.* Unique camshaft can be assembled in a variety of configurations. The design is for a dual overhead cam application. It allows for optimization of all valving events.

Courtesy: Clemson University

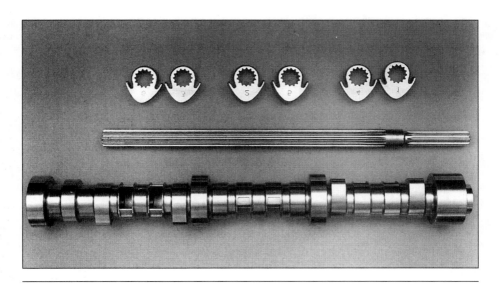

FIGURE 3.32 *Clemson Variable Camshaft.* Photo illustrates simple design. With this design, the entire camlobe rotates. A design utilizing split camlobes, in which movable camlobe halves rotate axially out of plane with their mating fixed halves, may be used to control dwell, phasing, and timing.

Courtesy: Clemson University

decreased because charge air can flow freely into the engine. Variable valve lift systems control engine speed by varying valve timing instead of throttling charge airflow. As a result, the engine runs under conditions that more closely emulate WOT.

Another VVA concept centers on electromagnetic valve lift. These systems are like the Electromagnetic Valve Actuator (EVA) developed by Aura Systems, El Segundo, California (Figure 3.33 and Table 3.4). Aura's EVA is configured as a complete drop-in module that replaces the entire valve and seat assembly, as well as the rest of the valvetrain. Valve lift is then controlled by the car's electronic control unit. Dwell, timing, and phasing are infinitely variable. Moreover, the operating schedule can be easily switched to accommodate different fuels. In combination with similarly controlled fuel metering, flexible fuel vehicles can operate in essentially a dedicated mode with different fuels. Similar valve systems have been developed at BMW and Renault. The electrical power required to operate an electromagnetic valve lift system at high engine speeds has been a continuing problem. Twelve-volt electrical systems can become overtaxed at high engine speeds. The 42-volt electrical system, however, will make adequate electrical power available for the first time.

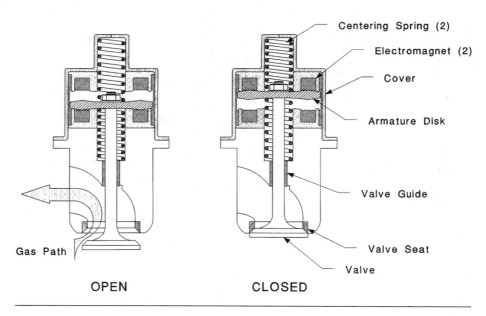

OPEN   CLOSED

FIGURE 3.33 Aura Systems Electromagnetic Valve Actuator (EVA)

www.aurasystems.com

TABLE 3.4 Aura Systems EVA Specifications and Performance

Specification	2-Valve Cylinder*	4-Valve Cylinder**
Mechanical		
Maximum rpm	6,250	7,500
Transition Time	2.42 msec	3.3 msec
Lift	7.62 mm	9.5 mm
Valve Opening Force	500 N	500 N
Valve Seating Force (adjustable)***	250 N	250 N
Touch-Down Velocity (setable)	0.15m/s	0.15 m/s
Diameter of EVA Body	62 mm	40 mm
Length (surface of head to top of body)	72 mm	85 mm
Valve Mass	90 g	40 g
Overall EVA Weight (includes valve)	0.85 kg	0.35 kg
Electrical		
Required Voltage	12 or 24 V Nominal	12 or 24 V Nominal
Current Draw @ 24 Volts:		
Initialization (on start of engine)	32 amp, 8 msec	22.8 amp, 5.5 msec
Transition (once per transition)	18.4 amp, 5.2 msec	13.1 amp, 5.5 msec
Holding (between lifts)	4.5 amp capture to release	3.2 amp capture to release
Power Consumption in Watts (including 50% alternator efficiency loss):	Ea. Valve (8 Valves)	Ea. Valve (16 Valves)
800 rpm	14.4 (115.2)	9.3 (148.9)
2,500 rpm	31.4 (251.2)	20.3 (325.2)
6,000 rpm	66.4 (531.2)	43.1 (689.3)
7,500 rpm	—	52.8 (845.0)

Source: Aura Systems, Inc.

*Verified by tests. Tested on a 2.3L Ford Ranger.

**Estimated.

***The force at which the valve is held against the valve seat.

Diesel Engine

The compression-ignition (CI) engine was conceived by Dr. Rudolph Diesel in an attempt to improve on the poor fuel efficiency of the spark-ignition engine. A paper on Diesel's theories was published in 1893 and stimulated additional research among German contemporaries. Early designs were cantankerous and some prototype engines actually exploded during tests. While experiments were being conducted in Germany on Diesel's design, an English engineer by the name of Herbert Akroyd-Stuart developed and patented a more successful version. Stuart's design utilized an intake of uncharged air that was ignited by injecting a liquid fuel directly into the combustion chamber, which is how the modern diesel engine operates. In a diesel engine, ignition takes place when the fuel is injected. Ignition timing is controlled by fuel delivery, rather than by a spark. Engine rpm is controlled by regulating injection duration. Air intake is unthrottled.

Diesel engines are either direct injection (DI) or indirect injection (IDI) designs. The DI engine is more fuel efficient, requires more exacting tolerances, is less tolerant of fuel inconsistencies, and is inherently more noisy and rough than an IDI engine. In the early 1980s, a typical diesel-powered automobile would consume about 30 percent less fuel than an equivalent gasoline-fueled car. Today, improvements in SI engine technology have reduced the gap to about 10 to 15 percent in the diesel's favor. Diesel engines are still more fuel efficient, and efficiency is likely to improve with new developments in turbocharging and low heat-rejection technologies.

Part of the diesel's greater fuel economy is caused by the slightly greater volumetric energy content of diesel fuel, but most of it comes from the inherently greater thermal efficiency of the cycle. The high compression ratio and the reduced pumping losses of the unthrottled intake are generally accepted as the primary contributors to the diesel's high fuel efficiency. Both power and fuel economy are significantly improved by turbocharging. Typical power and fuel consumption curves are shown in Figure 3.34.

Regulated emissions are inherently lower with the diesel engine. Without aftertreatment, diesel exhaust emissions are as low as those of a typical gasoline engine equipped with a three-way catalytic converter. The diesel's inherently clean combustion has therefore allowed diesels to operate without exhaust aftertreatment. The emissions profile is not all positive, however, and more stringent regulations are due. Diesel engines produce a large quantity of particulate emissions, primarily in the form of soot that is laden with liquid hydrocarbons. Animal studies indicate that some of these hydrocarbons may be carcinogenic. Emissions of aldehyde are also significantly greater than with a gasoline engine. Consequently, different exhaust species may require special emphasis in diesel emission controls.

A diesel engine is typically more costly to manufacture and more expensive to service. Also, a diesel's high-pressure fuel pump is more easily damaged by contaminated or water-laden fuel; however, proper filtration and routine maintenance can avoid problems attributed to fuel contamination, and higher manufacturing costs are ultimately returned to the customer in the form of increased engine life. A diesel engine will typically run 320,000 km (200,000 mi.) or more.

Regardless of its inherent advantages, diesel engines have yet to win favor with U.S. consumers. In Europe, the diesel engine is much more popular and

TYPICAL DIESEL POWER CURVE

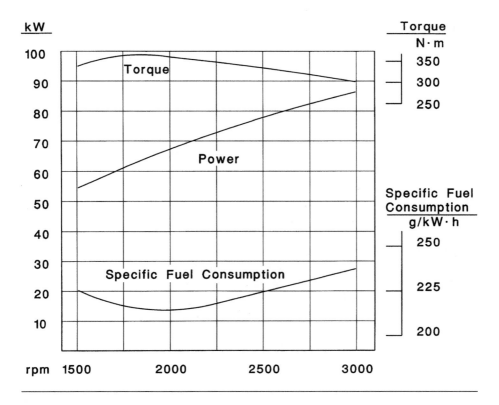

FIGURE 3.34 *Typical Diesel Power Curve*

far more prevalent. One-third of new cars sold in Belgium are equipped with diesel engines. In France, diesels account for nearly one-quarter of new-car sales, and in Germany, Spain, and the United Kingdom, diesel sales average 11, 14, and 11 percent, respectively. Although the diesel engine is inherently noisier and rougher than its gasoline counterpart, proper design can largely eliminate these problems. In addition, the engine also offers much opportunity for significant improvements in operating efficiency. Improvements center on refined fuel delivery systems, more precisely controlled combustion, and low heat-rejection technologies.

Supercharging

In the past, supercharging has been associated with increased power for high-performance cars, and often at the expense of fuel economy. But supercharging can also be used to increase engine efficiency and improve vehicle fuel economy. The difference in results depends on the orientation of the application. If a given vehicle is supercharged for increased power and performance, fuel consumption will also increase. If, however, the goal is to improve fuel economy, the designer can use supercharging to achieve equal power and performance from a smaller, lighter engine and thereby improve fuel economy. Fuel economy gains as high as 20 percent or more are typical for a supercharged diesel. Traditionally, the spark-ignition engine is less responsive to supercharger-induced improvements in fuel economy. One can, however, operate a small engine at relatively high loading and utilize the supercharger to increase output in order to meet peak demands, which will significantly improve fuel economy.

The supercharger operates by pressurizing the incoming air. Pressurization reduces the internal pumping losses associated with air intake, and it raises combustion chamber pressure for increased combustion efficiency. A supercharger therefore increases the power output of the engine. A supercharged engine of a given size burns more fuel and produces more power than a naturally aspirated engine. A supercharger also allows a smaller engine to produce equal power, and it improves the low-rpm torque characteristics of the engine.

The traditional supercharger is essentially a mechanically driven compressor or blower. A supercharger is normally belt-driven from the engine output shaft. A more widely used method of pressurizing charge air is turbocharging. The turbocharger works on the same principal, except it is driven by exhaust gases passing through a turbine assembly. The turbocharger is therefore a type of waste heat recovery system. Another exhaust heat recov-

TABLE 3.5 *Comparison of Supercharge Systems (+ signifies advantage over other systems)*

Characteristics	Exhaust Turbocharger	Comprex Wave Charger	Positive Displacement Supercharger
Nominal power output	++	+	
Torque at low engine rpm		++	+
Response		+	++
Improved fuel consumption in consumer driving cycles		++	+
Noise radiation	++		
Pollutant emissions	+	++	
Function with additional exhaust cleanup devices	+		++
Space requirement and free choice of placement	++		+
Technology close to production release	++	+	
Manufacturing and installation cost	+		++

Source: *Automobile Technology of the Future,* SAE R-107, Society of Automotive Engineers, Warrendale, PA, 1991.

ery system is the pressure-wave Comprex unit. Each type of system has advantages and disadvantages.

An interesting concept called air-injection supercharging was explored many years ago at General Motors Research Labs.[12] Instead of a conventional pump set to pressurize charge air, compressed air from high-pressure tanks was introduced through a third valve directly into the cylinders during the compression stroke. An engine-driven compressor then recharged the compressed air reservoirs. Tests of a converted 2.7L engine showed a 250 percent increase in power output at 2,000 rpm. With the compressor inoperative, fuel economy was 47 percent greater at 32 km/h (20 mph) and 25 percent greater at 97 km/h (60 mph). With the compressor engaged, gains dropped to about 35 percent and 15 percent, respectively. Disadvantages included the large volume of air required and the complications of the high-pressure delivery and metering system, which required an extra valve at each of the eight cylinders.

Low-Heat-Rejection Engines

The low-heat-rejection (adiabatic) diesel engine is based on the concept of retaining the waste heat that is now discarded through the cooling and exhaust systems. Experimental designs concentrate on the heat lost to the coolant system, leaving exhaust recovery to various forms of supercharging. With an adiabatic engine, the cooling system is eliminated and the engine is then insulated against radiation heat losses. Ceramic surfaces on pistons, valves, and cylinder heads are normally required to withstand the extremely high temperatures. Operating temperature of the cast iron increases from 200°C to about 875°C; even higher temperatures are possible. Theoretically, much of the heat that is normally lost through the coolant should be retained within the engine, resulting in a corresponding increase in thermal efficiency.

Several automobile manufacturers have built experimental adiabatic engines. In the early 1980s, an experimental Ford DI diesel, with a 93 percent reduction in cooling system heat loss, produced fuel economy gains of 2 to 25 percent (depending on engine load) and 38 to 56 percent lower particulate emissions in the light to medium load ranges.[13] Adiabatic technology is still embryonic. Fuel efficiency gains as great as 40 to 50 percent are theoretically possible if waste heat can be retained. Early experiments, however, indicate that much of the rejected heat is not necessarily retained by simply eliminating the cooling system. Instead, heat is redirected to the exhaust system, where it is rejected in the form of higher exhaust temperatures.

Two-Stroke Cycle Engine

Although some automobiles have been equipped with two-stroke engines, traditionally the two-stroke cycle has not been considered a satisfactory automobile power plant. The most limiting characteristics of the engine have been its high fuel consumption and high exhaust emissions. Irregular part-load combustion, relatively short life, and reduced low-rpm torque are other disadvantages. For a while, it looked as though its characteristically high emissions might actually force the two-stroke engine into oblivion. Today, electronic controls and new fuel delivery and scavenging techniques point to a brighter future.

Theoretically, the two-stroke cycle is more fuel efficient than the four-stroke cycle. This results primarily from the power stroke at each revolution, which essentially reduces friction losses by half. Another advantage includes improved part-load fuel consumption. Lowest specific fuel consumption occurs at approximately 50 percent load, which is much closer to the typical loads experienced during urban driving. The two-stroke engine also produces more power per given displacement, and it is lighter, mechanically simplified, and less costly to manufacture than a four-stroke engine of equal power. The engine's two primary deficits, poor fuel consumption and high exhaust emissions, have been its undoing. These shortfalls arise primarily from the difficulty in controlling the exchange of gases.

In its most basic form, the two-stroke engine has no valves, at least not in the traditional sense. Gases flow into and out of the cylinder when ports in the cylinder wall are uncovered as the piston nears the bottom of the stroke (Figure 3.35). By uncovering the exhaust port first, cylinder pressure drops, and much of the burned charge is expelled into the exhaust header. Slightly afterward, the intake port is uncovered. At this point, charge air, which has been pressurized in the crankcase by the downward-migrating piston, is injected into the cylinder. As the piston moves upward after bottom dead center, the ports are closed off and the fresh charge is trapped and compressed, where it is ignited near top dead center, and the process begins anew.

The task of clearing out exhaust gases and introducing a fresh charge takes place at essentially the same time, during approximately 250 to 270 degrees of crankshaft rotation. As a result, some of the exhaust gases mix with the fresh charge, and some of the charge air ends up being discharged through the exhaust port, unburned. Short-circuiting losses alone can amount to as much as 10 to 30 percent of the total fuel consumed. The engine's characteristically high emissions of unburned hydrocarbons results primarily from the unburned fuel that escapes into the exhaust header. Because the four-stroke engine dedicates one entire cycle to the exchange of gases, short-circuiting losses do not occur, and it is much easier to keep exhaust and intake gases from mixing.

The limited time available for exchange of gases is another problem. Lacking a separate stroke for gas exchange, one might assume that the two-stroke engine must complete the process in half the time. Actually, it must be accomplished in one-third or less of the time available in a four-stroke cycle engine. Engine efficiency is significantly improved when uncharged air is used for scavenging and the fuel is injected directly into the cylinder after the ports have been covered by the upward-migrating piston. This requires an injector nozzle pressure of approximately 100 bar, and events must take place over an extremely condensed time budget. Table 3.6 shows the time allotted to events in a DI two-stroke engine at 6,000 rpm.[14]

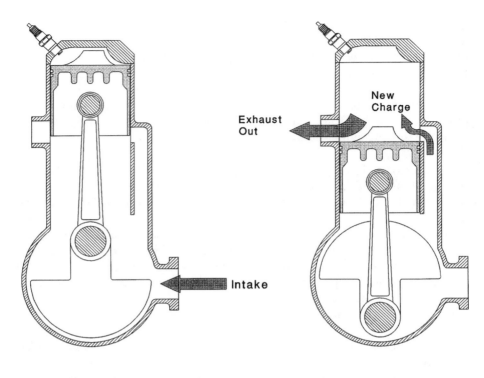

COMPRESSION STROKE   GAS EXCHANGE

FIGURE 3.35 Two-Stroke Cycle

TABLE 3.6 Direct Injection Two-Stroke Cycle Schedule of Events

Event	Time Available	Degrees Crank Angle
Exhaust and Induction	4 millisecond	144
Fuel Injection	1 millisecond	36
Mixing and Vaporization	2 millisecond	72
Combustion and Expansion	3 millisecond	108

Source: SAE Paper No. 861242, www.sae.org

Injection and vaporization require a spray velocity in the order of 100 meters per second, and droplets must be no larger than 10 to 20 microns immediately after leaving the nozzle. Complete vaporization and precise control of events were not possible before the advent of electronic controls and high-pressure DI systems. This technology is responsible for the resurgence of interest in the two-stroke cycle engine.

Active Thermo-Atmosphere Combustion

The active thermo-atmosphere combustion (ATAC) engine is somewhat of a hybrid design that combines attributes of the spark-ignition engine and the compression-ignition engine. Scientists in Japan discovered the process in the 1970s and coined the name active thermo-atmosphere combustion. Today, it is more often called homogeneous charge compression ignition (HCCI). Combustion in an HCCI engine is initiated by controlled auto- ignition. It is a different combustion process than occurs during either spark-ignition or diesel combustion. In a spark-ignition engine, combustion begins when a kernel is established by the spark plug. A flame front is thereby established, and the flame then propagates outward from the kernel and quenches against the cooler surfaces of the cylinder walls, piston, and cylinder head. Under this scenario, combustion begins and ends at a precisely defined point in time. During HCCI there is no traditional flame front, and the onset of combustion is much more gradual. Instead of a traditionally propagating flame front, HCCI is a more homogeneous process, with combustion initiating at multiple points throughout the charge and progressing at a slower rate.[15]

Once initiated, HCCI continues thereafter regardless of the occurrence or absence of an electric spark. The shift from spark-controlled combustion to HCCI is smooth. During HCCI, cycle-to-cycle P_{max} becomes much more uniform, and the engine runs more smoothly and quietly, and without knock. During light to moderate loads, HCCI stabilizes the lean-burn charge, essentially eliminates the misfires typical of the two-stroke cycle, and thereby significantly reduces fuel consumption. A diagram of cylinder pressure curves reveals a uniform rise and fall in pressure over multiple cycles that is unequaled by the traditional combustion process. Because HCCI is not dependent on spark ignition, however, once established it can produce a runaway condition. Variable compression ratio and control of charge air temperature are used to control HCCI combustion. HCCI occurs at high inlet temperatures and high compression ratios.

Because combustion is homogeneous and very lean mixtures are used, combustion temperatures are lower, which results in very low NO_x emissions. But low combustion temperatures also tend to produce greater amounts of unburned hydrocarbons. Part-load fuel economy, however, is greatly improved, and overall fuel economy is also greater. Fuel consumption in a vehicle powered by an HCCI engine can be as little as one-half that of an equivalent vehicle powered by a conventional spark-ignition engine. Because HCCI combustion is difficult to maintain at light loads and low engine speeds, an engine typically transitions between HCCI combustion and spark-ignition combustion, depending on the operating condition.

Transmission as a Tool for Reducing Fuel Consumption

Unfortunately, IC engines characteristically develop power in a way that is not very well matched to the requirements of an automobile's operating schedule (Figure 3.36). Modern automobile engines can deliver little power at very low rpm. Engine output climbs steadily from idle speed and typically reaches maximum levels in the range of 3,500–5,000 rpm. In general, the vehicle's demand for power is diametrically opposed to these output characteristics. A vehicle typically needs maximum power (for acceleration) near its minimum speed. Road load at cruising speeds may demand as little as 5 to10 percent of the engine's capability. Passing and negotiating grades impose high, but temporary, power demands with moderate to zero change in vehicle speed. The component that makes it possible for these two mismatched curves to work together is the transmission.

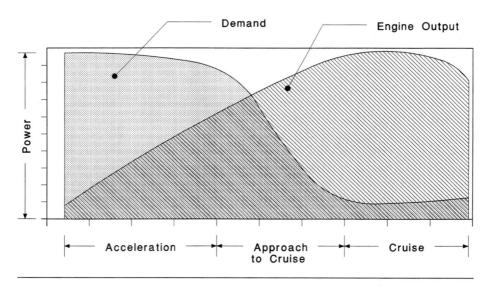

FIGURE 3.36 Vehicle Power Demand Compared to Engine Output Characteristics

In the case of a manual-shift transmission, the task of selecting an appropriate ratio is delegated to the driver. In the 1940s, the automatic transmission emerged and ultimately assumed the job of managing ratios, at least for most U.S. drivers. Early versions were sluggish, troublesome, and inefficient devices that could not match the performance of an adroit human at the controls of a manual-shift gearbox. Today, automatic transmissions are much more efficient, and they typically run 160,000 km (100,000 mi.) or more with very little service. In addition, the automatic transmission has actually become a tool for increasing vehicle performance and fuel economy.

The ability of the transmission to improve fuel economy results from the part-load fuel efficiency characteristics of the IC engine. When acceleration power is no longer required, the transmission can upshift to a ratio that will load the engine into its region of lowest brake specific fuel consumption. A properly designed shift schedule can have a significant effect on fuel economy. Unfortunately, the match between vehicle speed and engine speed is usually a compromised value because of the limited number of ratios that are feasible with a conventional multiratio transmission. Multiratio transmissions are normally limited to three, four, or five discrete ratios. Precise engine load management, however, requires a greater degree of ratio selectivity. A promising alternative is the continuously variable transmission.

Continuously Variable Transmissions

The continuously variable transmission (CVT) has been around as long as the automobile. Engineers have always recognized its theoretical advantage over the multiratio gearbox. A CVT enables the engine to run at its most fuel-efficient or most power-efficient speed while driving the vehicle at any speed desired. With a CVT, engine speed and vehicle speed are no longer connected by a series of discrete ratios. Instead, they can function independently across a wide and stepless band according to engine characteristics and performance requirements. The advantages of this infinite ratio selectivity are enormous. Most obvious in the IC engine application is that the engine can be loaded into its most fuel-efficient region at cruising speeds, then allowed to accelerate into its region of greatest output when peak power is needed, regardless of vehicle speed. Practical problems have consistently plagued the design, but the CVT is now coming of age.

There are essentially two types of continuously variable transmissions: belt CVTs and traction drives. A belt CVT utilizes a belt that connects two pulleys of changeable pitch diameters. The most familiar variety is the unlubricated composite V-belt type normally found on snowmobiles and all-terrain vehicles. Quincy-Lynn's Urbacar and Trimuter, as well as several European microcars, have utilized transmissions of this type. Perhaps the most notable automotive application is the DAF Variomatic developed by Van Doorne. Unfortunately, transmissions of this type work less well as engine size is increased. Excessive wear and the inability to transmit high torque loads have been consistent problems. In an effort to improve power transfer capability and reduce wear, engineers at Van Doorne developed an improved belt-drive CVT that utilizes a steel belt running in an oil bath. Several versions have been based on this principle, including the production electro continuously variable transmission (ECVT) used in the Subaru Justy.

Another type of CVT, the traction drive, operates differently. These designs employ rolling elements of various configurations and vary the ratio by moving the power transfer points (roller contact points) either closer to, or farther from, the axes of the input and output discs. Traction drives have also been plagued with problems, primarily as a result of inadequate lubricants. Better lubricants have recently been developed, and infinitely variable traction drives may soon replace the conventional multiratio automatic-shift transmission.

With the simple composite V-belt CVT, centrifugal force generated by engine speed causes weights to fly outward and force the movable face against the fixed face of the driver pulley. The greater the engine speed, the greater the force squeezing the two pulley faces together. This causes the belt to ride farther out on the rim as the gap between the fixed and movable faces closes (Figure 3.37). Concurrently, the belt is pulled deeper into the driven pulley, forcing its two faces apart. Torque-sensitive units employ a cam on the driven pulley that is designed to increase belt squeeze in response to increased torque. Torque increases when the vehicle encounters greater resistance or when the operator applies power. Consequently, the actual shift position is a result of the continuous balance between the centrifugal forces of the driver pulley and the counteracting forces of the driven pulley.

Transmission efficiency depends on how the unit is tuned for the application (Figure 3.38). If the belt is subjected to too much squeeze, friction increases and efficiency suffers. If there is too little squeeze, the belt slips and efficiency also suffers. Efficient operation, therefore, relies on a precise balance of forces throughout the operating range. Belt squeeze must be continuously matched to torque. This ideal operating profile can be difficult to maintain; however, when correctly tuned in a relatively low-horsepower application, efficiency can be quite high. More sophisticated versions were successfully used on the DAF in Europe.

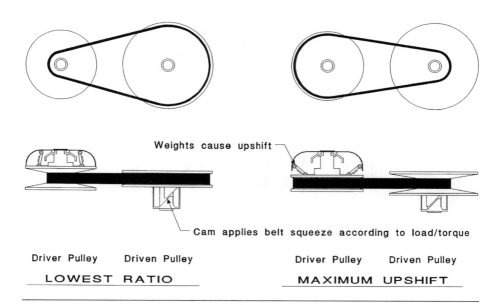

FIGURE 3.37 Salsbury Torque-Sensitive CVT

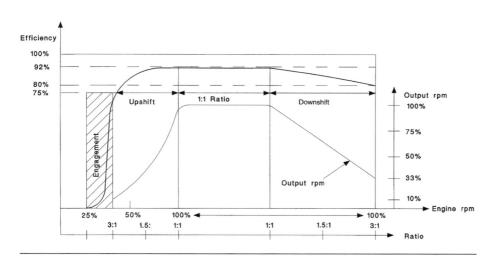

FIGURE 3.38 Salsbury CVT Efficiency Curve

FIGURE 3.39 Ford Continuously Variable Transmission

Courtesy: Ford Motor Company, www.ford.com

Van Doorne's lubricated-steel-belt CVT works on a similar principle. But instead of a composite belt, the design relies on a lubricated steel belt to transfer power and employs hydraulic pressure to provide belt squeeze and effect changes in shift position. Also, the unit can utilize electronic feedback and logic to analyze vehicle and road conditions and modify the shift schedule accordingly. In CVT terminology, the pulley and belt assembly comprises the variator. The variator is then combined with other components such as a fixed-ratio reverse gear, hydraulic pumps, governors and actuators, and an automatic clutch or a hydraulic torque converter. A steel-belt CVT is capable of transferring much greater power and with much-improved wear characteristics. This type of design also provides more precise control over belt squeeze and shift position. Unfortunately, parasitic losses to subsystems and the torque converter are essentially on par with those of a conventional automatic transmission. Because of the design's ability to take advantage of engine operating characteristics, however, vehicle fuel economy can be significantly improved. With an optimized shift schedule, the Subaru Justy ECVT achieved fuel economy that was 15 to 20 percent greater than the same vehicle equipped with a standard three-speed automatic transmission.[16] Ford's belt-type cut is shown in Figure 3.39.

Traction drives are also continuously lubricated by oil. These drives control input and output ratios and transfer power in a manner similar to gears, except the rolling elements have no teeth. Instead, they rely on the hydro-elastic properties of the lubricating fluid to transfer shear loads at the contact points between the rollers and discs. As one might imagine, maintaining a lubricant film while preventing roller slippage has been one of the greatest challenges. But new lubricants that provide a barrier between metal components, yet transfer high shearing forces, have been developed. New lubricants can transfer approximately 6 percent of the roller contact pressure in the shear direction, compared to about 1 percent for conventional transmission oil. Traction drives are simple, responsive, and compact.

Designs may be based either on the toroidal or the half-toroidal layout (Figure 3.40). Characteristically, half-toroidal variators exhibit lower spin losses; however, the axial load on rollers and input and output discs are extremely high. Consequently, excessive internal loads and reduced bearing life are common difficulties. A full-toroidal layout virtually eliminates axial loads on the rollers, and better lubricants have solved the historical disadvantage in load transfer capability.

The full-toroidal Torotrak transmission, development by British Technology Group with assistance from Ford Motor Company, is the most successful design to date. The Torotrak design utilizes a two-cavity layout, with the output disc in the center and an input disc at each end (Figure 3.41). Roller pressure is maintained by the hydraulic clamping action of the two input discs, which traps the rollers between them. A dual-cavity Torotrak variator has been tested at 94 percent efficiency. Half-toroidal designs are reported to operate at roughly 90 percent efficiency. Shift rate is extremely rapid and smooth. Rollers can migrate across the discs in as little as one-half revolution of the engine, and changes in ratio are virtually unfelt by the vehicle's occupants.

Another unique feature of the Torotrak unit is what they call the "geared neutral." The geared neutral allows the vehicle to remain at rest, to creep forward, or to pull away in a burst of acceleration utilizing an infinitely variable startup gear that circumvents the need for a traditional torque converter or clutch. Because of its infinitely variable ratio capability, from neutral to full

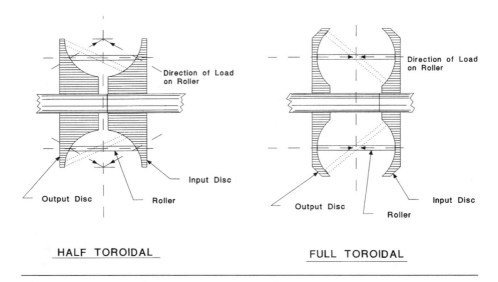

FIGURE 3.40 *Toroidal and Half-Toroidal Variator*

FIGURE 3.41 *Torotrak Variator*

Courtesy: Torotrak, www.torotrak.com

upshift in forward and reverse without having a discrete reverse gear, the company calls their unit an infinitely variable transmission (IVT), rather than a CVT. The heart of the system is a planetary gearset.

To visualize how the geared neutral works, consider the behavior of the gearset illustrated in Figure 3.42. When the sun gear is driven at speed Na, and the planet carrier is "grounded" with the planet gears free to rotate, then the annulus gear rotates in the opposite direction at a speed that is determined by the relative size of the gears. Likewise, if the planet carrier is free to rotate and the annulus is driven at the same speed as in View 1, the planet gears will rotate at the same speed as before, and the carrier will still remain stationary. If the speed of the annulus is reduced or increased, however, the planet carrier will start to rotate in either the forward or reverse direction so that the sum of the speeds of the epicyclic elements remains constant. By connecting the variator input shaft to the sun gear, the variator output disc to the annulus, and the vehicle wheels to the planet carrier, a slight change in variator ratio will begin to move the vehicle in either the forward or reverse direction. If the variator ratios are matched, the vehicle will not move, although it is still "in gear."

The geared neutral in combination with the variator provides a new level of freedom to power plant engineers. With Torotrak's IVT, the shape of the torque curve is no longer limited by vehicle dynamics, but instead is free to comply with the optimum characteristics of the engine. The engine and transmission form a much more fluid and synergistic package when their individual management schedules are more completely integrated. Vehicles can now be tuned for maximum performance, maximum fuel efficiency, and minimum emissions within a dynamic system of integrated controls and continuous feedback.

As of 1999, Torotrak had licensed its technology to nine manufacturers, including Ford, General Motors, Toyota, Koyo Seiki, and LG Machinery. The company expects an 80 percent share of the automobile transmission market by year 2010. A typical IVT layout is shown in Figure 3.43.

NEW TECHNOLOGY AND ALTERNATIVE CARS

Most of the options for improving vehicle fuel economy can be applied to cars of any size. Even the basic idea of vehicle mass reduction is already being applied to conventional multipassenger cars. The primary difference has to do with magnitude. Significantly smaller personal mobility products are naturally lighter and therefore require less energy. Much greater fuel

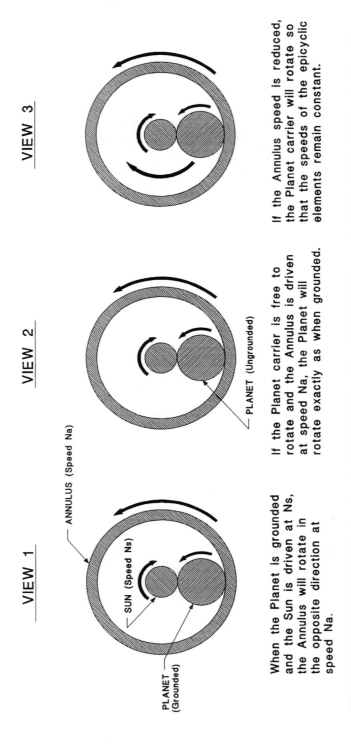

VIEW 1

VIEW 2

VIEW 3

ANNULUS (Speed Na)

SUN (Speed Ns)

PLANET (Ungrounded)

PLANET (Grounded)

When the Planet is grounded, and the Sun is driven at Ns, the Annulus will rotate in the opposite direction at speed Na.

If the Planet carrier is free to rotate and the Annulus is driven at speed Na, the Planet will rotate exactly as when grounded.

If the Annulus speed is reduced, the Planet carrier will rotate so that the speeds of the epicyclic elements remain constant.

FIGURE 3.42 Torotrak's Epicyclic Gearset Provides Geared Neutral

FIGURE 3.43 Torotrak CVT General Layout

Courtesy: Torotrak, www.torotrak.com

economy is therefore possible, regardless of the powertrain technology. A modern automotive powertrain can achieve fuel economy in excess of 100 mpg, provided mass is limited and the vehicle is aerodynamically clean. When framed in terms of a choice between reducing the work or doing an equivalent work more efficiently with better technology, the point can easily become obscured. The best approach is of course to do both.

Although vehicle downsizing is not normally considered a technology, in actual practice, development efforts with ULM designs will ultimately result in a technology of its own. Structural composites can significantly reduce vehicle mass; however, the higher loads typical of conventional high-mass vehicles can often tax a material's capability or make particular applications infeasible. Lightweight plastic composites are especially suited to the structural requirements of ULM vehicles. Overall vehicle repackaging can take advantage of new ideas in powertrain packaging. Combination starter/generators, electromagnetic valve lift, composite engine components, and miniature drives become more mechanically and/or economically feasible with smaller engines of fewer cylinders. At reduced power levels, traction drives and CVTs are simplified and less costly. One unit of mass reduction can translate into two or more units in the vehicle as a whole because of the trickle-down effects throughout the rest of the vehicle.

Different aspects of existing technologies may be emphasized when applied to new types of products. In the process of developing designs for significantly smaller vehicles, new technologies are likely to emerge, and existing technologies may produce new benefits when utilized in new ways. In this respect, a vast reservoir of existing technology is already awaiting a new orientation in systems engineering to create a new level of product benefits.

REFERENCES

1. R. L. Bechtold, "Ingredients of Fuel Economy," SAE Paper No. 790928.

2. Gino Sovran and Mark S. Bohn, "Formulae for the Tractive-Energy Requirements of Vehicles Driving the EPA Schedules," SAE Paper No. 810184.

3. Thomas D. Gillespie, *Fundamentals of Vehicle Dynamics* (Warrensdale, PA: SAE, 1992).

4. Charles E. Scheffler and George W. Niepoth, "Customer Fuel Economy Estimated from Engineering Tests," SAE Paper No. 650861.

5. Thomas P. Yasin, "The Analytical Basis of Automobile Coastdown Testing," SAE Paper No. 780334.

6. David M. Tenniswood and Helmut A. Graetzel, "Minimum Road Load for Electric Cars," SAE Paper No. 670177.

7. Walter Korff, *Designing Tomorrow's Cars* (New York: M-C Publications, 1980), p. 190.

8. Marcus C. Inman Hunter, *Rotary Valve Engines* (New York: John Wiley & Sons, 1946).

9. Peter W. Gabelish, Albany R. Vial, and Philip E. Irving, "Rotary Valves for Small Four-Cycle IC Engines," SAE Paper No. 891793.

10. Wolfgang Demmelbauer-Ebner et al., "Variable Valve Actuation," *Automotive Engineering*, October 1991.

11. Michael T. Nelson and Alvon C. Elrod, "Continuous-Camlobe Phasing: An Advanced Valve-Timing Approach," SAE Paper No. 870612.

12. "Air-Injection Supercharging," *Automotive Engineering*, July 1992.

13. W. R. Wade et al., "Fuel Economy Opportunities with an Uncooled DI Diesel Engine," SAE Paper No. 841286.

14. N. John Beck et al., "Electronic Fuel Injection for Two-Stroke Cycle Gasoline Engines," SAE Paper No. 861242.

15. Shigeru Onishi et al., "Active Thermo-Atmosphere Combustion (ATAC): A New Combustion Process for Internal Combustion Engines," SAE Paper No. 790501. Also see: Paul M. Najt et al., "Compression-Ignited Homogeneous Charge Combustion," SAE Paper No. 830264.

16. Yasuhito Saki, "The ECVT Electro Continuously Variable Transmission," SAE Paper No. 880481.

CHAPTER FOUR

ALTERNATIVE FUELS

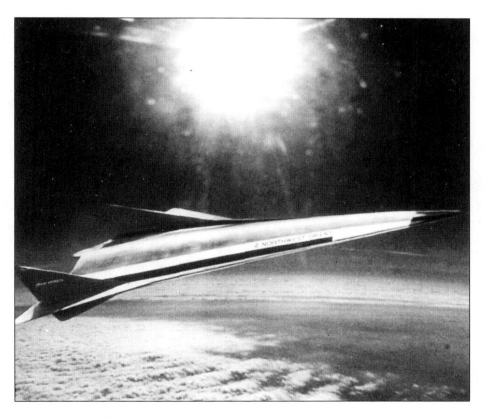

Courtesy: American Gas Association, www.aga.org

Following the 1973–1974 OPEC oil embargo, alternative fuels emerged as a popular and immediate answer to concerns about future motor fuel supplies. Vehicles that run on methanol, natural gas, and hydrogen have since regularly emerged from automobile manufacturers worldwide. In the popular view, the idea of switching to alternative fuels is usually seen as an either/or choice. At some point in the future, the transportation sector will convert en masse to some type of fuel other than gasoline. In reality, alternative fuels have already arrived in the form of reformulated gasoline in which various additives, including alcohol, reduce emissions and lower the fuel's petroleum content. But gasoline is still the world's primary source of transportation energy, and will probably remain so well into the 21st century.

If gasoline had not existed, we probably would have wanted to invent it. Gasoline is an excellent motor fuel, and the reciprocating gasoline engine runs well on it even when built to relatively primitive standards. An early study at Allied Signal showed that gasoline, although selected almost by accident at the beginning of the 20th century, is one of the best ambient-temperature liquid motor fuels available in terms of energy content and other qualities. Its undesirable attributes arise mainly from the political implications of petroleum monopolies and from its harmful emissions. But much progress has been made in emissions reduction.

Today's gasoline cars run 97 percent cleaner than before the advent of emission controls. The value of an alternative fuel, therefore, begins with its ability to solve the liabilities of gasoline while avoiding new ones of its own. There should be plentiful and preferably domestic supplies, and the fuel should be environmentally friendly. Further, a viable alternative fuel must be at least potentially inexpensive, nontoxic, and easily contained and dispensed. It should provide a vehicle range and power profile similar to that of gasoline, and it must be adaptable to the existing transportation fuel infrastructure.

Even if the perfect alternative were available, substituting another fuel for gasoline could have ramifications that extend far beyond the merits of the fuel. Large-scale change will impact the design, manufacture, and service of vehicles. A variety of petrochemicals, often considered byproducts of motor fuel production, would be affected. Ill-conceived change could create disruptions within the refining and supply infrastructure and place unwarranted financial demands across the spectrum of transportation, energy supply, and petrochemical industries. A forced or too-rapid shift to an alternative fuel may open windows of opportunity for entrepreneurial companies and create windows of vulnerability for companies that are heavily invested in existing technology. Ultimately, the decision to make a large-scale switch to any alternative fuel must be based on the economic soundness of the changeover.

The most promising alternatives to gasoline appear to be methanol, ethanol, natural gas, hydrogen, and electricity. Liquefied petroleum gas is also included in the discussion. Much of what we know about these fuels (or energy carriers) comes from controlled laboratory experiments, experimental vehicles, and comparatively limited in-the-field fleet experience. Consequently, it is difficult to predict with certainty the environmental and economic impact of large-scale use of any alternative fuel. The recent problems with groundwater contamination by the additive MTBE is just one illustration of unforeseen consequences resulting from a seemingly benign improvement in gasoline formulations.

ADVANTAGES AND DISADVANTAGES OF ALTERNATIVE FUELS

The following table presents the advantages and disadvantages of the most promising alternative fuels:

TABLE 4.1 *Advantages and Disadvantages of Alternative Fuels*

Fuel Type	Advantages	Disadvantages
Methanol	• Familiar liquid fuel • Vehicle development relatively advanced • Organic emissions (ozone precursors) will have lower reactivity than gasoline • Lower emissions of toxic pollutants, except formaldehyde • Engine efficiency should be greater • Abundant natural gas feedstock • Less flammable than gasoline • Can be made from coal or wood (as can gasoline), though at higher cost • Flexfuel "transition" vehicle available	• Range as much as half less, or larger fuel tanks • Would likely be imported from overseas • Formaldehyde emissions a potential problem, especially at higher mileage; requires improved controls • More toxic than gasoline • M100 has nonvisible flame, explosive in enclosed tanks • Costs likely somewhat higher than gasoline, especially during transition period • Cold-starts a problem for M100 • Greenhouse problem if made from coal

(continued)

TABLE 4.1 (continued)

Fuel Type	Advantages	Disadvantages
Ethanol	• Familiar liquid fuel • Organic emissions will have lower reactivity than gasoline emissions (but higher than methanol) • Lower emissions of toxic pollutants • Engine efficiency should be greater • Produced from domestic sources • Flexfuel "transition" vehicle available • Lower CO with gasohol (10 percent ethanol blend) • Enzyme-based production from wood being developed	• Much higher cost than gasoline • Food/fuel competition at high production levels • Supply is limited, especially if made from corn • Range as much as one-third less, or larger fuel tanks • Cold-starts a problem for E100
Natural Gas	• Though imported, likely North American source for moderate supply (1 mmbd or more gasoline displaced) • Excellent emission characteristics, except for potential of somewhat higher NO_x emissions • Gas is abundant worldwide • Can be made from coal	• Dedicated vehicles have remaining development needs • Retail fuel distribution system must be built • Range quite limited; need large fuel tanks with added costs, reduced space (LNG range not as limited, comparable to methanol) • Dual-fuel "transition" vehicle has moderate performance, space penalties • Slower refueling • Greenhouse problems if made from coal

(continued)

TABLE 4.1 *(continued)*

Fuel Type	Advantages	Disadvantages
Electric	• Fuel is domestically produced and widely available • Minimal vehicular emissions • Fuel capacity available (for nighttime recharging) • Big greenhouse advantage if powered by nuclear or solar • Wide variety of feedstocks in regular commercial use	• Range, power very limited • Much battery development required • Slow refueling • Batteries are heavy, bulky, have high replacement costs • Vehicle space conditioning difficult • Potential battery disposal problem • Emissions for power generation can be significant
Hydrogen	• Excellent emission characteristics: minimal hydrocarbons • Would be domestically produced • Big greenhouse advantage if derived from photovoltaic energy • Possible fuel-cell use	• Range very limited, need heavy, bulky fuel storage • Vehicle and total costs high • Extensive research and development effort required • Needs new infrastructure
Reformulated Gasoline	• No infrastructure change except refineries • Probable small to moderate emission reduction • Engine modifications not required • May be available for use by entire fleet, not just new vehicles	• Emission benefits remain highly uncertain • Costs uncertain, but will be significant • No energy security or greenhouse advantage

Source: U.S. Office of Technology Assessment

TABLE 4.2 *Energy Density Comparison of Select Alternative Fuels*

Energy Storage Medium (MJ/kg)	Energy Density*
Gasoline	48.24
Diesel	45.72
Methanol	22.68
Ethanol	29.52
Hydrogen (liquid)$_1$	41.84
Hydrogen (gas)$_2$	2.34
LPG Propane	46.3
LPG Butane	45.6
Methane	5.44
Pb/Acid Battery	0.19
Regen. Fuel Cell (H_2/Cl_2)**	0.44

*Based on higher heating value of fuels
**Includes weight of magnesium-based hydride storage

Table 4.2 shows the comparative energy density of alternative energy sources.[1]

ALCOHOL FUELS

The alcohol fuels are methanol and ethanol. Alcohol is an excellent motor fuel, and it can be made from a variety of feedstocks, including renewable biomasses. Ethyl alcohol, or ethanol, is normally made by fermenting a biomass, primarily corn. The other alcohol, methanol, is most economically made from natural gas. Ethanol is in many respects the superior motor fuel. Its major drawbacks are that it is more costly to produce than methanol, and when made from corn, it ends up as another competitor in the already overstressed food chain. But recent advances in making ethanol from wood and lignocellulosic materials may ultimately change the equation. Table 4.3 compares the attributes of methanol, ethanol, and gasoline.

ETHANOL (C_2H_5OH)

Ethanol is grain alcohol, like the "white lightning" that people made during Prohibition. The most economical way to produce ethanol is by fermenting starch or sugar crops. As a fuel, its volumetric energy density is slightly less

TABLE 4.3 Comparison of Methanol, Ethanol, and Gasoline

	Methanol	Ethanol	Gasoline
Oxygen Content, wt%	50.0	34.8	0
Boiling Point, K	338	351	308–483
Lower Heating Value, Mj/kg	19.9	26.8	42–44
Heat of Vaporization, Mj/kg	1.17	0.93	0.18
Stoichiometric Air-Fuel Mass Ratio	16.45:1	9.0:1	14.6:1
Specific Energy, Mj/kg per Air-Fuel Ratio	3.08	3.00	2.92
Research Octane Number	109	109	90–100

than two-thirds that of gasoline. Consequently, an ethanol fuel tank must be 1.5 times the volume of a gasoline tank to contain enough ethanol for equal range. This limitation is slightly offset when the engine is dedicated and tuned to take advantage of the higher octane rating of "neat" ethanol.

The most extensive experience with ethanol motor fuel has occurred in Brazil. In an effort to stem oil imports, in 1975 the Brazilian government initiated a conversion-to-ethanol program by providing disincentives for using gasoline. By 1985, alcohol-fueled cars accounted for 92 percent of new-car sales. Unfortunately, the Brazilian experience is a prime example of the unforeseen repercussions that can occur with a major shift to an alternative fuel. In the 1970s it made perfect sense to policymakers to utilize Brazil's vast sugar cane crops to support that nation's switch to domestically produced ethanol and away from ever more costly petroleum. Years later, after a massive, costly, and largely successful conversion effort, Brazilians were then confronted with new conditions: a world of cheap oil and expensive sugar (plus new oil finds of their own), which made it unprofitable to use their valuable crop of sugar cane to produce ethanol motor fuel. Today, after reversing its ethanol program, ethanol cars now account for only 1 percent of new-car sales in Brazil. Brazilian gasoline still contains 22 percent bioethanol, which has helped combat local smog problems and reduce the nation's greenhouse emissions.

Reservations about a switch to ethanol in other countries are based mainly on the economic unsoundness of ethanol as an alternative fuel and on the possible negative effects of dedicating huge crops to the production of motor fuel. With current technology, more energy is spent making bioethanol than is obtained from the fuel. The U.S. Office of Technology Assessment (OTA) has estimated that competition for grain and corn to produce large quantities of ethanol might raise that nation's food bill by billions of dollars annually. The OTA has also expressed concern about the environmental effects of increasing corn production to meet the demands of ethanol production. Corn is a very energy-intensive, agricultural-chemical-intensive, and erosive crop. Large-scale use of agricultural ethanol motor fuel could result in an overall negative effect on the environment and the economy (Table 4.4).

Depending on the production scenario, greenhouse gases may or may not be reduced with ethanol. Substantial quantities of CO_2 are produced during crop growing, harvesting, distillation, and other parts of the ethanol production cycle. The OTA concluded that "it is unlikely that ethanol production and use with current technology and fuel use patterns will create any significant greenhouse benefits." But if the motor fuel consumed by farm equipment were ethanol, and the electrical energy used to produce ethanol were derived from renewable resources, then the entire cycle would be virtually free of CO_2 loading. Although CO_2 is released during combustion, the carbon that is released was first absorbed from the environment during crop growing. So bioethanol motor fuel offers the potential of zero net emissions of atmospheric carbon.

Several programs are under way to develop technology for economically producing ethanol from wood and lignocellulosic materials. New technology could change ethanol's environmental/cost equation and eliminate the food-versus-fuel competition that has plagued the food-to-ethanol production cycle.

METHANOL (CH_3OH)

Methanol is a close chemical cousin to ethanol. Its primary advantage over ethanol is that it can be economically produced from abundant resources of natural gas. Methanol can also be produced from coal or wood, but at a higher cost. As a blend of 85 percent methanol and 15 percent gasoline (M85), it is a fuel to which manufacturers can adapt motor vehicles with minimal changes.

Flexible fuel vehicles (FFVs) that can run on either gasoline or gasoline/methanol blends are also easily designed. An FFV will provide consumers with a familiar vehicle, a familiar liquid fuel, and performance that could surpass their dedicated gasoline-fueled automobile. Such vehicles would also ease consumer resistance to an unfamiliar fuel. With FFVs, consumers could conceivably be unaware of, and even indifferent to, which fuel was actually being dispensed into their vehicles.

TABLE 4.4 Environmental Impacts of Agricultural Ethanol

Water	• Water use (irrigated only) that can conflict with other uses or cause groundwater mining
	• Leaching of salts and nutrients into surface and groundwaters (and runoff into surface waters), which can cause pollution of drinking water supplies for animals and humans, excessive alga growth in streams and ponds, damage to aquatic habitats, and odors
	• Flow of sediments into surface waters, causing increased turbidity, obstruction of streams, filling of reservoirs, destruction of aquatic habitat, and increase of flood potential
	• Flow of pesticides into surface and groundwaters, potential buildup in food chain causing both aquatic and terrestrial effects such as thinning of eggshells of birds
	• Thermal pollution of streams caused by land clearing on stream banks, loss of shade, and thus greater solar heating
Air	• Dust from decreased cover on land, operation of heavy farm machinery
	• Pesticides from aerial spraying or as a component of dust
	• Changed pollen count, human health effects
	• Exhaust emissions from farm machinery
Land	• Erosion and loss of topsoil, decreased cover, plowing, increased water flow because of lower retention; degrading of productivity
	• Displacement of alternative land uses: wilderness, wildlife, aesthetics, etc.
	• Change in water-retention capabilities of land, increased flooding potential
	• Buildup of pesticide residue in soil, potential damage to soil microbial populations
	• Increased soil salinity (especially from irrigated agriculture), degrading of soil productivity
	• Depletion of nutrients and organic matter from soil
Other	• Promotion of plant diseases by monoculture cropping practices
	• Occupational health and safety problems associated with pesticide residues and involvement in spraying operations

Source: U.S. Office of Technology Assessment.

Methanol also has disadvantages that make an early switch unlikely as long as gasoline supplies remain high and prices remain low, but the advent of direct methanol fuel cells could change this situation. The primary disadvantages of methanol include its low energy density (about one-half that of gasoline), a nonvisible flame with M100 (neat methanol), significant cold-starting diffi-culties, increased formaldehyde emissions, and the likelihood that signifi-cant supplies would come from oil-exporting nations. In addition, methanol is toxic and corrosive. Spills can damage clothes, shoes, and automobile paint, and prolonged skin contact can result in poisoning.

Researchers have made significant progress toward solving many of the technical problems associated with a methanol-based motor fuel, and metha-nol is also an excellent carrier fuel for the hydrogen needed to run fuel cells. Consequently, methanol, probably in the form of a gasoline/methanol blend for combustion vehicles and pure methanol for fuel-cell vehicles, may be the best option for an alternative fuel.

Source and Abundance

At least initially, methanol could be economically produced from the world's plentiful resources of natural gas (Figure 4.1). The United States has large reserves of natural gas, but even now some domestic gas needs are being supplied by imports. The present resource base of natural gas is, however, much broader and less regional than that of petroleum. As a result, the potential for political difficulties that arise from regional monopolies is not nearly as great over the near term. Ultimately, the rapid development of nat-ural gas resources will return market power to the holders of the largest reserves. The world's largest reserves of natural gas lie under the Middle Eastern Organization of Petroleum Exporting Countries (OPEC) nations.

Methanol can also be produced from coal. A method of economically con-verting coal to methanol could make several regions energy self-sufficient throughout the 21st century. In the absence of such technology, many of the world's industrialized countries will continue to import a significant portion of their energy needs. Methanol would then be shipped into ports in the same manner that petroleum is now imported.

Environmental and Safety Factors

Environmental and safety issues occur throughout the chain of production, delivery, and consumption of methanol. The most visible impact, and the most popular subject of discussion, occurs at the end of the chain where the fuel is dispensed and subsequently converted into gaseous and particulate exhaust emissions. In general, a methanol engine produces fewer harmful exhaust emissions than its gasoline-burning counterpart. The low combus-

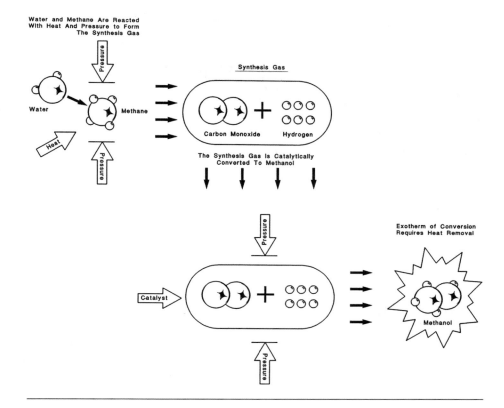

FIGURE 4.1 Methanol from Natural Gas

tion temperature of methanol can reduce NO_x emissions to nearly half that of gasoline. Unburned fuel emerges from the combustion chamber as methanol, instead of hydrocarbons, which reduces the challenge of managing emissions of that category. Emissions of carbon monoxide are almost identical to a typical gasoline engine at an equivalent fuel/air mixture. Methanol, however, is stable at much leaner mixtures than gasoline, and lean mixtures produce lower CO emissions. In the final analysis, emission levels depend largely on the particular design. Table 4.5 shows the results of studies on the subject of methanol emissions, including those of the U.S. Environmental Protection Agency (EPA) and the OTA.[2]

Methanol contains virtually no sulfur, and as a result sulfurous emissions are nonexistent. Also, an over-rich mixture, although bad for fuel economy, does not produce the black smoke typical of an over-rich gasoline charge. Emission of aldehydes, which gives the exhaust its characteristic odor, is

TABLE 4.5 *Methanol Emissions Compared to Ethanol and Gasoline Vehicles*

Study	Emission Change (% relative to gasoline vehicle emissions)		
	NMHC*	CO	NO$_x$
Methanol FFVs			
DeLuchi et al. (1988)	−33 to −4	−43 to +142	−40 to −80
Alson et al. (1989)	−40 to −30**	equal	equal
U.S. EPA (1989)	−30*	NA	NA
M100 Dedicated Vehicle			
U.S. EPA (1989)	−80**	NA	NA
Change et al. (1991)***	lower	equal	equal
M85 Dedicated Vehicles			
Deluchi et al. (1988)	−50 to +62	−70 to +187	−73 to +74
OTA (1990)	−40 to +20	equal	equal
Chang et al. (1991)***	lower	equal	equal
Sperling et al. (1991)	−50	0	0
Ethanol Vehicles			
OTA (1990)	similar to MVs	lower	equal
U.S. EPA (1990a)	similar to MVs	equal	equal
Chang et al. (1991)***	lower	equal	equal

Source: Institute of Transportation Studies, www.its.berkeley.edu

*NMHC = Nonmethane hydrocarbons

**Adjusted for ozone-forming potentials of HC

***Qualitative summary of various studies

much greater than in gasoline engines. Experiments with gasoline/ methanol blends demonstrate that aldehyde emissions increase in a linear relationship to the amount of methanol in the blend.[3] Formaldehyde accounts for up to 98 percent of the total aldehyde emissions from a methanol-burning engine. By comparison, only 31 to 54 percent of the total aldehydes produced by a gasoline engine is in the form of formaldehyde. Tests conducted under U.S. Federal Test Procedure conditions demonstrated that vehicles using neat methanol (M100) had aldehyde emissions 10 times higher and unburned fuel emissions equal to those of a gasoline-fueled vehicle.[4] Formaldehyde is a major constituent of smog and is also responsible for much of the associated eye irritation. Exhaust aftertreatment can remove much of the formaldehyde and up to 90 percent of the unburned methanol.[5]

Fuel evaporation during processing, transportation, and dispensing is also a significant contributor to the environmental impact of transportation fuels. Although methanol produces evaporative emissions, the fuel's photoreactive emissions are less than those of gasoline, which give methanol high marks in this area.

Methanol's toxicity has caused concerns over the relative safety of introducing the fuel into widespread use. Both methanol and gasoline are harmful if they are ingested or absorbed through the skin. Methanol, however, is much more readily absorbed and considerably more poisonous than gasoline. Ingestion of 60–120 cm^3 (2–4 oz.) of methanol is usually fatal to an adult. A fatal dosage for a three-year-old child is slightly more than 15 cm^3 (1/2 oz.)— an amount that can be easily ingested in a single swallow. The fact that methanol is tasteless exacerbates the potential hazard to young children, who might otherwise be repelled by the taste of gasoline.

Self-service methanol stations would also present an increased risk to consumers. Spilled methanol is absorbed through the skin much more rapidly than gasoline. Ingestion or absorption of even small amounts can cause blindness. The familiar act of rescuing a stranded motorist by siphoning fuel from another vehicle could turn out to be a deadly act of altruism. In general, however, hazards can be offset by appropriate product design and legislation. Anti-siphon screens for fuel tanks, service station shut-off valves set to prevent topping off, and the prohibition of methanol-fueled lawn mowers and trimmers along with their attendant small containers of fuel that are accessible to children would reduce the hazards associated with methanol motor fuel. Regardless of the precautionary measures, however, accidental poisonings will undoubtedly occur with large-scale use of the fuel.

Compatibility with Current IC Engine Design

Methanol is an ideal fuel for the Otto engine. Some changes in engine design are required for M100. M85 (containing 15 percent gasoline), however, alleviates many of the difficulties associated with a methanol motor fuel. The undesirable characteristics of methanol motor fuel include cold-starting difficulties, lubricant contamination problems, increased engine wear, and materials incompatibilities caused by methanol's corrosiveness.

Cold-Starting

Unlike gasoline, methanol does not vaporize well at lower temperatures. This characteristic results in increasingly difficult starting as the ambient temperature drops. At a temperature of approximately 12°C, an unmodified, carbureted, spark-ignition (SI) methanol engine will refuse to start. The cold-starting temperature limit can be extended into the range of 0–5°C by improved fuel atomization, applying heat for vaporization, and by mixing methanol with fuels of greater volatility. Also, a major contributor to successful cold-starting is higher cranking speeds (110 rpm). Toyota has reported cold-starts in the –20°C range with M85 in their lean-burn system.[6] Difficult cold-weather starting appears to be easily solvable.

Lubricant Contamination

The effects of engine lubricant contamination have been an ongoing problem with methanol motor fuel. During cold-starts, methanol quickly enters engine lubricating oil, and can rapidly reach significant levels. Although gasoline also enters engine oil, modern oil formulations can tolerate gasoline contamination without harm. Methanol, however, is not compatible with engine oil, and as a result contamination rapidly changes the oil's characteristics. Within 15 minutes of initial cranking, the lubricating oil has been transformed into an oil-methanol-water emulsion that can no longer adequately lubricate the engine. The oil remains in this condition until bulk temperature reaches approximately 70°C, at which time the contaminants have largely evaporated.[7]

This contamination/evaporation cycle may take place over a period of 20 minutes or longer, depending on how rapidly the oil reaches its methanol boil-off temperature. Unfortunately, short trips on the order of five to ten minutes result in a relatively continuous degradation of lubricating oil.

Increased Engine Wear

Increased engine wear has been a perplexing problem with methanol-fueled vehicles. Several studies have confirmed that methanol fuel significantly increases the rate at which engine components break down, especially along the upper portion of the cylinder wall and at the top ring. Initially it was believed that methanol was washing away lubricant in these areas and leaving the cylinder walls unprotected. Studies at Southwest Research Institute

(SRI), however, have indicated that increased wear is actually caused by the corrosive effects of acids formed during combustion.[8]

The SRI used a water-cooled cylindrical combustion apparatus, which was designed to cause condensates to form on the cylinder wall. The condensate was then collected and analyzed. It consisted primarily of water and methanol, along with small amounts of formaldehyde, formic acid, and rust-colored precipitate, which turned out to be iron formate, rather than iron oxide. In the presence of hydrogen peroxide (produced by methanol combustion), formic acid is transformed into performic acid, which is highly reactive to iron.

Performic acid is so reactive that SRI was unable to obtain a sample that actually contained the acid. All of the reactants, however, were present during combustion, and the byproduct of performic acid reaction with iron was also present. Experiments therefore strongly suggest that performic acid is the causative element of accelerated engine degradation in methanol-fueled engines. Special formulations of lubricating oil designed to protect the engine from performic acid are expected to resolve this difficulty.

Materials Compatibility

Many of the materials used in the fuel system of a gasoline-fueled automobile are incompatible with methanol. Methanol (and to a lesser degree, ethanol) is very corrosive to lead, magnesium, aluminum, and to the plastics and elastomers that are typically found in an automobile fuel system. In an existing engine, components such as the fuel tank, fuel pump diaphragm, carburetor, fuel gauge float, fuel pump, carburetor housing, and many seals, rings, and washers are made of materials that will rapidly degrade in the presence of methanol. Incompatibility problems rarely occur with blends containing small amounts of methanol (such as M15). The more methanol-rich blends such as M85 or M100 require fuel systems made of methanol-compatible materials.

Use in Fuel Cells

Methanol is one of the best ambient-temperature liquid fuels available as a carrier fuel for hydrogen. One liter of methanol actually contains more hydrogen than one liter of cryogenic liquid hydrogen. Because of its molecular structure, it is much easier to reform hydrogen from methanol than from any other hydrocarbon fuel. Moreover, direct methanol fuel cells operate on a methanol/water mix and do not need a reformer. So methanol is at the top of the list as an ambient-temperature liquid fuel for fuel cells.

Reservations about a methanol/fuel-cell transportation system centers mainly on safety and economic issues. Direct methanol fuel cells must be supplied with pure methanol, which implies costly upgrades to the fuel supply infrastructure—upgrades that would not be required with a methanol/gasoline blend. In addition, the additives that make methanol's flame visible act as a poison to the catalyst used in fuel cells. This dilemma, coupled with the highly toxic nature of pure methanol and the increased harmful emissions of methanol fuel cells, has led to the current debate over whether to center on gasoline fuel cells, methanol fuel cells, or work toward building a hydrogen supply infrastructure for pure hydrogen fuel cells.

LIQUEFIED PETROLEUM GASES

Liquefied petroleum gases (LPGs) consist of several chemical species that are both byproducts of petroleum and natural gas recovery and essential feedstocks for petrochemical production processes. LPG is also used as a fuel for lighters, barbecues, camp stoves, domestic heating, and as an alternative fuel for vehicles. LPG may be sold as either propane or butane. Spark-ignition engines run well on either gas, but propane is preferred because of its lower boiling point and higher octane rating. Both gases have a naturally high octane rating and a low content of sulfur, aromatics, and other contaminants. Vehicles that run on LPG produce few harmful emissions, and engines and lubricating oil remain cleaner. The main advantages of LPG are its low cost and clean-burning characteristics. In most countries, LPG is either not taxed or it is taxed at a lower rate than gasoline. Propane is most extensively used as a vehicle fuel in Japan and in many European countries. More than 270,000 LPG vehicles are on the roads in the United States alone.

Source, Composition, and Properties

Much of the world's LPG is produced in natural gas processing plants where propane and butane, along with other gases and contaminants, are removed from "wet" natural gas before it is compressed for delivery through pipelines. Not all natural gas fields produce recoverable amounts of LPG, but significant quantities are recovered from fields in the United States, France, Germany, the Middle East, and Australia. LPG is also recovered at petroleum refineries. Yields depend on the composition of wellhead natural gas and of the parent

TABLE 4.6 Properties of Commercial Propane and Butane

Product	Density Liquid kg/l	Density Vapor @ 0°C kg/m³	Boiling Point @ 1013 mbar (°C)	Ignition Temp. (°C)	Specific Heating Value MJ/kg	Theoretical Air Requirement kg/kg
Propane	0.51	2.0	−43	470	46.3	15.6
Butane	0.58	2.7	−10	365	45.6	15.4

crude and the refinery processes used to produce light-end petroleum products. Supplies are therefore linked to natural gas and petroleum reserves.

Commercial LPG consists of either propane (C_3H_8) or butane (C_4H_{10}), but rarely in pure form. Specifications for motor fuel LPG vary among countries. In the United States, motor fuel LPG (HD5) must be at least 90 percent propane and normally contains small amounts of ethane, propylene, butane, and other trace components. In Europe, LPG motor fuel varies from the predominantly propane blends in the United Kingdom, through a range of 50/50 propane/butane blends to 80/20 butane/propane blends in Italy. The actual properties of LPG depend on the properties of its chemical components. Approximate properties of commercial grade propane and butane are shown in Table 4.6.

Vehicle Emissions

Actual emissions depend on variables such as fuel formulations, driving cycle, and vehicle systems. In his book *Liquefied Petroleum Gases*, A.F. Williams reported that vehicles running on LPG can produce up to 80 percent less CO, 50 to 70 percent less HC, 50 percent less NO_x, and about 10 times less SO_x emissions (because of propane's low sulfur content).[9] The SRI tested three aftermarket LPG conversion kits with four different grades of propane to document emission levels. The SRI reported that emission levels varied only slightly according to the purity of the propane (test fuels ranged from 85 to 98 percent pure propane) but varied significantly according to the particular conversion kit. In comparison to operating on gasoline, the SRI's lowest emissions kit resulted in approximately 40 percent less HC, 50 percent less CO, 12 percent less CO_2, and 60 percent more NO_x emissions.[10]

The SRI also speciated HC emissions and calculated reactivity adjustment factors (RAFs) ranging from a high of 0.337 to a low of 0.292. RAFs indicate the ozone-forming potential of the individual HC species of exhaust gases and account for the lower reactivity of alternative fuels. (The RAF for a baseline gasoline vehicle is assumed as 1.) The RAF adopted by the California Air Resources Board (CARB) for LPG is presently 0.50, or about 60 percent greater than the average RAF calculated by the SRI for their LPG test vehicles. In the final analysis, LPG may be nearly as clean as natural gas.

Vehicle Fuel System

LPG is stored in a liquefied state in pressure vessels at approximately 1378 kPa above ambient (200 psig) in propane systems and at slightly lower pressures in butane systems. Because of the relatively low pressure required for liquefaction, fuel systems do not suffer from the same weight penalty as natural gas, and smaller tanks can contain more energy. In dedicated vehicles, LPG fuel tanks may be located in the space normally occupied by the gasoline tank. Otherwise, vessels are located in the luggage compartment and consequently reduce luggage capacity. Fuel is introduced to the engine in a gaseous state, which makes LPG fuel delivery systems similar to those of natural gas. A typical LPG fuel system is illustrated in Figure 4.2.

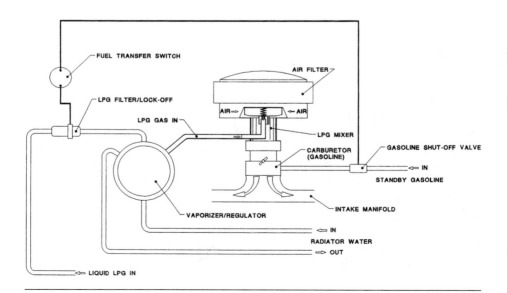

FIGURE 4.2 Typical LPG Fuel System Layout

Compatibility with IC Engines

Spark-ignition engines run well on either butane or propane, but propane is the preferred fuel. Although butane has an octane rating about equal to that of gasoline, propane's research octane number (RON) is on the order of 110–112. In addition, propane's lower boiling point translates into a higher tank pressure and more reliable fuel delivery. Butane blends containing at least 20 percent propane can ensure satisfactory fuel delivery in all but the coldest conditions.

In a dedicated propane vehicle, engine compression is normally increased to about 11:1, and plugs of a lower heat range are installed. Based on propane's volumetric energy content, fuel consumption should be approximately 28 percent greater than an equivalent gasoline vehicle; however, with the increased thermal efficiency of an optimized propane-fueled vehicle, actual loss in fuel economy may be as little as 15 percent. Improved firing regularity (smoother running) and longer engine and lubricating oil life are characteristic of LPG vehicles.

NATURAL GAS AS A TRANSPORTATION FUEL

Methane, the primary constituent of natural gas, is the lightest of all the hydrocarbon fuels. The methane molecule consists of one carbon atom and four hydrogen atoms (CH_4) (Figure 4.3). This simple and clean-burning gas is already one of the world's most successful alternative transportation fuels. Worldwide, nearly one million motor vehicles run on natural gas, and the momentum is growing for broader application. Approximately 102,000 natural gas vehicles exist in the United States, 36,000 in Canada, 300,000 in Italy, and 400,000 in Argentina. Twenty-two percent of all new transit bus orders are for natural gas vehicles.

Industry efforts are underway to gain a significantly greater portion of the transportation fleet market. Natural gas vehicles can be easily serviced by dedicated on-site refueling stations. Existing applications are based on well-proven compressed natural gas (CNG) technology, but home refueling systems for private automobiles are also viable. Today's systems use CNG technology. Experimental systems include those based on low-pressure adsorbent storage (ANG) and liquefied natural gas (LNG). The technology of low-pressure ANG was developed by the Atlanta Gas Light Adsorbent Research group (AGLAR). For some time, the AGLAR group has served as a conduit for human resources and funds from several participating utility companies, whose objective is to broaden the market for natural gas by solving the technical difficulties of high-pressure CNG systems.

FIGURE 4.3 The Simple Methane Molecule Has Broad Application. The methane molecule is made of one carbon atom and four hydrogen atoms (CH_4). Potential transportation applications are immense.

Courtesy: American Gas Association, www.aga.org

The prospect of using this abundant and low-cost gas for private automobiles is extremely appealing. Natural gas could help pave the way for the ultimate adoption of hydrogen. Compressed hydrogen/natural-gas blends can be delivered through the existing natural gas pipeline distribution network, and blends containing as little as 10 percent hydrogen produce a much cleaner exhaust than natural gas alone.

Source and Supply

Natural gas is one of the most plentiful fuels in the world (Figure 4.4). The United States is virtually energy sufficient in natural gas. According to the American Gas Association, proven domestic reserves within the lower 48

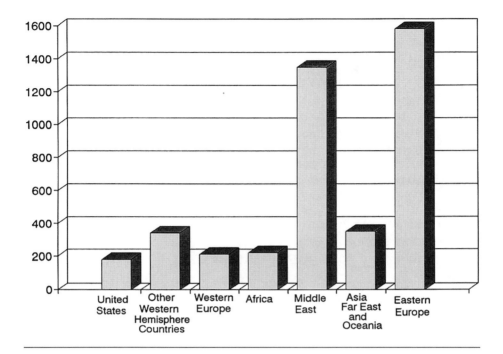

FIGURE 4.4 *World Natural Gas Reserves*

states alone can meet current U.S. consumption levels for at least another 50 years. Improved recovery methods could increase reserves to a 200-year supply. These estimates do not include Canadian reserves or the vast U.S. natural gas deposits in Alaska. On a worldwide scale, natural gas reserves are distributed among nations and continents much more evenly than those of petroleum. As a result, political pressures and market control by individual nations are far less likely to be a problem.

A large infrastructure for delivering natural gas to consumers is already in place throughout the industrialized world. The pipeline delivery network that supplies the United States is currently 25 percent underutilized. This excess capacity alone can deliver enough fuel to supply half of that nation's transportation energy needs if vehicles were converted to run on natural gas.

The cost of natural gas is almost half that of gasoline, on the basis of the cost per unit of energy delivered to the customer (Table 4.7). On a per-liter energy-equivalence basis, for example, the cost of natural gas calculates to

$0.11–$0.21 per liter ($0.40–$0.80 per gallon), depending on regional prices. Wellhead prices of natural gas are expected to remain more stable than prices of crude oil. The increased predictability of supplies and prices, and the extra dollars available to consumers as a result of a switch to natural gas, would have a positive effect on the economy.

TABLE 4.7 *Fuel/Energy Price Comparison in U.S. Dollars*

Energy Source	$/Common Unit	$/Gasoline Equivalent	$/Energy
Gasoline			
Unleaded (whlsale)	0.70/gal, 0.19/L	NA	6.06/mmBTU, 5.74/GJ
Premium (whlsale)	0.77/gal, 0.20/L		
Methanol (whlsale)	0.42/gal, 0.11/L	0.86/gal, 0.23/L	7.42/mmBTU, 7.03/GJ
Ethanol (whlsale)	1.25/gal, 0.33/L	1.88/gal, 0.50/L	16.30/mmBTU, 15.45/GJ
Natural Gas			
Core	0.46/therm*	0.54/gal, 0.14/L	4.64/mmBTU, 4.40/GJ
Noncore	0.39/therm*	0.45/gal, 0.12/L	3.89/mmBTU, 3.69/GJ
LPG	0.32/gal, 0.08/L	0.42/gal, 0.11/L	3.63/mmBTU, 3.44/GJ
Electricity			
Off-Peak	0.05/kWh	1.56/gal, 0.41/L	13.54/mmBTU, 12.83/GJ
Residential	0.11/kWh	3.57/gal, 0.94/L	30.94/mmBTU, 9.33/GJ
Hydrogen			
Steam Reformation	3.43/kSCF, 0.12/m^3	1.43/gal, 0.38/L	12.40/mmBTU, 11.75/GJ
Electrolysis	12.98/kSCF, 0.46/m^3	5.40/gal, 1.42/L	46.72/mmBTU, 44.29/GJ
H_2/CNG Blends**			
% by Energy Content			
5% H_2		0.50/gal, 0.13/L	4.32/mmBTU, 4.09/GJ
10% H_2		0.55/gal, 0.15/L	4.74/mmBTU, 4.49/GJ
15% H_2		0.60/gal, 0.18/L	5.17/mmBTU, 4.90/GJ
25% H_2		0.70/gal, 0.18/L	6.02/mmBTU, 7.73/GJ
50% H_2		0.94/gal, 0.25/L	8.15/mmBTU, 9.73/GJ
75% H_2		1.19/gal, 0.31/L	10.27/mmBTU, 9.73/GJ

Based on estimates by Hydrogen Consultants, Inc.

*Therm = 100 ft^3 (2.83 m^3) of atmospheric natural gas and approximately 100,000 BTU (105.5 MJ).

**Prices estimated by Hydrogen Consultants, Inc.

Environmental and Safety Factors

Natural gas is a very clean-burning fuel, and it may be one of the safest vehicle fuels available. Technically unsound movie scenes, however, have given natural gas a bad image with dramatic portrayals of suicides and explosions. Although both of the foregoing have occurred, they are not representative of the characteristics of the fuel. The hazard of breathing air that is heavily laden with natural gas is primarily a result of displaced oxygen, not the toxicity of the gas. As for an accidental explosion, the gas-to-oxygen ratio must be within a fairly narrow range at the time the ignition source is introduced, or the gas will not ignite. This will of course occur at some point around the edges of a gas cloud, or in some areas at some moments as a room fills with escaping gas.

The safety record of natural gas vehicles (NGVs) is well documented. Vehicles have been using natural gas as a fuel for nearly half a century in some areas of the world. Today, approximately 50,000 NGVs are in use in the United States and approximately 700,000 are being used worldwide. According to the Natural Gas Vehicle Coalition, a fire or explosion caused by a collision involving a natural gas–fueled vehicle has never been documented.

The remarkable safety record of NGVs is probably largely the result of the strength of the high-pressure storage vessels and the stringent certification and testing standards imposed on fuel system components. But development of low-pressure technology (adsorbent storage) could lead to thinner fuel vessels, less stringent standards, and possibly an increased risk of a ruptured fuel system. Such an event, however, is in many ways less hazardous than a ruptured gasoline fuel system. Unlike a liquid fuel, natural gas will not remain at the site in a highly flammable state. Instead, the escaping gas rapidly disperses into the air, where it can no longer ignite.

Like gasoline and methanol, natural gas exhaust emissions vary depending on the tradeoffs made between performance, fuel efficiency, and other factors; however, natural gas is basically a low-emissions fuel. In fact, exhaust emissions are among the lowest of all the alternative fuels, with the exception of NO_x. Emissions of NO_x can actually exceed those of gasoline, if the natural gas engine is not appropriately tuned.

Natural gas contains no sulfur, and it does not mix with oil. Consequently, there are no sulfuric emissions, and the lubricating oil, combustion chamber, and spark plugs remain cleaner in a natural gas–fueled engine. Its gaseous nature eliminates the cold-starting and warm-up problems associated with poor vaporization of liquid fuels and thereby avoids the characteristic mixture enrichment typical of liquid-fueled engines.

TABLE 4.8 Exhaust Emissions Comparison (g/km*, g/mi)

Fuel	Reactive Hydrocarbons		CO	NO$_x$
	Exhaust	Evaporative		
Gasoline	0.22, 0.35	0.02, 0.04	0.87, 1.4	0.41, 0.66
Methanol (M85)	0.16, 0.25	0.056, 0.09	0.62, 1.0	0.28, 0.45
Ethanol (ABBE)	0.63, 1.02	0.03, 0.05	1.12, 1.8	0.38, 0.61
Natural Gas	0.12, 0.19	0	0.06, 0.1	0.27, 0.44

Source: California Air Resources Board (CARB), www.arb.ca.gov
*CARB figures are compiled in g/mi. Equivalent g/km values inserted by the author.

The largest percentage of exhaust hydrocarbon emissions is in the form of methane, which is not photoreactive. Reactive hydrocarbon emissions are extremely low. A natural gas–fueled automobile will therefore contribute little to ozone formation. Methane, however, is a powerful greenhouse gas, and much is unknown about the long-term global effects of introducing additional methane emissions into the atmosphere. Emissions of NO$_x$ can be controlled with a reduction catalyst as long as the mixture is stoichiometric. Table 4.8 shows a comparison between emissions of gasoline, methanol, ethanol, and natural gas.

Total emissions throughout the cycle of production and use should be moderately lower with dedicated natural gas vehicles. Also, studies at the Gas Research Institute (GRI) indicate that blends of natural gas and small amounts of hydrogen (up to 10 percent) can reduce the already low emissions by as much as 40 percent.[11] Other studies indicate that virtually all natural gas/hydrogen blends exhibit a reduction in NO$_x$ and CO emissions, and that the rate of emissions reduction is greater than the rate at which the relative proportion of hydrogen is increased.[12] Chrysler Corp. completed a study comparing the operating costs and the environmental benefits between natural gas and electric vehicles. The study concluded that the cost to purchase and operate natural gas vehicles would be 16 percent lower than for electric vehicles, and emissions of oxides of nitrogen would also be significantly lower. These environmental impacts were based on a source-fuels mixture of 55 percent coal, 13 percent natural gas, 5 percent oil, and 27 percent other sources such as nuclear and hydroelectric.[13] Based on this mixture of fuels, the study concluded, however, that natural gas vehicles would have slightly higher hydrocarbon and carbon dioxide emissions, when compared to electric vehicles.

Assessment of the overall effect of natural gas fuel on air quality is a complex issue. Concerns have been expressed about the uncertainty of methane's potency as a greenhouse gas and the projected magnitude of increased atmospheric methane resulting from system leaks throughout an expanded infrastructure. At present, not enough is known to make an accurate forecast of the environmental effects of large-scale use of natural gas as a transportation fuel, but initial findings indicate that expanded use of natural gas will have an overall beneficial effect on the environment.

Compatibility with Current Motor Vehicle Design

Because of its high octane number, natural gas will not operate well in a compression-ignition engine, but it is well suited to spark-ignition engines. Converting a conventional gasoline engine to run on natural gas is relatively simple. Aftermarket conversion systems use a gas induction system installed upstream of the carburetor. These rather straightforward conversions result in engines that run smoothly but at slightly reduced power. As long as the timing is not advanced past the limits of gasoline, flexible fuel operation is also possible. In this case, the vehicle can be switched over to run on gasoline at the discretion of the operator.

If flexible fuel operation is unnecessary, the gasoline fuel system is removed and the engine is then optimized for natural gas. Optimization consists primarily of increasing the compression ratio to a value as high as 15:1 and advancing the timing several degrees past the tolerable setting for gasoline. These modifications let the engine take advantage of the 130 octane rating of natural gas. A dedicated engine runs with increased power and efficiency, comparable to or surpassing that of a dedicated gasoline engine.

Fuel Metering

Fuel may be introduced and metered by a mixer, which operates similar to a carburetor, or by injectors, which operate similar to the injectors of a liquid-fueled engine. Older carburetor-type systems use a venturi or orifice system in which the incoming air pushes a mechanical element to operate a gas control valve (Figure 4.5). Because a gaseous fuel does not require vaporization, it could appear that the demands on the fuel delivery system are less, but quite the opposite is true. The gaseous nature of the fuel requires a greater volumetric capacity in the fuel injection system, and precise and rapid metering become more challenging. The process is made more demanding by the variable composition of natural gas and by changes in air and gas temperatures, all of which affect the fuel/air ratio.

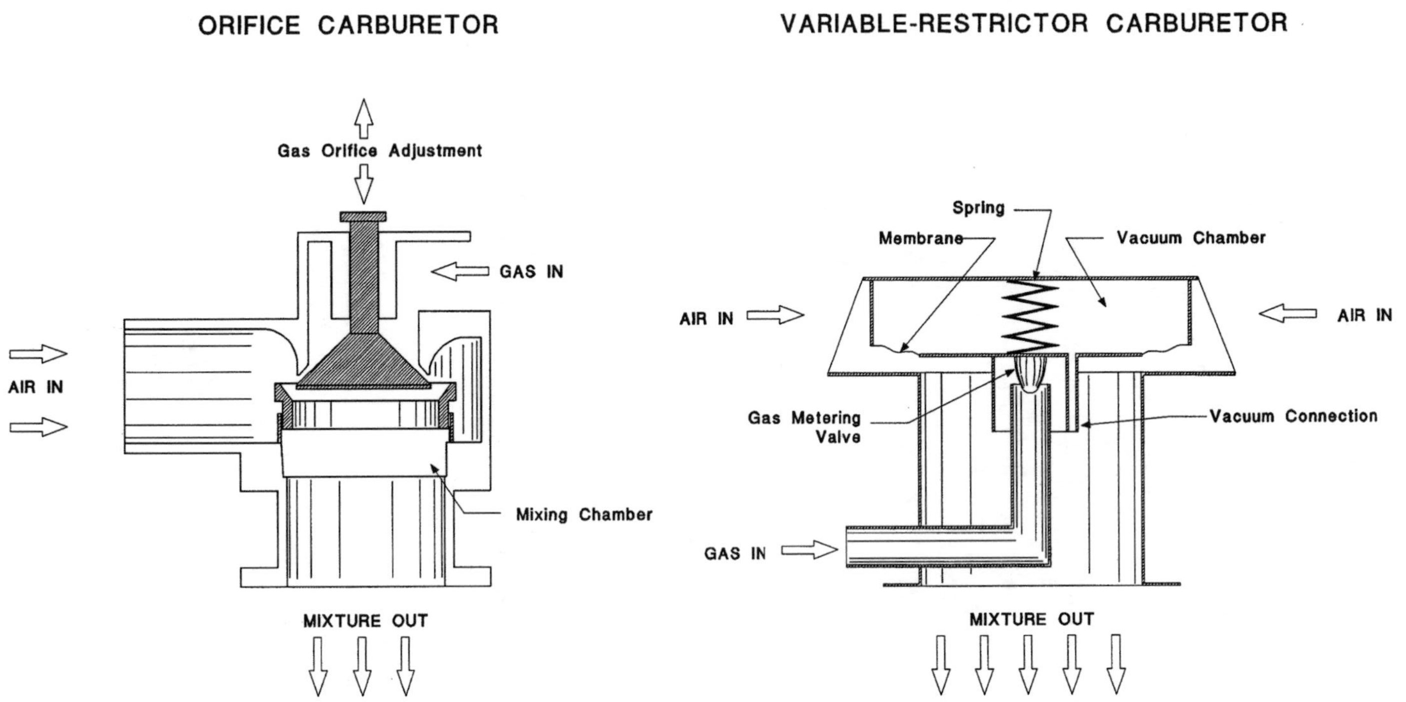

ORIFICE CARBURETOR

VARIABLE-RESTRICTOR CARBURETOR

FIGURE 4.5 Natural Gas Carburetors

Today's CNG vehicles use fuel injection systems. Modern natural gas injectors interface with automotive electronic engine controls and provide precise control over fuel metering. A system made by GFI Control Systems features a single-stage regulator, combined with an electronic controller and metering system with throttle body injection. It uses closed-loop control and electronic spark advance. Control strategy includes an adaptive feature to compensate for variations in gas composition and component variability.

Onboard Storage of Natural Gas

Natural gas for vehicular application is normally stored in a compressed state at pressures up to 20.7 MPa above ambient (3,000 psig). The gas may also be stored in a cryogenic state as a liquid (LNG) or by utilizing a low-pressure technology known as adsorbent storage (ANG). Actual fuel delivery is unchanged, regardless of the storage system.

Compressed Natural Gas

Of the nearly 190 million motor vehicles in the United States, only 70,000 run on natural gas, and these vehicles rely almost exclusively on CNG technology. A CNG system stores gas in heavy cylinders at pressures of up to 20.7 MPa above ambient (3,000 psig). The vehicle is then refueled with high-pressure gas from a specially equipped refueling station. Figure 4.6 shows a CNG Honda Civic.

The volume/range relationship of natural gas as compared to other fuels is easy to estimate by using an energy equivalence factor. It takes about 0.93 m³ of atmospheric natural gas to equal the energy available from 1 liter of gasoline (125 ft³/gal). To equal the energy in a liter of diesel fuel requires slightly more than 1 m³ of natural gas (135–140 ft³/gal). Consequently, 70 cubic meters of atmospheric natural gas is required to equal the range of a 75-liter (20-gallon) tank of gasoline. In order to contain this volume of CNG at 20.7 MPa above ambient (3,000 psig), the storage vessel must have a volume of approximately 0.28 m³ (10 ft³). By comparison, an energy-equivalent gasoline tank occupies only about one-fourth the space, and a tank of diesel fuel occupies only about one-fifth the space of a CNG container.

Weight of the fuel container per given energy content is also an important consideration. Gasoline and its steel container weigh approximately 0.96 kg/L (8 lb./gal). When empty, a high-pressure steel CNG vessel weighs approximately 0.96 kg per volumetric liter (60 lb./ft.³), which calculates to about 3.77 kg per liter equivalent (31.5 lb./gal) of gasoline when filled with

FIGURE 4.6 Honda Civic GX Natural Gas Vehicle. Developed in parallel with the standard Civic, the GX was created specifically as a CNG vehicle. The 2001 Civic GX is the third-generation CNG vehicle from Honda.

Courtesy: Honda Motor Company, www.honda.com

CNG. A fiber-reinforced aluminum CNG vessel will weigh about half that on a volume-equivalence basis. A similar fiber-reinforced stainless steel vessel can weigh as little as 0.32 kg/L (20 lb./ft.³); however, reduced weight has a price penalty, as can be seen in Table 4.9.

The volume, weight, and cost of energy stores define limitations that engineers must seek to minimize with the design of the vehicle. For example, when vehicle energy efficiency is improved, allocations to expensive and heavy vessels can be reduced. Additional economies can be gained by utilizing storage vessels for structural reinforcement. CNG vessels may be positioned to provide crash protection or to strengthen an otherwise underrated chassis; however, the periodic recertification required for high-pressure vessels complicates the challenges associated with integral fuel tanks. To obtain recertification, vessels must be removed from the vehicle and hydrostatically tested at 2.5 times their rated pressure. Consequently, tanks must be installed for quick and easy removal.

TABLE 4.9 Cost and Weight Comparison of CNG Storage Vessels

Vessel Type	Weight of Empty Vessel (lb/ft³, kg/L)	Charged Weight Per Energy Equivalent of Gasoline* (lb/gal, kg/L)	Cost of Vessel	
			According to Vessel Volume ($/ft³, $/L)	According to Gasoline Energy Equivalent** ($/gal, $/L)
Steel	60, 0.962	31.50, 3.77	90.00, 3.18	43.27, 11.42
Aluminum	48, 0.770	25.75, 3.08	200.00, 7.06	96.15, 25.37
Fiber-Reinforced Steel	56, 0.890	29.60, 3.54	175.00, 6.18	84.13, 22.20
Fiber w/Aluminum	37, 0.593	20.50, 2.45	155.00, 5.47	74.50, 19.66
Fiber w/Stainless Steel	20, 0.320	12.30, 1.47	325.00, 1.48	156.25, 41.23

*Includes approximately 90 g/L (5.6 lb/ft³) for weight of CNG at 20.7 MPa above ambient (3000 psig). Assumes a natural-gas/gasoline equivalence ratio of 0.93 m³/L (125 ft³/gal).
**Cost on a gasoline-energy-equivalence basis for high-pressure CNG vessel.

A CNG vehicle is normally refueled from a station set up to provide rapid delivery of high-pressure natural gas. A typical CNG refueling station consists of a compressor and storage vessels, from which the fuel is delivered to the vehicle. Rapid refueling is normally accomplished by a cascade delivery system. Cascade delivery utilizes the compressor to recharge a multiple of storage vessels. Fuel is then delivered to the vehicle by drawing in order from the vessel of lowest pressure to the one of highest pressure. A cascade system can refuel a typical CNG passenger car in as little as 10 minutes (Figure 4.7). The size of the compressor and the number of storage vessels are optimized for the number of daily refuelings, and the cost of the system varies accordingly. Using the Canadian experience with public natural gas refueling stations as a guideline, the average cost to retrofit a traditional service station with CNG compression and dispensing equipment is expected to be in the range of $320,000.[14]

Regardless of the technical feasibility of operating vehicles on compressed natural gas, an undeveloped distribution infrastructure for high-pressure gas will continue to limit the fuel to fleet operations that can amortize the expense of an on-site refueling station. In a chicken-or-the-egg dilemma, a consumer-oriented distribution system for high-pressure gas is unlikely to develop in the absence of vehicles to refuel, and natural gas–fueled vehicles will remain an unattractive option for consumers until adequate distribution is in place.

FIGURE 4.7 CNG Refueling Station. Refueling with compressed natural gas takes only slightly longer than refueling a conventional gasoline vehicle.

Courtesy: American Gas Association, www.aga.org

Liquefied Natural Gas

LNG offers the benefit of greater onboard energy stores. In comparison to CNG, LNG results in nearly triple the volumetric energy density. Compressed natural gas contains about one-fifth to one-quarter of the energy contained in an equal volume of diesel fuel. In comparison, a liter of cryogenic LNG contains 60 percent of the energy in a liter of diesel fuel. So a volume of fuel equal to that of conventional gasoline vehicles would supply enough energy for nearly the same range. Liquefied natural gas can be stored at low pressure in a vacuum-insulated vessel for as long as a week or more without losses to boil-off. In addition, the liquefaction process removes much of the impurities present in pipeline natural gas, which translates into a cleaner-burning fuel. The downside to LNG is a general lack of familiarity with cryogenic storage. Methane liquefies at $-162°C$.

LNG might be delivered by truck to local stations, where vehicles would be refueled. Another option is home liquefaction units, which could liquefy natural gas from an ordinary residential service and recharge vehicles overnight. General Pneumatics' Western Research Center has developed a Stirling cycle liquefier that could make such a system practical. The unit is roughly the size of a conventional air-conditioning unit and would sell for about $3,500. Cost of the natural gas and the electrical energy used to liquefy the gas and refuel the vehicle would total about $0.26 per liter ($1.00 per gallon).

Adsorbent Storage

Like LNG, perfected ANG storage technology has the potential to open vast new markets for natural gas in the transportation and consumer appliance markets. If development goals are realized, this technology of low-pressure natural gas storage could make it possible to refuel boats, recreational vehicles, lawn mowers, and automobiles from a device connected to a residential gas line.

Adsorbent storage is based on the ability of some materials to assimilate methane gas. The most efficient sorbents to date are made of activated carbon. When a natural gas vessel is filled with activated carbon, the amount of gas that can be stored at relatively low pressure is greatly enhanced (Figure 4.8). Capacity is greater up to a pressure of about 12.4 MPa (1800 psig). At 12.4 MPa above ambient (1800 psig), the presence of carbon actually begins to degrade the storage capacity of the container by occupying space that would otherwise be available for compressed gas. Carbon sorbency takes place at relatively low pressures. At higher pressures, an ANG vessel operates as a simple compression vessel.

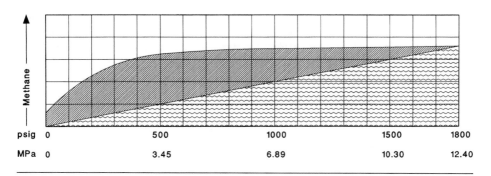

FIGURE 4.8 Adsorbed/Compressed Ratio Typical of Methane Adsorbed on Activated Charcoal

The comparatively low operating pressure of an ANG storage system makes it possible to utilize relatively inexpensive support equipment. Reduced operating pressure (normally 3.5–6.2 MPa above ambient) reduces system complexity, lowers hardware costs, and eliminates hazards that might otherwise be associated with a high-pressure system. For example, an ANG compressor may be configured as a simple two-stage unit, provided system pressure is limited to 3.5 MPa above ambient (500 psig). High-pressure CNG systems require a significantly more sophisticated, and therefore more costly, four-stage compressor. In addition, the lower pressure typical of ANG allows for less expensive components throughout the system, including the substitution of thinner cylinders, which do not require periodic recertification. This benefit reduces manufacturing and maintenance costs and provides the designer with additional options, including the freedom to design the storage tank as an integral and load-bearing part of the chassis.

Difficulties and Costs

The benefits of ANG are not without costs and challenges. Methane adsorption capacity is a function of the surface area of the sorbent and the specific number of adsorption sites on the carbon. Controlling the characteristics of the carbon to maximize its performance has been one of the most difficult challenges of ANG technology. According to the GRI, the carbon AX-21, previously manufactured by Anderson Development Company, has proven to be one of the most efficient carbons to date.[15]

Other problems include contamination of the carbon bed with the normal impurities of natural gas and managing the effects of temperature variations that result from rapid charge/discharge cycles. For example, during a rapid charge from atmospheric to 3.5 MPa above ambient (500 psig), the heat of adsorption can reduce the storage capacity by approximately 20 percent unless the sorbent bed is cooled. Later, heat may have to be applied to the bed to induce complete desorption. Contamination problems have been successfully eliminated by filtering the incoming gas. A proprietary Thermal Energy Storage (TES) system developed by the GRI successfully deals with exothermic and endothermic reactions to adsorption/desorption.

The final difficulty is the high cost of producing the carbon. The cost of carbon varies widely from a low of $3.30/kg ($1.50/lb.) to a high of $44/kg ($20/lb.). The cost of the carbon has no relationship to the material's effectiveness as a sorbent. Until recently, however, the most costly carbons have also been the most efficient methane sorbents, and therefore ANG could not economically compete with CNG technology. According to AGLAR's Doug Horne, the economic threshold is approximately $6.60/kg ($3/lb.) for the carbon sorbent, and that goal has been achieved.

Enhanced Capability ANG Storage

Studies at the Institute of Gas Technology (IGT) have shown that the capacity of an adsorbent storage system is significantly enhanced by chilling the gas before introducing it into the sorbent bed. According to the IGT experiments, at 3.5 MPa above ambient (500 psig), a cylinder packed with AX-21 carbon to a density of .4g/cm^3 had a storage capacity of 205 Vs/V at 227 K (−46°C) and 256 Vs/V at 200 K (−73°C).* By comparison, at 293 K (20°C) and 20.7 MPa above ambient (3000 psig), a standard CNG cylinder stores about 260 Vs/V. Prechilling the gas also eliminated the need for the GRI's TES system. Low-temperature ANG introduces the need, however, to apply additional heat (more than is normally required) for desorbtion. In general, heat to promote desorption is not viewed as problematic because both electrical and power system heat is normally available onboard vehicles.

Researchers at the IGT believe their low-temperature concept will reduce carbon requirements, as well as eliminate the need for a system to manage the heat of adsorption. They estimate that the total reduction in required storage volume will be 40 percent with a 227 K (−46°C) prechill and 52 percent with a prechill to 200 K (−73°C). Carbon requirements would drop by 27 and 41 percent, respectively, and storage system weight would be reduced by 59 and 67 percent, respectively. These substantial technical improvements are somewhat offset by greater costs associated with the fuel delivery system.

Creative Tank Designs for ANG Systems

The low operating pressure of an ANG system opens the possibility for tank designs that would otherwise be infeasible with high-pressure systems. High-pressure CNG containers must be made in cylindrical or spherical shapes in order to contain the extremely high pressure of the gas; however, these round shapes are not as space efficient as a square or rectangular vessel. The relatively low pressure of an ANG system opens the way for creative tank designs that are much more compact.

---

* Vs/V = Volumes of atmospheric gas per volume of tank capacity at pressure.

For example, a square or flat tank can be designed by stacking a series of essentially triangular cells. The efficiency gains of such a design are significant. A square tank will store 27 percent more fuel than a cylindrical tank of equal outside dimensions, and a flat tank presents even more possibilities. Using this simple approach, a flat tank might form a load-bearing floor and assume the role of the vehicle's frame. An integral frame/tank design could result in greatly increased fuel storage capability with little or no compromise in passenger or cargo area.

Home Refueling with Natural Gas

One of the most interesting prospects of adsorption storage is the ability to refuel at home from a standard residential gas service (Figure 4.9). A slow-fill residential system designed for overnight refueling would require signifi-

FIGURE 4.9 High-Pressure Home Refueling Station. A modern residential refueling system for CNG vehicles is currently manufactured by Fuelmaker Corporation of Salt Lake City, Utah. A low-pressure system would be slightly smaller and about half the cost.

Courtesy: FuelMaker Corporation, www.fuelmaker.ca

cantly smaller and less costly support equipment. Such a system would have no need for heat management because excess heat would dissipate over the relatively long recharge cycle. Prechill capability could be provided by a small refrigeration unit. A residential refueling appliance for an ANG system could be based on a relatively inexpensive compressor designed to charge directly into the vehicle's storage vessel. Vehicle fuel costs would then be included in the consumer's monthly utilities statement.

Hydrogen/Methane Blends

Hydrogen is often referred to as the ultimate alternative fuel, but a consumer fuel-delivery infrastructure is virtually nonexistent, and onboard storage presents a variety of technical problems. In contrast, the infrastructure for methane (natural gas) is already in place, requiring only that it be adapted to the transportation sector, and a hydrogen/methane blend could be delivered over the same pipeline distribution network. As a prelude to widespread use, hydrogen could be introduced to the transportation sector initially as an enhancer or additive to natural gas. Gaseous hydrogen/methane blends could provide a relatively low-cost, low-risk means of making a gradual transition to the so-called hydrogen economy.

Adding hydrogen to natural gas results in significantly reduced hydrocarbon and carbon monoxide exhaust emissions. Early tests show that harmful emissions are significantly reduced in blends containing as little as 10 percent hydrogen. Hydrogen is also adsorbed on activated carbon. The effect of hydrogen on sorbent efficiency with a predominately methane blend is shown in Figure 4.10.[16]

HYDROGEN AS A FUEL

Hydrogen is the lightest and simplest of all the elements, and yet it is capable of releasing tremendous energy when combined with oxygen and burned. Hydrogen may be used as a gaseous fuel in an IC engine, similar to the way in which natural gas is used. A more efficient way to utilize hydrogen's energy is in a fuel cell to produce electricity. Either way, when hydrogen is converted into energy, the byproduct is water. Consequently, hydrogen is normally considered a zero-pollution energy source, even though pollutants may be created during hydrogen production. Hydrogen is technically not a fuel, but like electricity, it is an energy carrier. Both electricity and hydrogen can be generated using virtually any primary form of energy.

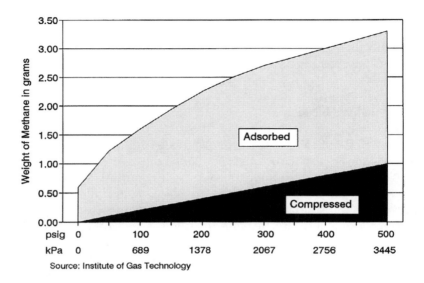

ADSORPTION OF METHANE ON AX-21 CARBON
40 cm³ Cylinder, 25 deg C, 0.4g/cm³ Packing Density

Source: Institute of Gas Technology

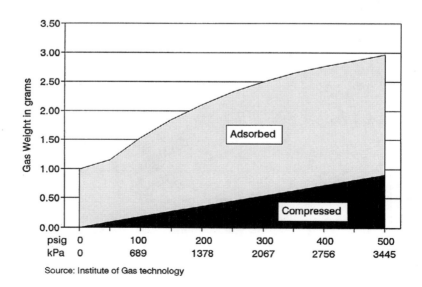

ADSORPTION OF 10% HYDROGEN/90% METHANE BLEND
40 cm³ Cylinder, 25 deg C, 0.4g/cm³ Packing Density

Source: Institute of Gas technology

FIGURE 4.10 *Effects of Hydrogen on Adsorption*

Source and Abundance

Hydrogen is the most abundant element in the universe. Although hydrogen barely exists on the earth's surface as a free element, it has combined with a variety of other elements and may be extracted in several ways. The methods of producing hydrogen can be grouped into two primary categories: (1) the chemical technologies that break down hydrocarbons into more simple molecules, and (2) electrolysis, which can use any source of electrical power to produce hydrogen from water. The most economical method is to split hydrogen from natural gas through a process called *steam reformation*. With improvements in photovoltaic and other solar-to-electricity technologies, electrolysis could provide a nearly endless supply of cheap hydrogen.

Worldwide production of hydrogen currently amounts to 20,860,000 t (23 million U.S. tons) annually. The most extensive use of hydrogen is in the manufacture of ammonia, which consumes two-thirds of the world's production. Oil refineries use one-fifth of the hydrogen, and the remainder goes to fertilizer, dye, and rocket fuel production.

Supply Infrastructure and Cost

Today, hydrogen is comparatively expensive, and no consumer infrastructure exists. Although hydrogen can be inexpensively reformed from natural gas in the quantities needed to supply transportation energy, delivery, dispensing, and storage expenses, both at the dispensing site and onboard the vehicle, add to the total expense of hydrogen fuel. Estimations of the end-user cost of hydrogen vary widely, depending on how the hydrogen is produced and delivered. Estimations of at-the-pump prices range from about on par with the cost of gasoline, on an energy-equivalent basis, to $6.60 or more per liter equivalent of gasoline ($25/gal). Table 4.10 provides a comparison of the production costs of hydrogen by source.[17] Table 4.11 provides a comparison of delivery costs by method.

Whether hydrogen is in a gaseous or liquid state determines much about the cost and means of delivery. Gaseous hydrogen may be delivered through pipelines just as natural gas is delivered today. Liquid hydrogen can be economically transported by truck. Either way, shipping/delivery costs are expected to be greater than those of gasoline, with truck shipment costing about 1.5 times more than gaseous pipeline delivery. The cost to dispense gaseous hydrogen at public refueling stations has been estimated at roughly equal to the cost of dispensing natural gas through similar facilities.[18]

TABLE 4.10 Hydrogen Production Costs

| Energy Source | Source Energy $ Cost ($/GJ)* | Energy Conv. Effic. (%) | Hydrogen Cost $/mmBTU ($/GJ)* | | Production Cost $/mmBTU ($/GJ)* |
			Energy Cost	Capital Cost	
Natural Gas	2.00/mmBTU (1.90)	65–75	2.80	2.80	5.60
			(2.65)	(2.65)	(5.31)
Coal	2.00/mmBTU (1.90)	<55	4.00–5.00	7.00–8.00	12.00
			(3.79–4.74)	(6.63–7.58)	(11.37)
Grid Electric	0.03/kWh**	63–81	11.00–14.00	8.80	20.00–23.00
			(10.43–13.27)	(8.34)	(18.96–21.80)

Source: California Energy Commission, www.energy.ca.gov
*Units in parentheses calculated by the author
**1 kWh = 3.6 MJ

TABLE 4.11 Hydrogen Delivery and Dispensing Costs

| Fuel Resource | Estimated Delivery Cost to Refueling Station | | Estimated Vehicle Dispensing Cost $/mmBTU ($/GJ)* |
	$/mmBTU ($/GJ)*	$/kWh**	
Gasoline	6.00–8.00	0.02–0.03	0.38–1.15
	(5.70–7.60)		(0.36–1.09)
Natural gas by pipeline	3.00–6.00	0.01–0.02	1.00–1.50
	(2.70–5.70)		(0.95–1.42)
Hydrogen gas by pipeline	10.00–15.00	0.034–0.05	1.25–1.75
	(9.48–14.22)		(1.18–1.66)
Liquid hydrogen by tanker	15.00–25.00	0.50–0.085	
	(14.22–23.69)		

Source: California Energy Commission, www.energy.ca.gov
*Units in parentheses calculated by the author
**1 kWh = 3.6 MJ

Another delivery option would be to reform natural gas into hydrogen at the dispensing site, rather than reform it at a remote facility and deliver it in gaseous form by pipeline. This would piggyback on the existing natural gas infrastructure and avoid the need for a second pipeline network. In the

United States, for example, natural gas already provides about one-quarter of that nation's energy needs, and it is delivered through 1.2 million miles of pipelines to all major metropolitan areas and many remote locations. The infrastructure cost of setting up natural gas reformers at existing service stations is estimated at about $200 to $500 per vehicle, assuming a large-scale shift to hydrogen vehicles (to provide the necessary consumer base). Considering the efficiency of direct hydrogen fuel cells, the cost of hydrogen "at the pump" will likely be comparable to that of gasoline on a per-kilometer basis.

Another option is to produce hydrogen by electrolysis from water using on-site residential refueling stations. This approach would circumvent the costs of delivering and dispensing the fuel through a conventional infrastructure. The cost of on-site production must then include the cost of the electrical energy used to make hydrogen from water.

Another concept focuses on the existing natural gas pipeline distribution network. The idea of delivering gaseous hydrogen through natural gas pipelines is feasible. In the late 19th century, gaseous hydrogen was delivered through pipelines in the United States, and this network was later converted to natural gas delivery. In Europe, some consumers were still receiving hydrogen through pipelines as recently as the 1950s. Converting the natural gas delivery network to hydrogen, however, would entail some system upgrades.

Pipes and fittings can become brittle and crack as hydrogen diffuses into the metal. The magnitude of the problem depends largely on the type of steel and welds used when the pipeline was installed and the pressure of the hydrogen in the pipeline. Seals and metering equipment would also have to be upgraded. Hydrogen flows about three times faster through a pipe than natural gas, but it also carries about three times less energy per volume of gas. Because compressors operate on volume, not on energy content, the capacity of existing compressor stations would be about three times underrated for a hydrogen delivery system, and the size and spacing of compressor stations in the natural gas system would not be optimized for hydrogen delivery. When all factors are considered, pipeline transmission costs might be about 50 percent greater than for natural gas. Because natural gas is plentiful, it would be difficult to justify switching to hydrogen, then reforming hydrogen from natural gas and thereby increasing the cost of energy.

In the final analysis, the biggest problem for a hydrogen supply infrastructure is not the technical feasibility of making and delivering hydrogen, but the fact that no customers exist. Conversely, the biggest obstacle in the way of a switch to hydrogen fuel is that no hydrogen supply infrastructure exists.

For a period of perhaps 20 years, both the petroleum motor fuel and hydro-gen supply infrastructures must exist side-by-side as the changeover takes place. So the key lies in developing the most economical and least disruptive way to make the transition to a hydrogen economy.

Storage Technologies

Hydrogen can be stored onboard a vehicle as a compressed gas, as a super-cooled liquid (cryogenic storage), as a solid hydride, or by adsorption on acti-vated carbon or carbon nanotubes. The most common and least costly method of storing hydrogen is in compressed form in high-pressure vessels similar to CNG cylinders. For cryogenic storage, hydrogen must be liquefied by reducing the temperature of the gas to the extremely low temperature of 20.4 K. A properly designed cryogenic container can limit the boil-off rate to between 2 and 5 percent per day at an internal pressure of 0.6 MPa above ambient (approximately 100 psig).

Both hydride and adsorption storage are relatively new technologies. The principle of hydride storage is based on the ability of certain materials to assimilate hydrogen under pressure to form a hydrogen compound (metal hydride). Hydrogen is later released by applying heat to the hydride. The hydride can store more hydrogen for a given volume than is possible with cryogenic storage; however, hydrides are expensive and heavy and tend to react sluggishly to changing energy demands. Adsorption storage on acti-vated carbon is essentially the same technology as ANG, but hydrogen adsorp-tion takes place at significantly lower temperatures.

Table 4.12 provides a comparison of different hydrogen storage technologies to a conventional tank of gasoline. Even if hydrogen were stored as a liquid, a hydrogen-fueled IC vehicle would have to carry four to five times the volume of fuel to equal the range of a gasoline-fueled IC vehicle. (Fuel volumetric requirements could be reduced to less than half that with a pure-hydrogen fuel-cell vehicle.) Hydrogen has high energy content per unit of mass, but even in liquid form, its density is significantly lower than that of gasoline.

Another variation of hydrogen adsorption on carbon is the discovery in the early 1990s that certain carbon structures, called *nanotubes* and *nanofibers*, have an exceptionally high affinity for hydrogen. No one knows exactly why carbon nanomaterials are good at storing hydrogen, but it is believed to have something to do with the structure of the nanomaterials' surfaces. Gas mole-

TABLE 4.12   Gasoline and Hydrogen System Attribute Comparison

Storage Technology	Fuel Weight and Volume Fraction* (WF/VF)	Mass Energy Density (kWh/Kg)**	Volumetric Energy Density (kWh/l)**	Infra-structure	Propulsion System Inter-face***	Notes 1) Advantages 2) Disadvantages
Gasoline Tank	WF = 70% VF = 80%	9.0	8.1	Low-pressure tanks and pumps	No	1) Simple system; low cost 2) Crash safety
Compressed Gas Cylinder	WF = 1–5% VF = 60%	0.4 to 2.0	0.4 to 0.6	High-pressure tanks and compressor	No	1) Simple system; low cost 2) High pressure; safety; low capacity
Liquid/Slush Tank	WF = 10–25% VF = 55–72%	3.9 to 9.9	1.5 to 2	Medium-pressure tanks and compressor; cryogenic refrigeration	Yes	1) Mass capacity; volumetric capacity 2) Evaporative loss; low temperature; safety
Hydride Tank	WF = 1.5–4%	0.6 to 1.6	0.6 to 1.6	Low-pressure tanks and compressor	Yes	1) Crash safety; volumetric capacity 2) Poisoning; loss of capacity; heavy
Activated Carbon Adsorption Tank	WF = 4–8%	1.6 to 3.2	0.6 to 0.95	Medium-pressure tanks and compressor; refrigeration equipment	NA	1) Lightweight; crash safety 2) Refrigeration temperature; evaporative loss; high volume

Source: California Energy Commission, www.energy.ca.gov

*Fuel Weight fraction = Fuel Weight/(Fuel Weight + Storage System). Fuel Volume fraction = Fuel Volume/(Loaded Storage System Volume).

**Hydrogen higher heating value = 39.4 kWh/kg. Assuming a 2.5 to 1 efficiency ratio in favor of a fuel-cell vehicle, to achieve the same energy density as a gasoline tank, the hydrogen energy storage system must have a mass energy density of 3.6 kWh/kg and a volumetric energy density of 3.24 kWh/L. (1 kWh = 3.6 MJ.)

***Propulsion System Interface refers to the need for the storage system to be connected to the fuel cell for thermal or other reasons.

cules fit into pores in these surfaces, although exactly why they prefer some pores to others is still unclear.

Carbon nanotubes are analogous to lightweight cylinders about the diameter of several hydrogen molecules. Gaseous hydrogen is drawn up into these carbon tubes like water is drawn up into a drinking straw. Researchers are working to fabricate lightweight bundles of aligned nanotubes, which will create a structure that would work like a hydrogen sponge. Using nanotube bundles that have the capability of storing 6.5 percent of their weight in hydrogen, a 45-kg (98-lb.) fuel tank could hold enough hydrogen for about a 560-km (350-mile) range in a conventional automobile. A fuel-cell vehicle might achieve twice that range. The cost of producing carbon nanomaterials could be as little as $4.40 per kilogram ($2 per pound) in volume production. Nanotubes, however, are still at the stage of laboratory bench-top experimentation.

Compatibility with IC Engine

With little modification, a standard IC engine will run well on hydrogen. The three primary problems associated with hydrogen fuel include (1) reduced power in comparison to gasoline, (2) a tendency to flashback into the intake manifold, and (3) embrittlement of the iron components of the combustion chamber.

Reduced power results from the reduced volumetric energy of a stoichiometric charge of hydrogen. Several causes have been suggested to explain the phenomena of flashbacks. Flashbacks occur when hydrogen is mixed with air and then carried into the intake port as a combustible charge. Early experimenters were uncertain about the cause of flashbacks. The effectiveness of using water injection and cooled exhaust gas recirculation in reducing the tendency to flashback indicates that the primary cause is the presence of hot spots remaining at the cylinder intake port. Japanese researchers have successfully demonstrated a design for a direct injection system that does not deliver the hydrogen until ignition is desired.[19] Additional work also suggests that two-cycle engine characteristics might be especially complementary to the flame characteristics of hydrogen.[20]

No aftertreatment is required for a hydrogen-fueled IC engine. Emissions consist primarily of steam, with trace amounts of hydrocarbons from engine lubricating oil. The thermal efficiency of a hydrogen IC engine is slightly greater than that of the same engine optimized for gasoline.

Fuel Cell

Although fuel cells are reviewed in Chapter Five, a discussion of hydrogen would be incomplete without mention of this highly efficient conversion system. Fuel cells are the most energy-efficient way to convert hydrogen into

TABLE 4.13 *Overall Vehicle Efficiency and Emissions by Power System Type*

Vehicle Type	Emissions	% Energy Efficiency
Gasoline IC Engine	Low	12–15
Hydrogen IC Engine	Ultra-Low	12–16
Fuel-Cell Electric Vehicle	Zero	30–40
Fuel-Cell/Battery-Electric Vehicle	Zero	25–35

energy. Through a reactive process, rather than combustion, fuel cells convert hydrogen directly into electricity at an efficiency of 40 to 60 percent. Depending on the type, a fuel cell can operate up to 2.5 times more efficiently than an IC engine, and fuel-cell vehicles could be refueled as quickly as conventional automobiles, once a hydrogen supply infrastructure is in place. Table 4.13 compares the overall efficiency of a given vehicle configured for various power systems.

The primary technical challenges in automotive applications have centered on the fuel cell's low specific power, slow startup, and slow response to peak demands. Because of slower demand response, batteries or ultra-capacitors are normally used to supply power for peak demands. A fuel-cell vehicle is therefore configured as a hybrid-electric vehicle. Costs have also been prohibitively high. At its present state of development, the cost to manufacture a fuel cell must be reduced about 10-fold in order to compete with conventional power systems.

ELECTRIC POWER

Similar to hydrogen, electricity is not a fuel; however, its potential for replacing oil as a transportation fuel is significant. Although the transportation sector is more than 95 percent dependent on petroleum, only about 9 percent of the world's electrical energy comes from oil. Electrical energy is produced from a variety of alternative fuels and energy sources. Electric vehicles (EVs) are, in essence, vehicles that run on the source fuels used by utilities companies to generate electrical energy. The U.S. Energy Information Administration lists the following breakdown of electrical energy sources, along with projections through 2020 (Table 4.14).[21]

In 1996, total world energy consumption amounted to 396.3 EJ (375.5 quads), and oil provided 39 percent of it, or about 153.8 EJ (145.7 quads).

TABLE 4.14 World Source Fuels for Electrical Generation with Projections through Year 2020 [ExaJoules (10^{18})*, Quadrillion (10^{15}) BTUs (Quads) Per Year]

Electrical Energy Source	1996 ExaJoules, Quads	2010 ExaJoules, Quads	2020 ExaJoules, Quads
Oil	14.0, 13.3	17.3, 16.4	20.7, 19.6
Natural Gas	8.1, 7.7	14.6, 13.8	24.5, 23.2
Coal	55.7, 52.8	70.6, 66.9	82.6, 78.3
Nuclear Power	25.4, 24.1	26.6, 25.2	22.9, 21.7
Renewables	32.4, 30.7	44.2, 41.9	52.5, 49.7
Total (ExaJoules, Quads)**	153.3, 143.4	201.7, 191.1	239.6, 227.0

Source: U.S. Energy Information Administration, www.eia.doe.gov
*EAI figures are given in Quadrillion BTUs. ExaJoules calculated by the author.
**ExaJoule individual figures may not result in the indicated totals because of rounding.

Although transportation consumed only 27.2 percent of the total, it consumed roughly half of the world's oil, and nearly 60 percent of the oil consumed by transportation goes to automobiles and light-duty trucks. So a switch to the grid system for personal transportation energy would displace nearly one-third of the world's total oil consumption, even if vehicle energy efficiency remained unchanged.

Although the EV is an efficient consumer of energy, the entire energy chain must be considered when making comparisons. Considering the overall energy chain, EVs may be 30 to 50 percent more energy efficient, in terms of primary energy consumption, than their gasoline-fueled counterparts. So EVs first reduced petroleum demand by switching to the source fuels used to generate electricity, and second by consuming less primary energy.

There are other benefits to EVs as well. When the production of energy is assigned to a centralized facility, benefits of scale and consolidation also emerge. Producing energy becomes far more simplified and efficient. System attributes such as weight and size are no longer critical. The cost effectiveness of a fixed generating system can be evaluated over the long term, rather than the short-term life of an automobile energy system. Long-term amortization can justify investment in efficiency and emissions control that perhaps could not be justified with the relatively short service life of a vehicle.

Storage technology, which is often seen as a major hurdle for natural gas and hydrogen vehicles, is not especially relevant to fixed-site generation. Natural gas, for example, is delivered to fixed sites through pipelines as it is consumed. Hydroelectric, solar, and wind generation are also viable resources for fixed-site generation. In contrast, these energy sources are virtually impossible to adapt to mobile systems, and fixed sites are more easily converted to different fuels. Professional technicians manage system change-overs, upgrades, and routine maintenance, and the sheer size of the operation provides the economics to justify these operations. Such is not the case with individual vehicles, each with their own self-contained and somewhat unique energy conversion system.

Electric power might therefore be viewed as the ultimate alternative fuel. The environmental and economic profile of electrical power reverts to the characteristics of the fuels used for its generation. The industry's heavy reliance on coal may result in greater emissions of sulfur oxides as more coal is burned to produce more electrical energy for automobiles. Even so, studies have shown that electric vehicles are still cleaner than conventional gasoline cars. The actual reduction in harmful emissions varies according to the emissions profile of the generating plant (see Chapter Five).

Today, electric vehicles suffer from the same limitation that stifled their development a century ago, but these limitations are quickly being resolved. Solutions center on new batteries, hybrid power systems, and smaller, cheaper fuel cells. Battery-electric vehicles may ultimately assume a different transportation role (e.g., urban cars and stations cars). These vehicles of inherently lower performance and range requirements translate into batteries of lower mass and lower per-vehicle costs.

ALTERNATIVE FUELS IN THE 21ST CENTURY

Today's automobiles, as well as consumer attitudes and expectations, are natural byproducts of the low cost and great abundance associated with their petroleum-based propulsion system. Transportation energy in the form of motor fuel gasoline was abundant and inexpensive throughout the 20th century. Driving habits and preferences in automobiles have been nurtured by a seemingly unending supply of low-cost motor fuel. In 1999, for example, U.S. consumers could purchase a gallon of gasoline for $1.50 ($0.39/L) and receive enough energy to propel 1,600 kg (3,500 lb.) worth of hardware and passengers over a distance of 40 km (25 mi.) and at speeds of up to 160 km/h (100 mph). The same $1.50 would buy a loaf of bread, two soft drinks, or one cheap cigar. In the United States, gasoline is normally less expensive than bottled water.

Petroleum was the energy bargain of the 20th century, but it is unlikely to remain on sale through even the first quarter of the 21st century. Ultimately, the world will depend on OPEC petroleum for much of its motor fuel, the price will spiral upward, and petroleum will no longer be an economically feasible source of transportation energy. Alternative fuels will be phased in when the economics of the switch weigh in favor of a move away from petroleum.

References

1. Vernon P. Roan and Thomas A. Barber, "The New Breed of Hybrid Vehicles," SAE Paper No. 810270 ("Electric and Hybrid Vehicle Progress," SAE P-91).

2. Quanlu Wang et al., "Emission Impacts, Life-Cycle Cost Changes, and Emission Control Cost-Effectiveness of Methanol-, Ethanol-, Liquid Petroleum Gas-, Compressed Natural Gas-, and Electricity-fueled Vehicles," Institute of Transportation Studies of the University of California, Davis (May 1993).

3. A. Coinage, H. Menrad, and W. Bernhardt, "Alcohol Fuels in Automobiles," Alcohol Fuels Conference, Institute of Chemical Engineering, Sydney, Australia, August 9–11, 1978.

4. "Aldehyde and Unburned Fuel Emissions from Developmental Methanol-Fueled 2.5L Vehicles," SAE Paper No. 872051.

5. H. G. Adelman, D. G. Andrews, and R. S. Devoto, "Exhaust Emissions for a Methanol-Fueled Automobile," SAE Paper No. 72093.

6. "Unassisted Cold Starts to −29°C and Steady-State Tests of a Direct-Injection Stratified-Charge (DISC) Engine Operated on Neat Alcohols," SAE Paper No. 872066.

7. "Entry and Retention of Methanol Fuel in Engine Oil," SAE Paper No. 880040.

8. "Understanding the Mechanism of Cylinder Bore and Ring Wear in Methanol Fueled SI Engines," SAE Paper No. 861591.

9. A. F. Williams and W. L. Lom, *Liquified Petroleum Gases* (Chichester, England: Ellis Horwood Limited, 1982), p. 295.

10. Edward Bass et al., "LPG Conversion and HC Emissions Speciation of a Light-Duty Vehicle," SAE Paper No. 932745 (1993).

11. Personal conversation with Chris Blazak, "Gas Research Institute."

12. Gregory J. Egan, "Near-Term Introduction of Clean Hydrogen Vehicles Via H_2-CNG Blends," Paper presented at the Fourth Canadian Hydrogen Workshop, Toronto, Canada, November 1–2, 1989.

13. "Alternative Fuels Update," *Automotive Engineering*, May 1993, p. 8.

14. "A Comparative Analysis of Alternative Fuel Infrastructure Requirements," SAE Paper No. 892065.

15. "Economic Analysis of Low-Pressure Natural Gas Vehicle Storage Technology," Task 3 Topical Report (March 1989–April 1990), Gas Research Institute.

16. "Gaseous Fueled Vehicles: A Role For Hydrogen," Paper presented at the National Hydrogen Association's 2nd annual U.S. Hydrogen Meeting, Washington, D.C., March 13–15, 1991.

17. "Hydrogen-Fueled Vehicles Technology Assessment Report," California Energy Commission (June 29, 1991 draft).

18. *Ibid.*

19. S. Furuhama and Y. Kobayashi, "Development of a Hot-Surface-Ignition Hydrogen Injection Two-Stroke Engine," *International Journal of Hydrogen Energy*, Vol. 9, No. 33, pp. 205–213, 1984.

20. S. Furuhama and T. Fukuhama, "High-Output Power Hydrogen Engine with High-Pressure Fuel Injection, Hot Surface Ignition and Turbocharging," *International Journal of Hydrogen Energy*, Vol. 11, No. 6, pp. 399–407, 1986.

21. "International Energy Outlook, 1999," U.S. Energy Information Administration, DOE/EIA-0484(99).

CHAPTER FIVE

ELECTRIC AND HYBRID VEHICLES

Courtesy: Ford Motor Company, www.ford.com

The energy produced by the breaking down of the atom is a very poor kind of thing. Anyone who expects a source of power from the transformation of these atoms is talking moonshine.

—Ernest Rutherford, physicist

Electric cars have been around since the inception of the automobile, but in the early race for dominance, the internal combustion (IC) engine quickly won out as the best power source for cars. Although the electric power system was superior in many respects, as a source of energy, the battery was no match for the high energy content, ease of handling, and cheap and abundant supplies of gasoline. In the eyes of most IC engine folks, early electric vehicle (EV) proponents were just "talking moonshine." Today, nearly a century after the EV was forced into near oblivion, it appears that EVs in some form will actually become the ultimate winner. As automobile populations soar and cities become increasingly choked with combustion byproducts, the IC engine may ultimately become the victim of its own success. Automobiles have to become cleaner, more energy efficient, and less dependent on petroleum motor fuel, and electric vehicles appear to be the best option.

Electric vehicles are normally divided into two categories: battery-electric vehicles (BEVs) and hybrid-electric vehicles (HEVs). Each vehicle type offers a unique set of advantages and disadvantages. In this discussion, fuel-cell vehicles (FCVs) are placed within the HEV category because onboard energy stores are normally needed to make up for the fuel cell's slower response to peak power demands. Recent advancements in rapid-responding fuel cells make it possible to design FCVs having significantly less energy storage capability, or perhaps even none at all. So FCVs may emerge as vehicles that only remotely resemble those we now refer to as HEVs.

BATTERY-ELECTRIC VEHICLES

Today, emphasis within automotive circles has turned toward hybrids. But if one reviews industry thinking over the past 25 years or so, alternative technology has typically experienced the ebb and flow of contemporary opinion. Series hybrids entered the arena some 25 years ago as the most likely alternative to conventional vehicles. Emphasis soon switched to parallel hybrids, then back to series hybrids, then hybrids dropped out of favor altogether. In the 1980s, hardly anyone within the industry believed that hybrids had much of a future. Today, hybrid vehicles are moving onto the production line.

Automotive fuel cells have also encountered a few curves in the road. In the early 1980s, fuel cells for automotive application were only a fanciful idea, and as recently as 1996, the idea of cutting fuel cells from the Partnership for a New Generation of Vehicles (PNGV) was being considered, mainly because nearly everyone "knew" they could not be production-ready by the 2004 target date. Today, fuel cells have emerged as the most promising alternative power system technology in sight.

In the author's opinion, BEVs have a large role to play in the 21st century. The greatest obstacle to a clear vision of the future role of BEVs comes from the natural tendency to fit battery-powered vehicles into today's paradigm of personal mobility. But the personal mobility system could become quite unlike a simple projection of the system that exists today. On a conceptual level, the idea of a single power source for everything from hairdryers to urban mobility vehicles has a poetic simplicity about it that is difficult to resist.

The challenge of designing a practical battery-electric automobile centers mainly on ways to accommodate or improve on the limited specific energy of the battery. A lead-acid battery pack typically weighs 450 kg (990 lb.) or more, is far more costly than a liquid fuel system, takes far longer to replenish with energy, yet delivers less than 1 percent of the energy available from an equal mass of gasoline. Consider the technical challenge of designing a practical automobile having a 450-kg (990-lb.) fuel tank of less than four liters (1.05 gal.) capacity, and requiring four to eight hours to refuel. To make up for this extreme technical disadvantage, an emphasis on energy conservation and vehicle efficiency in BEV design is essential. Even with better batteries, a BEV is more than just a standard car powered by a different energy source. It is a new type of vehicle that demands a new level of emphasis on energy efficiency, and perhaps even a new applications and marketing perspective as well.

Aside from its shortfall in onboard energy stores, the battery-electric car is a superb transportation device. BEVs have no exhaust emissions, they are virtually maintenance-free (at least, potentially so), and they can be "refueled" at home using the existing infrastructure. With BEVs, fuel is converted into usable energy at a remote facility instead of on board the automobile. This shift in concept simplifies the tasks of maximizing conversion efficiency, controlling emissions, and utilizing alternative fuels. Just to control emissions, for example, each conventional vehicle (CV) requires its own mobile environmental protection system, along with the individual inspections and maintenance required to keep the system in good operating condition. With nearly 500 million automobiles in the world (a figure that could double in 20 years), the cost and complication of installing and maintaining individual vehicle systems is enormous. If emissions were controlled at the comparatively small number of centralized generating facilities, the job would be many times more manageable and much more cost effective.

Large stationary generating sites have inherently greater efficiency than the smaller power plants appropriate for mobile applications. In addition, fixed sites can easily utilize a variety of alternative fuels. Natural gas, which is considered an excellent alternative transportation fuel, already produces about 16 percent of the world's electrical energy. By year 2020, one-fourth of electrical energy is expected to come from natural gas. Coal produces 37 percent of the world's electrical energy, 16 percent comes from nuclear power, and 22 percent comes from a variety of renewable sources. Only 9 percent of electrical energy is derived from petroleum. Consequently, a switch to BEVs manifests as a switch to alternative energy sources.

Historically, the limitations of the battery have prevented this otherwise ideal vehicle from becoming a serious contender in the personal transportation arena. Today, after a decade of intensified development efforts, better batteries are on the horizon. But high manufacturing costs still present difficult challenges, perhaps even greater than those of developing the technology. Even with the presumed economic viability of traditional lead-acid batteries, carmakers now lose money on every BEV they manufacture. Ultimately, however, electric cars could be very inexpensive vehicles to manufacture.

BATTERY-ELECTRIC CARS: THEIR ENERGY SOURCE AND EMISSIONS

Several studies indicate that a switch to BEVs can significantly reduce harmful emissions, and to a lesser degree, reduce the total energy consumed by the world's fleet of private automobiles. Electric cars also have the potential to substantially reduce the world's consumption of petroleum. About 95 percent of transportation's energy now comes from petroleum motor fuels. A BEV typically consumes about 95 percent less petroleum than an equivalent CV. Because BEVs are powered by the source fuels used to generate electricity, they are the ultimate alternative fuel vehicles.

France, Belgium, Hungary, and Sweden all derive a large portion of their electrical power from nuclear generating plants. In these countries, electric cars are largely nuclear powered. In the United States, approximately 55 percent of electrical energy comes from coal, which makes U.S. BEVs predomi-

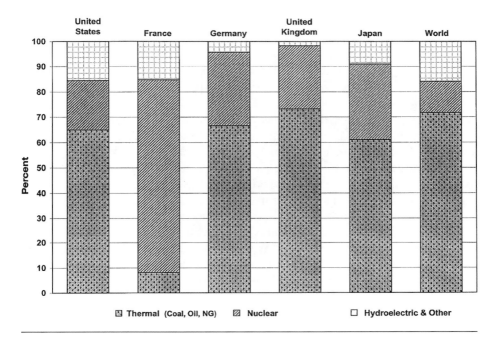

FIGURE 5.1 Source of Electric Power

nantly coal-powered cars.* Worldwide, fossil fuels, primarily coal and natural gas, still account for 62 percent of all electrical generating capacity; however, only a small percentage of electrical energy comes from petroleum. Regardless of the regional mix of source fuels, a switch to battery-electric cars manifests as a dramatic move away from petroleum.

In order to estimate the potential impact of BEVs on primary energy consumption, the entire energy chain, as well as the differences between the two energy systems, must be considered. With CVs, feedstocks, primarily petroleum, are refined into motor fuels and delivered to local service stations. Fuel is then dispensed into vehicles, where it is converted into power while the vehicle is under way. With BEVs, the conversion of fuel into power takes place at a centralized generating facility. Power is then delivered by electrical transmission lines and ends up at individual wall outlets ready to recharge EV batteries. Later, batteries are discharged as the vehicle is driven.

---

* Admittedly, the idea of coal-powered cars is not likely to warm the hearts of many environmentalists.

Various losses occur, and emissions are produced throughout both systems before power is ultimately delivered to the drive wheels. Consequently, the impact of BEVs on primary energy consumption, as well as on the resulting emissions, must account for efficiencies and emissions throughout the entire system or comparisons will not be accurate.

When energy is produced, the cycle begins with the process of recovering petroleum, mining coal, extracting natural gas, or producing biomass feed-stocks for fuel. Efficiency throughout the cycle, from feedstock recovery to conversion into usable energy, varies according to the particular technology. Extraction, mining, and recovery efficiencies are already quite high, and future improvements may be limited. Generating-plant average efficiency (conversion efficiency), however, is expected to improve over time as older, less efficient plants are retired and replaced with newer designs. According to estimates by the Institute of Transportation Studies (ITS), the systemwide efficiency of natural gas–fueled plants may increase by as much as 20 percent and coal-fired plants by 10 percent, between 1991 and 2011.[1]

When petroleum is the source fuel for electrical power, about 26.7 percent of the energy contained in petroleum arrives at the wall-plug as electrical energy. In the United States, only 5.9 percent of generating capacity comes from petroleum. Trends point to increased use of coal and natural gas, both in the United States and in the rest of the world. The overall energy efficiency of these two source fuels is 29.9 and 26.8 percent, respectively. Southern California Edison, the largest producer of electrical energy in the world, projected that the increased load of BEVs will be supported primarily by electrical energy produced from natural gas. By year 2010, 60.4 percent of U.S. generating capacity will come from coal, 14.1 percent will come from natural gas, 14 percent will come from nuclear plants, and 8 percent will be derived from renewable sources (e.g., geothermal, wind, solar, hydro). Only 3.5 percent will be produced from petroleum. Because of increased conversion efficiencies and the trend toward more efficient source fuels, battery-electric cars will become more energy efficient regardless of the particular vehicle technology. In a report for the ITS at Berkeley, California, Wang and DeLuchi estimated the overall energy efficiencies for seven energy-producing chains in year 1995 (as shown in Figure 5.2).[2]

Compared to wall-plug electricity, approximately double the amount of primary energy arrives at the service station as liquid motor fuel; however, wall-plug electricity has already been converted into usable power, whereas liquid motor fuel has not. In order to compare the consumption of primary energy between the two vehicle types, vehicle efficiency must be included. Automobile energy efficiency is highly variable among particular vehicles. Moreover,

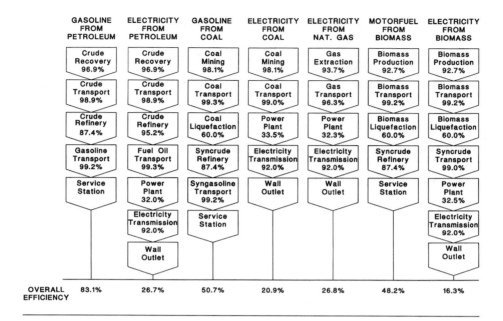

FIGURE 5.2 Overall Efficiency for Seven Energy-Production Chains

improvements that might be expected from advances in technology are difficult to accurately project. Even though IC engine technology is relatively mature, fuel consumption still varies widely among different CVs. BEV technology is embryonic, and energy consumption also varies significantly among different BEVs. As a result, estimates of future BEV energy efficiency tend to be more divergent than similar projections of CV efficiency.

In general, however, researchers agree that BEVs will consume less primary energy. Using their estimated 1995 U.S. source fuels mix and BEV technology as a basis, Wang and Deluchi estimated that EVs in the United States will consume 13 to 29 percent more primary energy than IC vehicles if petroleum, natural gas, or biomass is the source fuel, and 30 to 35 percent less energy if coal is the primary energy source.[3] In another report on the impact of BEVs in the Southern California Edison district, Wang and Sperling estimated that near-term BEVs will consume 10.3 percent less energy than an equal mix of IC vehicles.[4]

Although methods vary, simple comparisons between BEV and CV energy consumption can be made by assuming charger/battery efficiency for the BEV and converting the CV's motor fuel consumption into its energy equivalent. If

TABLE 5.1 Electric and Conventional Vehicle Energy Consumption Comparison

Vehicle	Weight (kg, lb)	Driving Cycle	kWh/mi (from Battery)	kWh/mi (from wallplug)	km/L, mpg	kJ/km, Btu/mi
EV1	1321, 2910	City*	0.180	0.298	—	66, 1020
Honda EV Plus	1632, 3594	City*	0.49	0.81		1818, 2772
Ford Ranger Electric		City*	0.38	0.63	—	1414, 2156
Chevrolet S10 Electric		City*	0.45	0.75	—	1684, 2567
Dodge Caravan Electric		City*	0.67	1.1	—	2469, 3764
U.S. CAFE[†]	1362, 3000[‡]	FUDS*	—	—	12.7, 28.8	2733, 4167[§]
Geo Metro		FUDS*	—	—	19.4, 44	1863, 2840[§]
Hypothetical ULM Vehicle[#]	450, 990	City**	—	—	32, 75	1049, 1600[§]

*FUDS is the Federal Urban Driving Schedule, which is used by the EPA to test urban fuel economy.
[†]Average fuel economy of new passenger cars sold in the United States in 1998.
[‡]Average weight of new passenger cars sold in the United States in 1998.
[§]Based on 125,000 btu/gal of gasoline.
[#]See Chapter Two for energy consumption of hypothetical 450-kg (990-lb.) ULM car.
**Estimated for urban driving cycle.

charger efficiency is assumed as 80 percent, and battery efficiency as 75 percent, then wall-plug energy consumption will be 1.66 times the BEV's on-the-road energy consumption. Table 5.1 shows results calculated for several different vehicles.

Results must then be adjusted to account for the upstream losses that occur when electrical energy and motor fuel are produced. A comparison of vehicle primary energy consumption for the most common source fuels is shown in Table 5.2. It is important to note that U.S. corporate average fuel economy (CAFE) figures do not reflect the average fuel economy of the existing fleet. Although U.S. CAFE for new cars sold in 1995 was 24.9 mpg, fleet-average fuel economy for the same year was 8.6 km/L (20.3 mpg) in the United States, compared to 9.5 km/L (22.6 mpg) in Japan, 11.8 km/L (27.7 mpg) in France, 10.8 km/L (25.8 mpg) in the United Kingdom, and 9.5 km/L (23 mpg) in Germany.[5] The Geo Metro and the hypothetical ULM vehicle are included to illustrate the effect of improved CV fuel economy. Conversion/

TABLE 5.2 Primary Energy Consumption by Vehicle Type in kJ/km

Vehicle	Gas. from Oil (83.1%)	Elect. from Oil (26.7%)	Gas. from Coal (50.7%)	Elect. from coal (29.9%)	Elect. from NG (26.8%)	Gas. from Biomass (48.2%)	Elect. from Biomass (16.3%)
GM'S EV1	—	2506	—	2237	2496	—	4104
Honda EV Plus	—	6809	—	6080	6784	—	11,153
Ford Ranger Electric	—	5296	—	4729	5276	—	8675
Geo Metro	2242	—	3675	—	—	3865	—
U.S. CAFE	3289	—	5391	—	—	5670	—
Hypothetical ULM Vehicle	1262	—	2069	—	—	2176	—

(Parentheses indicate percentage of primary energy delivered either to wall-plug as electricity or to service station as motor fuel.) Btu/mi = kJ/km × 1.609 × 0.9478.

production efficiencies in Table 5.2 are taken from Figure 5.2. As a comparison, Wang and DeLuchi estimated subcompact car petroleum-to-gasoline primary energy consumption at 5440 btu/mi (3,567 kJ/km).

Many factors affect comparisons between BEV and CV energy consumption. The impact of vehicle technology and source fuels mix stands out in Table 5.2. GM's EV-1 using electrical energy produced from coal is approximately 32 percent more energy efficient than the U.S. CAFE vehicle using petroleum motor fuel.[6] When EV-1 is compared to the conventionally powered ULM vehicle, however, the ULM vehicle is about 44 percent more energy efficient. Comparing Ford's electric Ranger to the U.S. CAFE figures results in about 44 percent more primary energy used by the Ranger, but the electric Ranger/U.S. CAFE comparison is biased against the Ranger because the conventional Ranger is rated at only 20 mpg (about 3,935 kJ/km). This translates into 4,735 kJ/km of primary energy using gasoline from oil, or roughly identical to the electric version using electricity from coal.

BEVs will become more energy efficient over time. Today's BEVs, for example, are significantly heavier than might be expected of future vehicles. Ford's Ranger is much heavier than its gasoline counterpart and is therefore doing more work for the same amount of energy. Although the extra work goes to transporting battery mass, with little direct consumer benefit, it does shift energy away from petroleum and it reduces overall emissions. As better batteries become available, existing weight penalties will be reduced and energy consumption will drop. It is also important to note that the Ranger is a converted gasoline vehicle, whereas GM's EV-1 was engineered from the ground up as a battery-electric vehicle. The EV-1 uses less than half the energy per vehicle-kilometer as the electric Ranger.

Sometime in the third decade of the 21st century, petroleum recovery efficiencies are expected to begin declining as petroleum supplies diminish and the remaining deposits become more difficult to recover. So even if BEVs do not save primary energy, switching away from petroleum motor fuels would ultimately have significant economic and energy security benefits. Transferring the job of emissions control to regional generation facilities could also be enormously beneficial.

Regardless of the diversity in estimations of primary energy savings, research indicates that BEVs will produce far fewer harmful emissions and significantly reduce the world's dependence on petroleum motor fuels. Wang and DeLuchi concluded that in many U.S. regions, electric cars would reduce transportation petroleum consumption by more than 90 percent. Even in worst-case areas, such as cities like New York that rely heavily on petroleum as a source fuel, BEVs will reduce petroleum use by 63 to 65 percent. In Chicago, Houston, and Los Angeles, subcompact car petroleum consumption, when replaced by comparable BEVs, will drop by 98.0, 98.2, and 91.9 percent, respectively.[7] Most of the reductions come from replacing petroleum motor fuel with the source fuels used to generate electricity, rather than reducing primary energy consumption.

Several studies have compared air pollution between electric cars and conventional IC vehicles. Initially it was hypothesized that large-scale use of battery-electric cars may do little more than replace tailpipe emissions with the emissions from large generating plants. If energy consumption is similar between the two vehicle types, then generating electrical energy for BEVs might simply shift emissions to a different source, with little effect on total air pollution; however, several studies in the United States and Europe have shown that large reductions in hydrocarbon (HC) and carbon monoxide (CO) emissions can be expected if CVs are replaced by BEVs. A study at the Electric Power Research Institute (EPRI) estimated that in U.S. urban areas, substituting BEVs for CVs will reduce nonmethane organic gases by 98 percent, nitrogen oxides (NO_x) by 92 percent, and CO by 99 percent. Additionally, the EPRI estimates that BEVs will produce only half the carbon dioxide (CO_2) of CVs.[8] The International Energy Agency estimates similar reductions in CO_2 emissions in most OECD countries, with variations depending on local source fuels mix.[9]

In another study, Wang and Santini surveyed six driving cycles in four U.S. cities and found that HC and CO emissions are consistently reduced by approximately 97 percent, regardless of the driving cycle or the local source fuels. This is because conventional vehicles produce large amounts of HC and CO emissions per unit of energy, whereas generating plants produce limited amounts. High HC and CO emissions result largely from cold-starts and short trips that do not allow individual vehicles to become fully warmed up. They also concluded that NO_x emissions may be reduced or increased, depending on the local mix of source fuels. Sulfur oxides (SO_x) and particulate matter (PM) may increase with expanded use of BEVs because of much smaller SO_x and PM emissions of conventional IC vehicles.[10] Greater SO_x and PM emissions are primarily caused by the use of coal for generating electricity. Emissions in this category are substantially lower in regions that rely more on natural gas, and nearly eliminated in regions that rely primarily on nuclear power.

Studying the effect on greenhouse gases, DeLuchi concluded that with the existing source fuels mix, BEVs are likely to have a mixed effect on CO_2 emissions, a primary constituent of the gases responsible for global warming. Compared to CVs using reformulated gasoline, electric vehicles using coal as a primary energy source produce significantly more CO_2 emissions. Natural gas results in slightly less CO_2 emissions, and nuclear power reduces emissions by 98 percent, when comparing BEVs to CVs. Because most (but not all) greenhouse gas emissions are proportional to energy consumption, DeLuchi concludes that increased energy efficiency is the key technical variable in reducing greenhouse gases produced by transportation. According to DeLuchi, the largest long-term reduction in greenhouse gases will come from substituting nonfossil energy sources such as biomass or solar power.[11]

ELECTRIC VEHICLE ONBOARD ENERGY FLOW

Battery-electric cars operate with a closed energy system. All of the energy is self-contained, and in general, the supply cannot be replenished by adding some type of fuel. Within this closed system, the energy/work loop can operate with nearly equal efficiency in both directions, either converting electrical energy into work or converting vehicle kinetic energy back into electrical energy. By comparison, conventional IC vehicles operate with an open energy system. When additional energy is needed, more fuel is added. Vehicles in this category process energy in one direction only, by transforming fuel into work. They cannot do the reverse and transform work (in the form of kinetic energy) into fuel. The inherent economy of the closed BEV system, along

TABLE 5.3 *Vehicle Average Operating Efficiency*

	Electric	Spark Ignition
Motor or Engine	80%	23%
Drivetrain	82%	78%
Battery and Charger	70%	—
Overall Vehicle	46%	18%

with its equally inherent scarcity, encourages conservation through onboard energy management and improved vehicle design. Table 5.3 provides a comparison of the onboard energy allocation between the two vehicle types.

Mechanical and electrical losses occur throughout the BEV power system, and just as with the IC vehicle, losses vary in magnitude according to system design and operating conditions. Also, batteries have different charge/discharge efficiencies, and efficiencies vary according to load. The Jet Propulsion Laboratory (JPL) tested several state-of-the-art electric cars under sponsorship of the U.S. Department of Energy (DOE) to determine the energy flow onboard an electric car. Dynamometer tests of the Electric Test Vehicle (ETV-1) equipped with a chopper-controlled, separately excited direct current (dc) motor, showed losses at various steady-state speeds, as shown in Figure 5.3.

Regenerative Braking

During deceleration, the brakes of a conventional vehicle convert kinetic energy into heat and discharge the energy into the air. Regenerative braking is designed to reclaim this otherwise wasted energy by generating a charge to the battery during deceleration. Under the most ideal conditions, the efficiency of the loop, from battery to kinetic energy and from kinetic energy back to the battery, is no greater than 50 percent. Theoretically, up to half of the energy required to overcome the inertia of acceleration can be recovered by using regenerative braking on deceleration. An early study by McConachie indicated that a 50 percent recovery efficiency would result in slightly more than a 50 percent increase in range over a city driving cycle (Table 5.4).[12] The 50 percent increase in range is based on the assumption that regeneration will be effective in all instances and that it completely replaces the vehicle's conventional braking system.

Different driving schedules indicate different recovery potentials. Over a particular driving schedule, the actual increase in range depends on the practi-

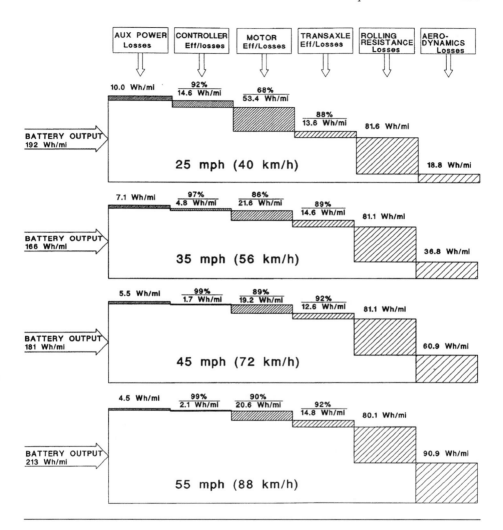

FIGURE 5.3 ETV-1 Energy Flow

cal limitations of an optimized regenerative braking system. Several studies with experimental regeneration systems have shown widely divergent results. During tests at the JPL using the General Electric-/Chrysler-built ETV-1, approximately 42 percent of the kinetic energy stored in the vehicle actually made it back to the battery terminals (Figure 5.4).[13] Energy accepted by the battery is then reduced by charging efficiency—another characteristic that varies between battery types, temperature, state of charge, and many other variables. In actual practice, regenerative braking normally increases vehicle

TABLE 5.4 Vehicle Operating Schedule over a City Driving Cycle (1500-kg Vehicle with 4.1-Liter Engine, Brisbane Metropolitan Area)

Test Vehicle	C1	C2	C3	C4
Trip Duration (minutes)	63	78	67	59
Average Speed (km/h)	18.2	14.9	17.2	19.6
Maximum Speed (km/h)	66	62	65	62
No. of Idle Periods	60	98	81	57
Percent of Time Idling	33.9	35.8	33.6	32.3
No. of Brake Applications	142	166	168	141
Percent Time Braking	22.8	16.9	20.9	21.4
Overall Energy Used (kJ/km)	670	756	745	720
Rolling Resistance (kJ/km)	242	232	237	242
Energy Dissipated (mJ)	8.26	10.1	9.8	9.23
% Range Increase with 50% Regeneration Efficiency	47	53	52	50

Note: Test covered a 19.33-km city route, which included moderate grade variations, a speed limit of 60 km/h, and no freeway driving.

range on the order of 15 percent; however, GM reports that in tests with the EV-1, regeneration produces a 25 percent increase in range.

Typically, regeneration systems are not designed to reclaim all of the energy available on deceleration. Designers tend to configure systems with emphasis on the upper speed ranges. Both mechanically and electronically, it is difficult to maintain regeneration at very low vehicle speeds. Also, batteries vary in their ability to accept high transient inputs, and they become less efficient at higher loads. As charge rate climbs, internal resistance goes up and energy is increasingly converted into waste heat.

Modern energy recovery is significantly improved over early systems. Initially, regenerative braking was controlled by the accelerator pedal. Suddenly releasing the pedal caused maximum regeneration, which also created high current transients and often loaded the drive wheels past the limits of adhesion. Drivability was a problem, especially if drive-wheel traction was lost. With modern systems, regeneration that exceeds the equivalent of conventional compression braking is controlled by the brake pedal. Increased brake pedal pressure also increases regeneration current, as well as vehicle braking effect. Hydraulic brakes are modulated to balance braking forces between front and rear wheels and to prevent loss of traction.

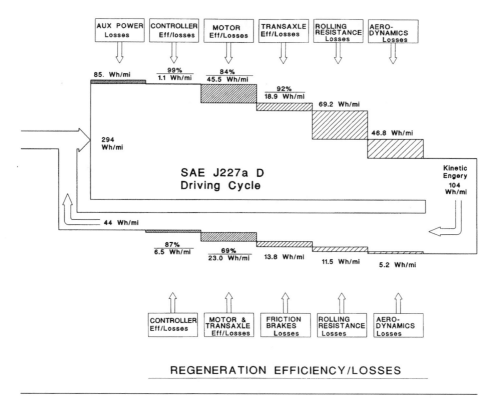

FIGURE 5.4 ETV-1 Energy Flow with Regenerative Braking

MECHANICAL OVERVIEW

Battery-electric cars are much more simplified, mechanically, and significantly more complex, electronically. The electric motor has few moving components (essentially, the rotor/armature and bearings). In general, the transmission is more simplified as well. BEV transmissions typically have only two discrete ratios and no reverse. The vehicle is powered in reverse by electronically reversing the rotation of the motor. With a few exceptions, batteries are passive systems with no moving parts. Processes take place on an electrochemical level, rather than mechanically. It is therefore tempting to regard the BEV as a relatively simple machine, which it is not. Complexity that may be lacking on a mechanical level is offset by increased complexity on an electrical level. Controlling the motor, interfacing vehicle systems, and operator inputs according to operating conditions, and managing the flow of energy within the system, are complex and demanding tasks. Proper system management requires continuous feedback from a variety of subsystems and a multitude of real-time computations, which can only be accomplished with sophisticated hardware/software architecture.

Electric Motor

The electric motor is the heart of the electric vehicle's tractive power system. It functions as a sort of bidirectional conduit between mechanical and electrical energy, converting electrical energy into mechanical work and converting mechanical work into electrical energy. In one form or another, the machine utilizes a rotating set of windings, which interact electromagnetically with a stationary set of windings. When electrical potential is applied, the windings generate magnetic fields, which cause the armature or rotor to rotate and thereby develop power. An electric motor will run indefinitely with little or no maintenance, and it produces no harmful emissions. It develops power smoothly and quietly and can achieve an efficiency of 85 to 95 percent.

Electric motors exhibit different operating characteristics depending on their design. Also, motors are designed to run on either direct current (dc) or alternating current (ac). Traditionally, dc motors are used in applications requiring variable speed and frequent acceleration and deceleration of large-inertia loads. Normally, a dc machine is heavier, smaller, produces more power for its size, and is more expensive than an ac motor. In the past, dc motors were used in most electric vehicle applications because of their high starting torque and their ability to run on direct current from the battery. Because of improved power-conditioning devices, ac induction motors have gained favor in recent years. Induction motors are less costly to manufacture, they are lighter, and they use no commutator brushes, so less maintenance is required.

Direct current motors are designed as series, shunt, or permanent magnet machines. Classifications refer to the way in which the field is energized. Traditionally, the most common EV motor has been the series machine (Figure 5.5). In a series machine, the field is connected in series with the armature. The series motor develops high stall torque. At a given voltage, torque and speed are inversely related and power is constant, except to the degree that it is reduced by efficiency losses along the output curve. Consequently, the series motor is often referred to as a "constant horsepower" machine. It is used in applications requiring high pullaway torque and a broad torque band. The series motor has no inherent speed limitation, and in an

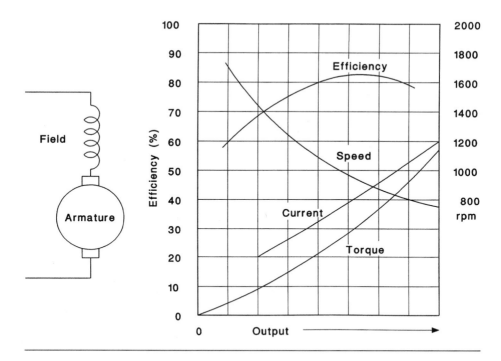

FIGURE 5.5 The Series Motor

unloaded condition it will continue to accelerate until bearing drag equals torque, or until it destroys itself.

A shunt machine has field windings that are connected in parallel with the armature (Figure 5.6). The motor is essentially a constant-speed, variable-output machine. When load is applied, it delivers power at slightly reduced rpm, and current increases in proportion to load. Unlike a series motor, the shunt motor develops very little stall torque. In applications that require greater pullout torque, a shunt motor may be equipped with series windings to improve low-rpm torque characteristics. Such a motor is then referred to as a *compound machine*.

With a shunt or compound machine, the field may be separately excited to provide independent control of the field and armature circuits. The relationship between the field and armature current determines the base speed of the motor. Increasing field current and/or reducing armature current reduces the motor's base speed. Conversely, decreasing field current and/or

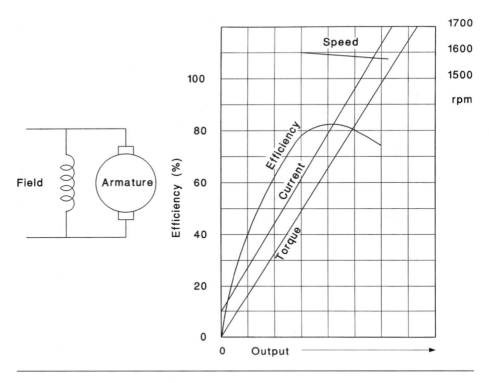

FIGURE 5.6 The Shunt and Compound Motor

increasing armature current results in a higher base speed. When the motor is running at less than base speed, both torque and current increase and the motor will accelerate until it reaches full rpm. When motor rpm is greater than base speed, counter electromotive force (emf) reverses the flow of current until motor rpm is reduced to base speed.

Controlling the relatively low-current field is a simple method of controlling the speed of the motor. Depending on the design of the machine, a 3:1 speed ratio can be achieved with field control alone; however, motoring efficiency suffers as the field is weakened. In EV applications, a shunt motor is often equipped with independent control of both the armature and field circuits. This provides maximum control of the machine's motoring/generating characteristics throughout a broad rpm range.

Permanent magnet motors use permanent magnets to establish the field, which eliminates the need for field windings. This slightly improves motoring efficiency, but at the expense of greater manufacturing cost. Zytek Electric Vehicles, Ltd. manufactures a line of high-performance liquid-cooled permanent magnet EV motors ranging up to 100 kW continuous duty. The small 50-kW motor runs up to 15,000 rpm and weighs just 13 kg (29 lb.). These motors are small enough to work as hub motors, which is how they were used in the Dodge Intrepid ESX.

Alternating Current for Electric Cars

The advantages of ac for vehicle motive power have been recognized for many years. Motors designed for ac are lighter, less expensive, can be more efficient, and require less maintenance because they lack the commutator brushes common to dc motors; however, the battery supplies dc, which must be converted into ac in order to run an ac machine. Until recently, dc-to-ac inverters were large, complicated, inefficient, and expensive devices, which placed ac systems at a disadvantage. Today, advances in high-power transistors and microprocessors have made ac power conditioning practical. As a result, ac systems appear to be the wave of the future, although proponents still exist in both camps.

The advantage of an ac system centers on the use of a three-phase induction motor, which has greater specific power and is less costly to manufacture than an equivalent dc motor. Greater specific power results primarily from the higher operating speeds possible with an ac machine. A dc motor is limited in rpm because of the limitations of commutation. With no commutator, an induction motor can easily operate at speeds in excess of 10,000 rpm. Alternating current motors can be air-cooled or liquid-cooled, and they offer good flexibility in packaging. They can vary considerably in length-to-diameter ratio without undue penalties in performance. At the extremes, very long rotors tend to become unstable at lower speeds, and very short rotors are less efficient. Two-pole motors tend to be favored over four-pole designs because they operate at twice the rpm per given frequency. When current is chopped at twice the frequency, switching losses increase.

Compared to dc systems, the power conditioning and control required for ac results in a much more complicated system. But inverters normally contain components that can be interlaced to serve as wall-plug chargers, which economizes on hardware and somewhat offsets the increased costs. Another consideration has to do with battery cell voltage. Battery packs for ac systems are normally configured for higher voltages (more than 200 V). Higher operating voltages enable the system to operate at reduced current levels, which reduces inverter switching losses. Battery systems, however, require more cells to produce higher voltages, which tends to increase costs, reduce bat-

TABLE 5.5 Electric Motor/System Attributes

Motor Developer/Type	Motor Power Density kg/kw*	L/kw	Voltage V	Electronics Power Density kg/kw*	L/kw	Trans. Y or N	System Peak Power kw
General Electric D.C. Sep. Exc. (ETV-1)	3.0	0.97	108	1.5	1.7	N	30
General Electric A.C. Induction (ETX-1)	1.2	0.13	192	1.3	1.4	Y	40
General Electric A.C. PM Sychr. (ETX-II)	0.98	0.15	192	0.69	0.91	Y	52
General Electric A.C. Induction (MEVP)	0.60	0.14	340	0.55	0.45	N	56
Cocconi Eng. A.C. Induction (EV1)	0.5	0.17	320	0.35	0.66**	N	90–100
Pentastar D.C. Sep Exc. (TEVan)	1.5	0.45	176	0.80	1.8**	Y	50

Source: SAE Paper No. 920447, www.sae.org

*All specific power and volume values based on maximum motor power.

**Includes dc-dc converter and battery charger.

tery performance, and decrease system reliability. Today, either ac or dc is an appropriate choice for an electric car. In the future, however, ac systems are likely to become more prevalent, and dc systems may become outmoded. Table 5.5 compares power system attributes among several early ac and dc systems.

Controller

The controller is the device that controls motoring speed, and thereby the speed of the vehicle. In a sense, EV controllers "throttle" the electric motor by modulating and conditioning the current. Controllers for dc machines are relatively simple, compared to the complex devices necessary for ac machines. Techniques of controlling dc motoring speed include resistance control, voltage stepping (selectively connecting batteries in series or parallel), and converting the current into high-frequency pulses using a pulse-width modulator, or "chopper" controller. Controllers for ac machines are more complex. Power-conditioning units for ac systems normally include a three-phase inverter, a pulse-width modulator, and microprocessor waveform conditioning and real-time feedback and signal processing.

Inexpensive dc speed controllers, such as the devices used in golf carts, are typically based on resistance control or voltage stepping; however, neither system is adequate for an electric automobile. Today, electronic control is considered the only practical means of speed control for an electric car. Moreover, the role of modern controllers has been expanded to include functions having to do with onboard energy management, regeneration, and battery charging.

Pulse-Width Modulator

Pulse-width modulators are the most efficient and reliable electronic controllers available for dc machines. These devices operate by rapidly turning the motor on and off. Current is literally "chopped" into segments of either full battery voltage or zero voltage. This high-frequency chopped current results in an average voltage that is less than battery terminal voltage. The actual value of applied voltage depends on the length of the on and off cycles and the number of cycles per second. Motoring speed is controlled by varying the duration and frequency of the on/off bands. Control is smooth and stepless and usually accompanied by an audible hum. Pulse-width modulators are commonly referred to as "chopper controllers" or simply "choppers" because of the chopped nature of the current.

There are primarily two types of choppers: the silicon controlled rectifier (SCR) and the transistor chopper. SCR choppers are larger, more expensive, and emit a louder hum than transistor-based devices. Operating frequency is a maximum of 1,000 pulses per second for SCR units and up to 20,000 pulses per second for transistor choppers. Before high-power transistors were developed, SCR choppers were used in high-power applications. Today, the transistor chopper has largely replaced the older SCR units. Depending on the output, operating efficiency of a transistor chopper extends from a low of about 85 percent to a maximum of 99 percent (Figure 5.7). A relay is often installed to bypass the chopper at full voltage to eliminate internal losses.

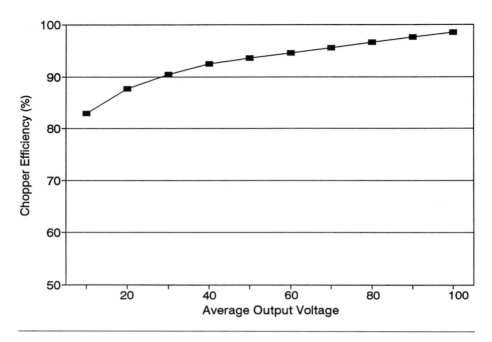

FIGURE 5.7 Chopper Efficiency by Load

Controllers for Alternating Current

Alternating current reverses direction many times per second to produce a sinusoidal wave form. A three-phase system is essentially the same as three ordinary single-phase systems, with the three single-phase waveforms shifted out of phase by one-third of the cycle. Power conditioners/controllers for ac motors change dc current from the battery into quasi-sinusoidal ac current with the appropriate frequency and magnitude for proper system performance. Inverters vary in type and control strategy, as well as in the details of implementation. In general, the systems are based on a chopped current in conjunction with microprocessor signal processing as needed to control the magnitude, frequency, and slip-angle of the waveform. Greater chopping speeds and internal processing result in greater losses to power conditioning and larger, heavier, and more costly control units. Ac controllers also tend to be noisier than their dc counterparts. Only in the last few years have components become available that make ac controllers power efficient and cost effective.

Motor as a Subsystem of the Transmission

With the advent of practical continuously variable transmissions (CVTs), the transmission could ultimately become the dominant subsystem in conventional IC engine powertrains. This conceptual switch in system dominance also has merit for electric cars. The idea of configuring a BEV power system for transmission dominance was first proposed to the author by the late Foster Salsbury, the founder of Salsbury Industries. This approach centers on the ability of a CVT to control vehicle speed independently of motoring speed.

Experiments in the late 1970s at the author's former design firm, Quincy-Lynn Enterprises, demonstrated the feasibility of using a CVT to control vehicle speed by varying the transmission's shift position. This control scheme eliminates the costly high-power electronic controller, and it provides torque multiplication in the process. In the CVT-equipped Quincy-Lynn car, the compound-wound dc motor was supplied with full system voltage and ran continuously at a steady-state speed. Vehicle speed was controlled by varying the shift position of the transmission. To accomplish this task, the transmission was configured to shift according to a signal supplied by the accelerator pedal. Transmission shift position tracked the degree of depression of the accelerator pedal and thereby controlled vehicle speed.

Quincy-Lynn's prototype system contained relatively little logic. When the accelerator pedal was depressed to 50 percent of its total travel, the transmission assumed a 50 percent upshift position, which accelerated the vehicle to 50 percent of its maximum speed. When the pedal was fully depressed, the transmission fully upshifted and accelerated the vehicle to maximum speed. The ratio spread was wide enough to smoothly control vehicle speed from 16 km/h to 97 km/h (10 to 60 mph) while the motor remained at a steady-state rpm. Below 16 km/h (10 mph), the transmission was allowed to slip, just as the clutch of a manual transmission would be slipped during startup. Transmission slip was also controlled by the accelerator pedal. Releasing the pedal below a preestablished speed disengaged the transmission.

Regenerative braking was a natural byproduct. Because transmission shift position tracked the depression of the pedal, reducing pedal pressure caused the transmission to downshift. This slowed the vehicle and accelerated the motor, which increased counter emf and reversed the flow of current. The amount of braking force and the magnitude of regeneration current was controlled by the pedal position in relation to vehicle speed. Drivability of the system was excellent. Speed control was natural, and there was no sense that anything unusual was taking place in the drivetrain.

FIGURE 5.8 Quincy Lynn Enterprises' CVT-Controlled Electric Vehicle. This Quincy-Lynn Enterprises' experimental BEV used an electronically controlled CVT to control vehicle speed. Motor ran at a steady-state speed. The device that made this possible was the electromatic drive transmission, which was developed jointly by Daryl Hillman and Foster Salsbury.

This first-generation Quincy-Lynn BEV was merely a proof-of-concept prototype designed to prove out the feasibility of CVT speed control (Figure 5.8). Several improvements were slated for a second-generation vehicle, but further development was not funded. For example, belt squeeze in the proof-of-concept transmission was controlled by springs in the driven pulley, which made belt squeeze dependent on shift inputs rather than torque requirements. Belt squeeze should be controlled at both the drive and driven pulleys and regulated according to torque in order to achieve maximum transmission efficiency. Too little belt squeeze results in slippage, and too much squeeze results in greater losses to friction. A field-control loop using low-power electronics would have broadened the vehicle's speed range and provided significantly better control of regeneration. By relying only on counter emf, regeneration was limited to the upper speed ranges, and the system's full potential for reclaiming energy was unrealized. In the prototype, the motor ran continuously, even when the vehicle was at rest. The second-generation system would have been configured to shut off the motor when the vehicle came to a stop. Free-running load of the motor was approximately 850 W.

Maintaining high transmission efficiency during shift and at high reduction ratios is the primary engineering challenge of a CVT speed control system. The electromatic drive transmission was dynamometer tested at 96 percent efficiency at full upshift, but data were not obtained for transient conditions and at high reduction ratios. Literature contains widely divergent reports of efficiencies during shift and at high ratios. In general, efficiency suffers during both conditions; however, it is difficult to estimate the degree to which improperly tuned units may have affected test results. A nonlubricated vee-belt CVT can be quite efficient throughout the range of operating modes, provided that belt squeeze is appropriately matched to torque. Electronically controlled torodial and lubricated steel-belt drives can also be used as EV speed controllers. Additional information on CVT drives is provided in Chapter Three.

Integrated Electronic Controls

Like its IC engine counterpart, the electric tractive system has reached a level of sophistication that demands microprocessor logic. Once microprocessors are introduced, multitasking and systems integration become possible. A state-of-the-art electronic control unit (ECU) can process feedback from various vehicle systems, sense and analyze operating conditions, then execute driver inputs in ways that balance system response and optimize performance. The ECU can manage the regenerative braking system, control the transmission shift schedule, and integrate a variety of electrical and mechanical functions that may be required to keep the vehicle precisely tuned to operating conditions and driver inputs. A fluid and integrated shift schedule can provide torque multiplication in response to demand, keep the motor in its region of maximum efficiency, and manage the flow of current into and out of the battery. Improved drivability is also a primary goal. The electric vehicle should emulate the feel and response of the familiar IC vehicle. A modern EV will incorporate drive-by-wire, brake-by-wire, and steer-by-wire processes, which will virtually eliminate mechanical connections between driver command modules and vehicle systems.

Powertrain control tasks must be done in real time and involve simultaneous and complex calculations. As events and status are sampled, data may be deposited in memory and shared with other subsystems. Some tasks must perform unique calculations in order to determine the appropriate torque, braking, and shift schedules. Other tasks can then use the information from the schedules to make appropriate outputs to solenoids, clutches, and operator interface systems. Battery state-of-charge will affect shift schedule biases and regeneration response to brake pedal pressures. Control parameters for shift schedule, motor current and frequency in ac systems, or field/armature current and ratio in dc applications depend on the complex relationship between vehicle speed, transmission ratio, brake input, vehicle acceleration/deceleration responses, and battery state-of-charge. During regeneration, the system must sense loss of traction and modulate the hydraulic brakes, balancing hydraulic braking with regenerative forces, and balancing both against vehicle conditions. In addition to powertrain control tasks, the ECU may also provide status, warning, and diagnostic information to the driver. It might also delay responses to certain driver inputs such as forced downshift or reversing at higher speeds, and provide default detection, systems override, or even systems shutdown when appropriate conditions exist.

In the process of achieving maximum energy efficiency and improved drivability, the mechanically simple electric automobile necessarily becomes an inherently complex machine. Hybrid vehicles place an even greater demand on system computers and software engineering. Only through appropriate hardware and software architecture can battery-electric and hybrid-electric vehicles achieve superior drivability and performance.

ONBOARD ENERGY STORAGE

Since the automobile's inception, the absence of an appropriate energy storage medium has been the Achilles heel of battery-electric cars. The electrochemical battery has been the traditional energy storage device. Energy can also be stored mechanically using a spinning flywheel. Flywheels of high specific energy imply very high rotor speeds, which introduces challenges regarding gyroscopic forces, materials capabilities, bearings, and system costs. Expectations for near- and midterm energy storage devices have therefore centered on improving the secondary (rechargeable) battery.

In September 1991, the U.S. Department of Energy, along with Ford, General Motors, and Chrysler, joined forces to develop advanced battery technologies for future electric cars. Together under the U.S. Advanced Battery Consortium (USABC), vast resources were committed to development efforts.

Similar projects began in Europe and Japan. Today's nickel metal hydride, lithium ion, and lithium polymer batteries were developed through these research and development efforts. Nickel metal hydride batteries are considered a transitional technology, which will ultimately give way to lithium ion or lithium polymer batteries.

An automotive energy storage system must meet many criteria, and often requirements are at odds with each other. The energy source must have high specific energy and high specific power, but in general, a battery's specific energy and its specific power are inversely related. When one is increased, the other decreases. Battery materials must be plentiful, inexpensive, nontoxic, and either recyclable or easily disposable without harm to the environment. Recharge time and maintenance must be low, and life must be high, preferably equal to the life of the vehicle. Unfortunately, the chemical changes that produce current also decompose battery materials. Self-discharge is another attribute typical of secondary batteries. A vehicle battery might rest unused for a period of days or even weeks, and power must be instantly available on startup. Yet all batteries experience some degree of self-discharge.

Many of the presumed requirements for battery performance were largely dictated by the characteristics of conventional liquid-fueled power systems, and it may be unrealistic to expect batteries to match them. In order to establish realistic development goals, the USABC identified the following primary criteria for advanced electric vehicle batteries, as shown in Table 5.6.

In general, battery systems fall into three classifications: the aqueous systems, the ambient-temperature lithium systems, and the high-temperature systems. Lead-acid batteries, various nickel systems, and flow batteries are aqueous systems. A zinc-bromine battery, for example, is an aqueous-flow system that received considerable attention early on; however, the mechanical complications of pumps and the safety hazards of bromine were two major concerns. Ambient-temperature lithium batteries have shown promising results. In their favor is their characteristic high cell voltage and low weight. High-temperature systems include lithium-metal sulfide, sodium-metal chloride, and sodium-sulfur batteries. Batteries in this class have high specific energy, but their high operating temperatures and corrosive solutions created difficult technical and safety challenges.

When comparing battery performance, it is important to consider a particular battery's state of development and the application for which it is designed. Technology is rapidly improving, and what is true today may be much different in a few years or even in a few months. Also, much of the information on advanced batteries is established under controlled and ideal

TABLE 5.6 USABC Advanced Battery Criteria

Parameter	Midterm Criteria	Commercialization Criteria	Long-Term Criteria
Price	@$150/kWh	<$150/kWh ($75 desired)	<$100/kWh
Cycle Life	600 @ 80% DoD	1000 @ 80% DoD 1600 @ 50% DoD 2670 @ 30% DoD	1000 @ 80% DoD
Range @ Life (Urban Miles)	100,000	100,000	100,000
Calendar Life	5 years	10 years	10 years
Power Density	250 W/L	460 W/L	600 W/L
Energy Density	135 Wh/L	230 Wh/L	300 Wh/L
Specific Power	150 W/kg (200 desired)	300 W/kg	400 W/L
Specific Energy	80 Wh/kg (100 desired)	150 Wh/kg	200 Wh/kg
Regenerative Specific Power	75 W/kg	150 W/kg	200 W/kg
End of Life (EOL)	20% of rated power and capacity specification	20% of rated power and capacity specification	20% of rated power and capacity specification
Operating Performance	−30° C to +65° C	20% Loss at extreme of −40° C and +50° C (10% desired)	−40° C to +85° C
Normal Charge	6 hrs, 20–100% SOC	6 hrs, 20–100% SOC (4 hrs desired)	3–6 hrs, 20-100% SOC
High Rate Charge	@15 min, 40–80% SOC	<30 min @ 150 W/kg, 20–70% SOC (20 min @ 270 W/kg desired)	<15 min, 40–80% SOC
Efficiency at EOL	75%	80%	80%
Off-Tether Pack Energy Loss	Thermal loss <3.2 W/kWh (<15% in 48 hours) Self discharge <15%/48 hrs	3 days: <15% with no performance loss. 12 days: Cumulative Loss <25% with some performance loss @ extreme temperature limits	Thermal loss <3.2 W/kWh (<15% in 48 hours) Self discharge <15%/month

DoD = Depth of Discharge
SOC = State of Charge

conditions, and therefore may not hold up in a real-world application. Even within a battery category, much about its performance depends on the trade-offs that designers have made between specific energy, specific power, and cycle life. Test results are affected by charge/discharge cycling rates, with slower rates yielding greater total energy and longer life. New charger technologies affect the life and charge rates of batteries. In addition, manufacturing costs are often difficult to project with an experimental system. So many factors affect test data, and many factors other than test data are critical to a battery's viability as a consumer product. A battery of exceptionally high specific energy may still be infeasible as a consumer product because of other negative attributes. The filters on the way from the laboratory to the marketplace are multilayered.

In the field, the characteristics of a vehicle's battery system are much more inconsistent than any of its other components or subsystems. Battery performance typically depends on age, charging procedures, previous discharge history, discharge rates, and even controller characteristics. For example, the specific energy of a lead-acid battery may be degraded by 30 percent or more when subjected to periods of high discharge, and the pulsed discharge of chopper controllers has a detrimental effect on specific energy. Battery performance is also subject to the complex and sometimes cantankerous nature of the chemical processes involved.

Attempts to estimate the manufacturing costs of an advanced battery are complicated by the lack of experience in producing the high volumes typical of automotive applications. Experimental batteries may appear advantaged or disadvantaged by early cost estimations. Cost projections are necessarily based on highly variable estimations of materials and processes. Projections often must rely on estimations based on undeveloped technologies and production processes, which implies a high degree of uncertainty.

Table 5.7 compares attributes of a number of battery couples. Today, the most promising production-ready EV battery is the nickel metal hydride battery. Over the longer term, lithium-ion and lithium polymer batteries appear to offer the best option, but a variety of other battery couples are also included in the following review.

Lead-Acid Battery (Pb/Acid)

Lead-acid batteries have been around since 1854 and essentially have matured along with the automobile industry. Like an incumbent politician, the lead-acid battery has the advantage of being an established entity. Familiarity in

TABLE 5.7 *Battery System Comparison*

Battery Type	Specific Energy Wh/kg	Specific Power W/kg	Cycle Life	Recycle % of Materials	Energy Efficiency
Lead-Acid*	40	300	750	97%	65%
Aluminum-Air	200	150	—	75%	35%
Lithium-Iron Disulfide	>130	>120	1000	50%	—
Lithium-Polymer	>180	300	1000	50%	—
Lithium-Ion	>140	300	1200	—	—
Nickel-Cadmium	56	200	2000	99%	65%
Nickel-Iron	55	130	1500	99%	60%
Nickel-Metal-Hydride	>80	200	1000	100%	90%
Nickel-Zinc	80	150	200	—	65%
Sodium-Sulfur	100	120	500	50%	85%
Zinc-Air	120	120	135	75%	60%
Zinc-Bromine	70	100	500	—	65%

*The C-3 specific energy rating for the Horizon No. 1H85 battery is used. Specific power is a mean between the fully charged capability and 20% state-of-charge capability. The Horizon lead-acid EV battery has the highest ratings in the industry.

product technology breeds a level of confidence that can come only with experience. For much the same reasons, the lead-acid battery is also relatively inexpensive. Having been in high-volume production for nearly 100 years, development, tools, and facilities costs have been amortized, and manufacturing techniques and economies are well established. A new battery technology must fight an uphill battle on all fronts, first to demonstrate its technical superiority, then to instill confidence that unforeseen problems will not develop in the field, and finally to compete with an established product during startup.

Technical advantages of the lead-acid battery include its simple manufacturing processes and its relatively high specific power. Lead-acid batteries are also easily recycled and the infrastructure is in place. Disadvantages include the relative scarcity of lead as a raw material, the battery's vulnerability to damage caused by sulfation and deep discharge, and its low specific energy.

Although lead-acid batteries are usually maligned because of their limited specific energy, maintenance problems have most frustrated EV fleet owners. This situation often surprises people because electric cars are assumed to be low-maintenance vehicles. Theoretically, EVs should require very little maintenance, but early vehicles have been largely experimental and often have not received the investment in engineering typical of most automotive products.

For maximum life, batteries must be conditioned by discharging to approximately 80 percent depth of discharge (DoD) for the first 10 to 15 cycles. If they are not properly "formed," cycle life and specific energy may suffer. Batteries must be periodically "equalized" by a very low charge rate maintained for a period after the batteries are fully charged. If batteries are not charged to gassing voltage, specific energy can drop because of electrolyte stratification. Overcharging can damage batteries as well.

Batteries self-discharge and they must be regularly supplied with a maintenance charge to prevent damage. This problem is not limited to lead-acid batteries. All batteries self-discharge at varying rates. At Quincy-Lynn Enterprises we would often store prototype electric cars for extended periods unattended. Batteries were usually dead when a vehicle emerged from a long hibernation. Moreover, they could not be revived by recharging. The problem was sulfation, which begins to occur when a lead-acid battery is unused for approximately 30 days. In general, maintenance problems can be easily managed by technical solutions, as long as the vehicle is regularly used and recharged with a properly designed charger.

The most tenacious problem of the lead-acid battery is its low specific energy. Within limits, the specific energy of a lead-acid battery may be increased by increasing the area-to-volume ratio of plate active materials. Essentially, thinner plates increase the battery's ability to store and deliver energy. When the surface-to-volume ratio of active material is increased in order to increase specific energy, however, the positive plate breaks down more rapidly and battery life suffers. Consequently, an inverse relationship exists between specific energy and service life (Figure 5.9). One method of increasing the surface-to-volume ratio of active plate material is by utilizing the tubular or gauntlet plate design. The tubular plate is designed to increase plate integrity at higher surface-to-volume ratios, and thereby allow batteries of greater specific energy while still retaining an acceptable cycle life. The tradeoff here is one of cost. Tubular plate batteries are more costly to manufacture.

One variant is the flow-through battery, which utilizes forced electrolyte circulation to improve battery performance. Circulating the electrolyte increases active material utilization by allowing the electrolyte to access the material's micropores. Circulation also eliminates specific gravity and heat gradients,

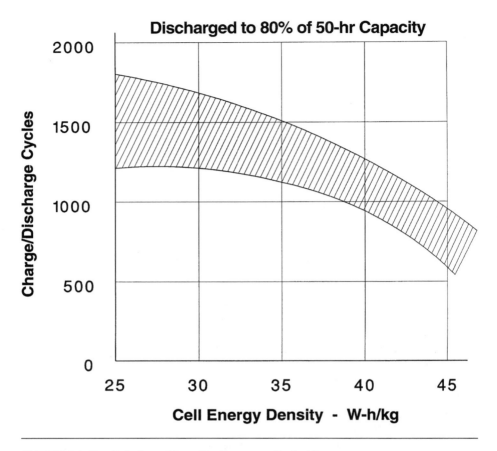

FIGURE 5.9 The Relation of Specific Energy to Cycle Life

and minimizes the overcharge required to keep batteries equalized. In general, however, flow-through batteries suffer from reduced service life, primarily because of the more aggressive electrochemical action promoted by electrolyte circulation. Nevertheless, specific energy and specific power are significantly improved. Experiments at Johnson Controls, Inc. demonstrated increases of 67 percent in specific energy and 54 percent in specific power as a result of the flow-through design.[14] Advantages of the flow-through battery include the following:

- Specific energy is significantly increased.
- Improves performance over the Simplified Federal Urban Driving Schedule (SFUDS).

- Makes the battery less sensitive to peak discharge rates.
- Provides the ability for thermal management.
- Improves active material utilization, which reduces demand on natural resources.

One of the most innovative lead-acid designs is the Horizon battery by Electrosource, Inc. It was developed by a group of aerospace engineers who adopted technology from strategic countermeasure devices. Plates are made of a co-extruded proprietary lead filament. Lead wire is extruded around a thin, glass fiber filament under extreme pressure. Strength comes from the glass-filament core, which eliminates the need for antimony, calcium, or other alloys normally required to stabilize plate materials. The resulting high-density, fine-grain lead also results in extended plate life. Wire is woven into a grid, which is coated with a proprietary paste and assembled into bipolar electrical plates (Figure 5.10). Plates are positioned horizontally, rather than vertically, in the finished modules.

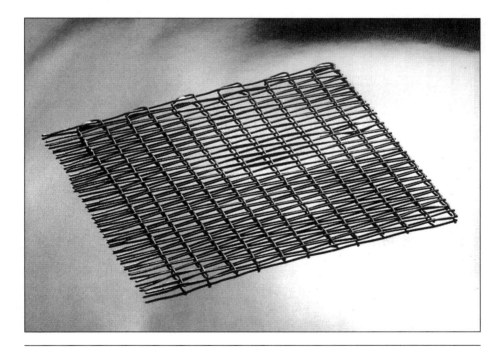

FIGURE 5.10 *Woven Co-Extruded Lead Wire Grid.* Advanced grid design creates a stronger structure without increasing costs and improves specific power and specific energy.

Courtesy: Electrosource Inc., www.electrosource.com

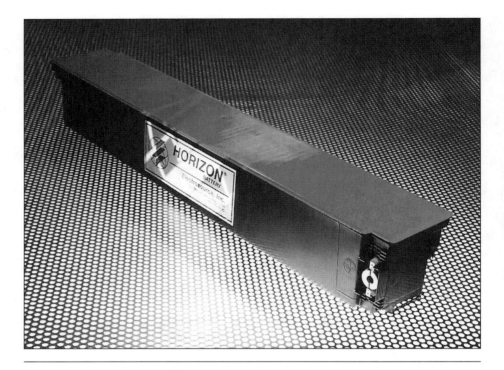

FIGURE 5.11 Horizon Battery by Electrosource, Inc. The Horizon battery has the highest rating of any lead-acid battery in the industry. The company uses a proprietary plate construction for superior life and improved specific energy.

Courtesy: Electrosource Inc., www.electrosource.com

Cycle life of production batteries averages 750 cycles (C-1 rate to 80% DoD) or up to 900 cycles maximum. Additionally, the battery can tolerate an eight-minute recharge to 50 percent capacity or a half-hour recharge to 99 percent capacity. Specific energy at the C-3 discharge rate is about 40 Wh/kg. Specific power is 400 W/kg at 100 percent charge, and 200 W/kg at 20 percent state of charge. Horizon batteries weigh 25 to 50 percent less than conventional lead-acid batteries (Figure 5.11).

Aluminum-Air (Al/Air)

Aluminum-air batteries are the first in a series of systems that use a metal anode and an atmospheric oxygen cathode. Iron-air, zinc-air, lithium-air, and magnesium-air are just a few. Metal-air batteries tend to exhibit high specific energy and low specific power. Theoretically, the specific energy of an Al-air battery is three times higher than zinc-air and seventeen times higher than a lead-acid battery. The breakdown of aluminum that occurs during the electrochemical reaction cannot be reversed, so the Al-air battery cannot be electrically recharged. Consequently, the device is not a true secondary battery. Aluminum-air batteries are essentially "refueled" by adding aluminum, and therefore they operate more like fuel cells. To recharge the battery, the aqueous alkaline electrolyte is replenished with water, and the waste aluminum hydroxide is removed. Approximately every fourth time the system is refurbished, the aluminum anode must also be replaced.

An Al-air battery is a very mechanical device (Figure 5.12). Incoming air is scrubbed of CO_2 and dehumidified before entering the battery. Pumps circulate electrolyte between the cells and a separate reservoir. A crystallizer

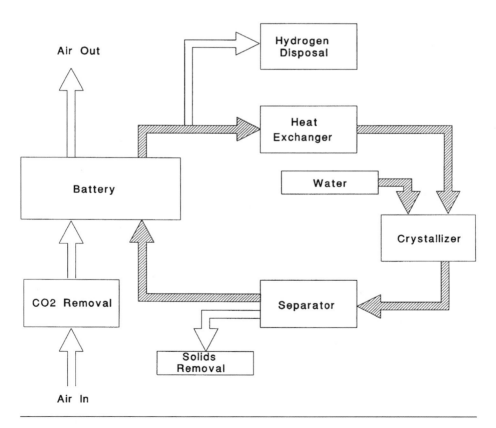

FIGURE 5.12 Aluminum-Air Flow System

removes soluble byproducts from the electrolyte to create an aluminum hydroxide slurry. Concentrated aluminum hydroxide is then washed into another receptacle for later removal. The system is complex and efficiency is low. Although aluminum is a plentiful metal, the mining and smelting process is very energy intensive.

Lithium-Iron Sulfide (Li/FeS)

Lithium is the lightest of all metals and has one of the highest electrochemical potentials. A perfected lithium-based battery could solve the range shortfalls that have plagued BEVs from the outset. Both lithium-iron monosulfide and lithium-iron disulfide are high-temperature batteries. The lithium-iron monosulfide battery utilizes a lithium alloy anode and an iron monosulfide cathode. The electrolyte is a molten lithium chloride and potassium chloride salt solution. Operating temperature is about 450°C. Lithium-iron disulfide is a similar high-temperature battery that utilizes a positive electrode of lithium-iron disulfide. This battery has a molten electrolyte of lithium chloride, lithium bromide, and potassium bromide. It operates at a slightly lower temperature (400°C), the electrolyte is more corrosive, and it has about twice the energy density of a lithium-iron monosulfide battery.

Lithium-Ion (Li/Ion)

Primary Li/ion batteries have been around for years. In order to make a rechargeable Li/ion battery, however, a new technology had to be developed. The breakthrough came when engineers at Sony discovered that carbon materials could be substituted for the lithium metals used in the non-rechargeable batteries. Rechargeable lithium-ion batteries use a carbon electrode that serves as a host for lithium ions.

Lithium-ion EV batteries have also been under development by VARTA, SAFT, and others. Many consider Li/ion to be the most promising rechargeable battery technology in sight. It has both high specific energy and high volumetric energy density, and it tolerates rapid charge and discharge rates. Cycle life is also among the highest of all the battery types. These attributes are important to varying degrees in both BEV and hybrid applications.

The anode of a Li/ion battery is carbon, and the electrolyte is $LiPF_6$ in an organic carbonate solution. The cathode may be $LiCoO_2$, $LiNiO_2$, or $LiMnO_2$ or an alloy containing various amounts of cobalt, nickel, or manganese. Cobalt is expensive but may be recyclable.

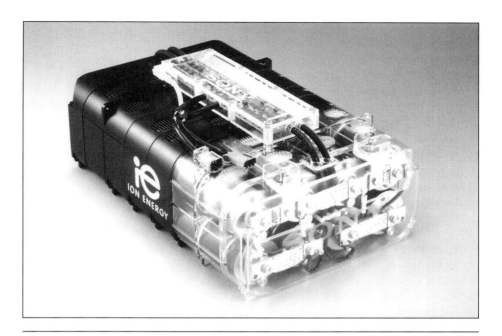

FIGURE 5.13 Nissan Altra EV Lithium-Ion Battery Module

Courtesy: Nissan Motor Co., www.nissandriven.com

The Li/ion system is flexible and adaptable to design variations needed to fit the vehicle's mission. A battery with higher specific power, Li/ion batteries would be used for hybrid applications. Batteries of lower specific power and greater specific energy would be used for BEV applications. By varying the chemistry, SAFT produces one EV battery with 140 Wh/kg specific energy and 400 W/kg specific power, and another battery with 100 Wh/kg specific energy and 950 W/kg specific power.

Nissan's Altra EV uses a 345-volt system supplied by 12 lithium-ion battery modules developed jointly between Nissan and Sony (Figure 5.13). Altra's range is on the order of 80 miles city and 100 miles highway. Batteries are air-cooled.

Lithium Polymer Battery (LPB)

The lithium polymer battery (LPB) was initially developed jointly by 3M Company and Canada's Hydro-Quebec. Argonne National Laboratory also worked on LPBs. The LPB relies on thin-film technology, with composite films that are only 100 microns thick. It is a solid-state battery made up of individual cells that are wound and shaped as needed. Individual cells are then packaged into modules, which are configured according to the application. The electrolyte consists of a solid plastic film within the original laminate.

A typical lithium polymer BEV battery pack might weigh on the order of 224 kg (500 lbs.) and contain energy stores as high as 45 kW/h. In comparison, EV-1's lead-acid battery pack weighs more than 480 kg (1,000 lbs.) and provides 16 kW/h of energy stores. So the LPB stores nearly three times the energy at half the mass of today's lead-acid batteries, which translates into a BEV with a 320- to 480-km (200- to 300-mi.) range. Cycle life is high, and the battery should be relatively cheap to build in production quantities.

LPBs can be easily configured for high specific power or high specific energy, which makes them suitable for either BEV or HEV applications. Preliminary performance of EV cells, modules, and packs tested under USABC cycling protocols shows that LPBs already meet performance characteristics necessary for a competitive BEV. Ford's TH!NK City has been equipped with LPBs for field testing.

Because of lithium's highly reactive nature, there was initial concern over the safety of a lithium battery. Lithium-ion batteries have been known to ignite and burn, but these characteristics can be handled through proper module design. Engineers at 3M crushed and drove nails through prototype LPB modules without harmful effects.

Nickel-Cadmium (Ni/Cd)

Nickel-cadmium batteries were developed in 1901. Ni/Cd batteries for EVs are currently available from SAFT (Figure 5.14). The SAFT EV battery has a cathode made of sinistered nickel chemically impregnated with a mix of hydroxides. The anode is a plastic-bonded cadmium composite. The electrolyte is a solution of potassium and lithium hydroxide. Open-circuit cell voltage is 1.35 V. SAFT's STM5-200 6-volt module is slightly smaller than a standard lead-acid traction battery, and at 25.5 kg, about the same weight. Specific energy is 52 Wh/kg and specific power 210 W/kg. Life can be as high as 2,000 cycles. SAFT estimates that Ni/Cd-powered EVs will run 200,000 km before batteries must be replaced.

The Ni/Cd battery is superior to the lead-acid battery in all areas except price. SAFT's STM5-200 EV battery retails for $1,235 each. Chrysler bought them for their TEVan, which uses 35 modules (monoblocks), and paid $1,100 each for a total of $38,500 per vehicle. Over the life of the batteries,

FIGURE 5.14 SAFT Nickel-Cadmium EV Battery Module

replacement cost calculates to $0.19 per kilometer ($0.31 per mile), which equates to approximately $6.19 per gallon of gasoline at 20 mpg. Costs, however, reflect low production volumes.

Technical problems have to do with the scarcity and toxicity of cadmium, and to a lesser degree with the battery's memory effect. If Ni/Cd batteries are not completely discharged or fully recharged, they tend to remember state-of-charge extremes and then behave as though they have less capacity. In electric car applications, batteries may take on the characteristics represented by the most common trip distance; however, they can be reconditioned by running them through several complete charge/discharge cycles. The relative scarcity of raw materials is not so easily overcome. Although nickel is abundant, cadmium is not. In addition, cadmium is very toxic and carcinogenic, and disposal is strictly regulated.

Nickel-Iron Battery (Ni/Fe)

Thomas Edison invented the nickel-iron battery at the beginning of the 20th century. These batteries have a negative electrode of iron and a positive

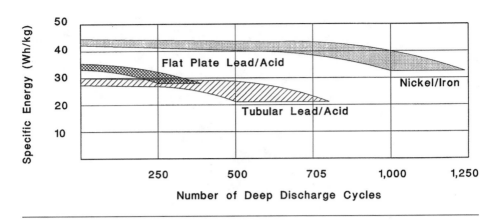

FIGURE 5.15 *Specific Energy Versus Cycle Life*

electrode of nickel oxide. The electrolyte is potassium hydroxide. Advantages include a smaller, lighter package with greater specific energy than the lead-acid battery. Long service life is also characteristic (Figure 5.15). A nickel-iron battery can weight 20 percent less, yet store as much as 50 percent more energy and have up to four times the service life of a similar lead-acid battery. Volumetric energy density is about 113 Wh/L. Batteries are packaged in 6-volt modules, similar in size to that of traditional lead-acid modules. Nickel-iron batteries are produced by Eagle-Picher Industries.

Disadvantages include higher initial costs and a relatively higher self-discharge rate. Eagle-Picher produces only 1,500 nickel-iron traction batteries per year. These batteries are essentially handmade. As a result, the single-unit price of their NIF200 module is $1,300. According to Darrel Ideker, Plant Manager at E-P's Joplin, Missouri facility, if production were increased to just 10,000 batteries per year, per-unit cost would drop to approximately $400. Accounting for increased service life, battery cost-per-mile could drop to that of a lead-acid battery. As for self-discharge, the NIF200 module will lose about 13 percent of its charge in seven days. With regular use, self-discharge does not present a problem.

Nickel Metal Hydride (NiMH)

Nickel metal hydride batteries have been under development since the 1970s at Ovonic Battery Co. in the United States and at SAFT in France. NiMH EV batteries are also manufactured by VARTA and Panasonic. Batteries can be recharged to 60 percent capacity in about 15 minutes. Specific energy is twice that of a lead-acid battery or about 80 Wh/kg. Batteries are compact, with a volumetric energy density as high as 215 Wh/L. Both NiMH and Ni/Cd batteries use an aqueous alkaline electrolyte and a nickel hydroxide cathode. NiMH batteries, however, do not use a cadmium anode. Instead, they use a metal alloy anode. The nominal cell voltage for NiMH batteries is 1.2 volts.

Early on, it was believed that NiMH batteries lacked high specific power. What was not understood is that existing cells had been developed for specific applications where specific energy was more important than specific power. The relationship of cell materials, primarily vanadium, nickel, titanium, zirconium, and additional transition metals, can be altered to tailor battery characteristics to the application. Cells developed specifically for EV application can achieve a peak power of 175 W/kg at 80 percent DoD. Excellent performance at continuously high discharge rates is also characteristic. The battery is tolerant of extremes in temperature, with more than 85 percent capacity available at –20°C. Cycle life is also high: batteries have been tested up to 1,000 cycles at 100 percent DoD.

Initially, a high rate of self-discharge was a problem with NiMH batteries. Batteries typically lost about 50 percent of their charge when left at room temperature for 30 days. Tests on extended C-size cells indicated a 15 to 20 percent self-discharge in seven days.[15] Advanced separator materials can increase charge retention to about 93 percent over a 30-day period.[16] Batteries are totally sealed, abuse resistant, and maintenance free. Based on present EPA protocols, NiMH batteries can be discarded in landfills as nonhazardous and nonpolluting waste. Figure 5.16 shows the architecture of the Ovonic EV battery.

Under the high demands typical of EV applications, NiMH batteries must be cooled to avoid overheating. General Motors has concentrated on the Ovonic design, which is an air-cooled battery available in two module sizes of 12 volts (Figure 5.17). Chrysler used the 12-volt SAFT module, which is equipped with built-in liquid cooling jackets (Figure 5.18). NiMH EV batteries are now in limited production for EV applications.

GM and Ovonic joined forces to manufacture NiMH EV batteries in 1994 under the company name of GM-Ovonic L.L.C. Ultimately, GM sold their interest in the company to Energy Conversion Devices, Inc. The ECD Ovonics website is located at www.ovonic.com.

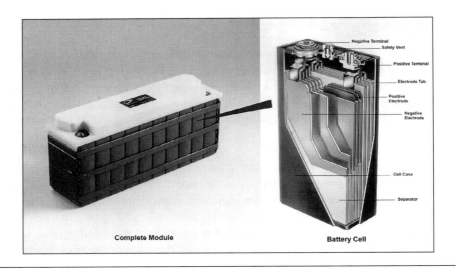

FIGURE 5.16 Ovonic NiMH Battery

Courtesy: Ovonic Battery Company, www.ovonic.com/enstor.html

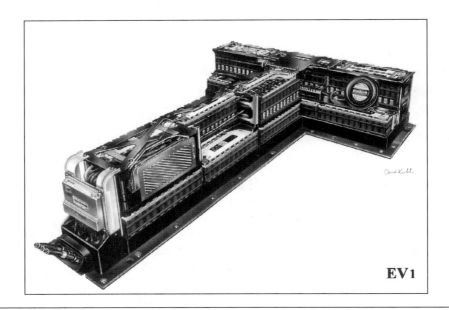

FIGURE 5.17 EVI Battery Layout with NiMH Batteries. GM's EV-1 uses the air-cooled Ovonic NiMH battery. Batteries are arranged in a "T" layout.

Courtesy: General Motors Corporation, www.gm.com

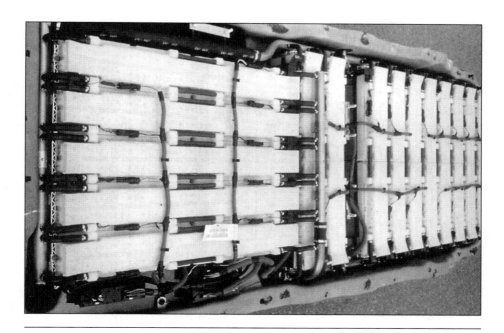

FIGURE 5.18 Chrysler's EPIC Battery Pack with SAFT Liquid-Cooled NiMH Batteries.
Chrysler's EPIC van is equipped with SAFT liquid-cooled NiMH modules, which are housed in a tub suspended under the vehicle's floor. Coolant jackets are molded into the battery case. Batteries are then interconnected to create the integral flow-through cooling system.

Courtesy: DaimlerChrylser, www.daimlerchrysler.com

Nickel-Zinc (Ni/Zn)

Nickel-zinc batteries have existed for nearly 100 years. This battery couple has a theoretical specific energy of about twice that of a lead-acid battery. It has excellent specific power and can deliver power even at extremely low temperatures. A lead-acid battery is virtually dead at –40°C, while the Ni/Zn battery can still deliver 40 to 60 percent of its 20°C capacity. A typical Ni/Zn cell has a positive electrode of nickel oxide, a negative electrode of zinc, and a potassium hydroxide electrolyte. Materials are abundant and inexpensive, and manufacturing processes are relatively straightforward.

The primary difficulty with the Ni/Zn couple is plate instability and a propensity to grow dendrites during recharge. During recharge, zinc is redeposited unevenly or in the wrong places. As a result, plates begin to change shape and composition. In addition, dendrites grow between the positive and negative plates and ultimately cause the cells to develop shorts. Life is typically limited to about 100 to 150 cycles.

Attempts to resolve the problem of plate instability have centered on reformulated and flowing electrolytes and improved separator materials. Work at Electrochimica Corp. with a reformulated chemistry for the zinc electrode-electrolyte system shows much-improved life with little compromise in the performance of the nickel electrode. Experiments at Electrochimica with prismatic cells incorporating their stabilized chemistry have demonstrated cell life of 600 to 800 cycles with 60 percent retention of original capacity.[17]

Sodium-Sulfur (Na/S)

Sodium-sulfur batteries were developed in the 1960s by Ford Motor Company. Later, Chloride Silent Power and ABB in Europe took over development. Several automobile manufacturers have built experimental vehicles powered by Na/S batteries. Na/S batteries have high specific energy as well as high specific power. Volumetric energy density is about 128 Wh/L. On the basis of weight, Na/S batteries have about three times the specific energy of a lead-acid battery.

Na/S batteries have two liquid electrodes separated by an ionically conductive ceramic electrolyte. Molten sodium is the anode, and molten sulfur-sodium polysulfide functions as the cathode. Individual cells are cylindrical and consist of the tubular ceramic electrolyte, which is filled with molten sodium on the inside and surrounded by molten sulfur on the outside. The cell's aluminum case is hermetically sealed and unvented. Approximately 360 individual cells are then assembled into strings and housed in a thermal enclosure designed to keep temperature within the operating range. Temperature must be maintained above 300°C for the battery to operate. Operating temperature is typically around 350°C. Load generates heat and increases internal temperature. Batteries must therefore be heated for startup, insulated to avoid heat loss under light or zero-load conditions, and cooled at higher loads to avoid excessive heat buildup. The cell stack is normally cooled by circulating air through the enclosure. Open-circuit cell voltage at 40 to 100 percent charge is 2.072 V. Below 40 percent, the state-of-charge cell voltage slopes downward to a 1.72 V cutoff, which is considered the fully discharged voltage.[18]

High manufacturing costs, low service life, and safety issues have been the primary problems of the Na/S battery. Although both sodium and sulfur are low-cost elements, high costs result from the attendant demands of sealed cells and their thermally controlled environment; the necessary multiple cell interconnects, bypasses, and fuses; and the safety measures associated with high energy density and corrosive, high-temperature materials. Safety issues appeared to be manageable through adequate fusing, improved electrical insulation, and crush-resistant cells and enclosures. Adequate life is also in sight. In 1977, cell life was approximately 70 cycles. Today, mean time to cell failure is approximately 1,000 cycles. When cells are assembled into batteries, life declines because of the combined failure rate of individual cells and subsystem components. Battery architecture and failure management strategies have increased battery life closer to that of aggregate cell life. High manufacturing costs remain a problem.

Vanadium Redox Flow Battery

This battery couple is not presently considered a candidate for EV application. It is included, however, because of its unique charged-liquid refueling capability. Development of the vanadium redox battery began in 1985 at the University of New South Wales in Australia. Like aluminum-air and zinc-air systems, a vanadium redox battery can be recharged by adding an active material. In essence, the battery can be "refueled." Unlike the other battery couples, the electrical charge is actually carried by the liquid electrolyte. Positive and negative electrolyte solutions are kept separated by a thin membrane as they flow through the battery's half-cells (Figure 5.19). Electrolyte can be recharged and stored fully charged outside the vehicle, then dispensed as needed for an "instant recharge." Recharging can also be done conventionally without changing electrolyte. Energy efficiency is extremely high: about 90 percent, or 87 to 88 percent when pumping losses are considered. Pumping losses occur because the electrolyte is pumped through the cells from separate reservoirs. Another important characteristic is that discharge rate has little effect on specific energy. Also, the battery can be completely discharged without damage. To increase energy stores, larger electrolyte reservoirs may be installed. The actual cell stack, however, remains unchanged.

Because the charge/discharge cycle does not involve solid-phase changes, there is no shedding or shorting, and electrolyte life is virtually infinite. The system can tolerate high charge rates, so recharging can be done in a fraction of the time required to recharge most EV batteries. Onboard recharge time is about 45 to 60 minutes. When electrolytes are dumped into a high-capacity stationary site, they can be electrically recharged in about 5 to 10

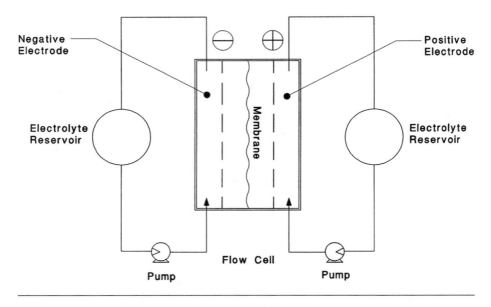

FIGURE 5.19 *Vanadium Redox Flow Schematic*

minutes. The ability to refuel with charged electrolytes and recharge depleted electrolytes during off-peak hours makes the system ideal for load-leveling purposes. Load leveling improves the efficiency of electrical generating facilities and thereby reduces the cost of generating electrical power.

The disadvantage of the vanadium battery for EV application is its relatively low specific energy compared to some of the other chemical battery couples. Specific energy is about 25 Wh/kg. Increasing the concentration of vanadium in solution from 2 molar to 3 molar or higher increases specific energy to about 40 Wh/kg, or about the same as a lead-acid battery. In situ chemical regeneration of the positive electrolyte while the car is running allows a 90 percent volumetric reduction of the positive electrolyte and a doubling of the negative electrolyte. This increases specific energy to about 80 Wh/kg or roughly equal to that of an NiMH battery.

Smaller systems appropriate for electric vehicle application are estimated to cost significantly more per kW/h than larger stationary industrial applications. Because the system is relatively new, much is still unknown about the economies of producing vanadium redox batteries in large quantities. Developers are optimistic that the cost may ultimately match that of the lead-acid battery. (See licensee website at www.vanadiumbattery.com.)

Zinc-Air Battery (Zn/Air)

Zinc-air batteries have been around for some time. Their primary application has been in hearing aids and other small electronic devices where high energy density at low current is important. Initially, zinc-air batteries were not considered rechargeable because recharging causes the growth of dendrites that ultimately short out cells; however, Dreisbach Automotive, Inc. (DEMI) in Santa Barbara, California, developed techniques for resolving the problems of dendrite growth and produced a rechargeable zinc-air EV battery system. In 1991, a Zn/Air-powered Honda CRX won the Inaugural Solar and Electric 500 race held at Phoenix International Raceway by leading the pack for two hours on a single charge (Figure 5.20). Table 5.8 shows the steady-state energy consumption and range of a Honda CRX powered by the system developed at DEMI.

FIGURE 5.20 Honda CRX Powered by Zinc-Air Battery. Demi/APS zinc-air electric-powered Honda CRX. Inset: Battery stack is loaded into the car by an APS technician. The rear seat is removed and the car becomes a two-seater.

Courtesy: Arizona Public Service Company, www.aps.com

TABLE 5.8 DEMI Honda CRX Range and Energy Consumption at Various Speeds

Nominal Speed		Energy Consumption		Range	
mph	km/h	Wh/mi	Wh/km	miles	km
30	48	124	77	322	518
35	56	145	90	275	442
45	72	186	116	215	346
55	88	228	142	175	282
65	105	270	168	148	238
75	121	311	193	128	206

Source: SAE Paper No. 911633, www.sae.org

Specific energy of a zinc-air battery can be varied to fit the application. Early versions had a specific energy of roughly 180 Wh/kg; however, specific power was marginal. As a result, power for the converted Honda had to be supplemented by an Ni/Cd battery pack weighing 159 kg (350 lb.). In more recent designs, some of the specific energy is sacrificed in favor of increased specific power. When specific energy is reduced to 120 Wh/kg, specific power increases to 120 W/kg, or nearly equal to that of a lead-acid battery. With increased specific power, the supplemental Ni/Cd battery pack is no longer required for most automotive applications. If greater acceleration is desired, power can still be supplemented by an Ni/Cd battery pack. Full recharge of the zinc-air battery takes about six hours, and an Ni/Cd pack can be continuously float-charged by the zinc-air system.

Zinc-air cells use a gas-permeable air cathode made of carbon, plastics, and proprietary catalysts. The anode is a zinc paste that is very similar to common flashlight anodes. The electrolyte is potassium hydroxide in a water gel, which is also similar to flashlight batteries. Air is typically ducted in from the front of the vehicle, routed through the battery, and then discharged at the rear. Airflow both cools the battery and provides the necessary reactive oxygen. Humidity and CO_2 content must be controlled within specific limits. Consequently, incoming air is electronically monitored, scrubbed of CO_2, and conditioned for humidity before it enters the battery.

Cycle life is not especially high (135 cycles); however, when specific energy is considered, zinc-air vehicles can run 40,000 km (25,000 mi.) before batteries need replacing. Zinc is plentiful, environmentally friendly, and costs are low.

DEMI estimated the cost of zinc-air batteries modeled after their most recent experimental battery at about $50 kW/h. Efficiency over a charge/discharge cycle is approximately 60 percent, figuring about 80 percent efficiency on discharge and about 75 percent on charge.

Zinc-Bromine (Zn/Br)

Zinc-bromine is an aqueous-flow battery system that has been under development for several years. Development began at Exxon Research & Engineering in the 1970s and has continued at Johnson Controls, Inc. (JCI) and at SEA in Austria. According to JCI, the advantages of their bipolar zinc-bromine battery are high specific energy and relatively low manufacturing costs. High specific energy is primarily the result of the nearly all-plastic battery stack. The battery consists of thin, conductive, carbon-filled plastic bipolar electrodes, thin microporous separators, and monopolar terminal electrodes. The liquid electrolyte is an aqueous solution of zinc bromide, bromine complexing agents, and optional supporting salts. Salts may be added to increase the electrolyte conductivity. Materials are abundant and inexpensive.

Deep discharge does not affect battery life, and self-discharge is minimal. One of the liabilities is the need to circulate the electrolyte, which is used to control battery temperature and prevent dendrite growth. JCI has built experimental batteries in the 65- to 80-Wh/kg range and projects manufacturing costs of approximately $55/kWh in high volumes. The battery operates at ambient temperature, and a service life of 1,000 to 2,000 cycles is expected with additional development.

The primary drawback of the zinc-bromine system is the highly corrosive and toxic nature of electrolyte solutions. Liquid electrolyte can cause severe burns, and inhalation of the vapor can be fatal. Consequently, there are concerns about the potential hazards of zinc-bromine batteries in the event of a collision.

Kinetic Energy Storage

Kinetic storage is another promising technology. Flywheels store energy mechanically instead of electrochemically. The idea of using flywheels to store energy for vehicles has been around for many years. Forty years ago, buses in Switzerland were powered by large flywheels, which were recharged at each stop. Near the end of his career, William Lear, of Lear Jet fame, became interested in kinetic storage for transportation applications. In order to increase the energy stored in a flywheel, either the speed, the mass, or both may be increased. A smaller mass spinning at higher speeds, however, results in greater specific energy. Modern high-speed flywheels operate in a vacuum at speeds of up to 100,000 rpm or more. The rotor is suspended in magnetic bearings. Energy is exchanged by electronic induction, rather than mechanically, in order to avoid the difficulties of mechanically interfacing with the high-speed rotor.

The greatest challenges in flywheel design have been to control disturbances that can destroy the rotor and to develop materials and designs that can tolerate the high stresses resulting from high spin speeds. Because the greatest velocity is at the rotor's rim, concentrating the mass at the rim results in greater stored energy; however, rim-type anisotropic flywheels can quickly exceed the structural limitations of materials because of unequal distribution of radial stresses. When the mass is concentrated at the rim, stress increases toward the hub, and the cross-section of the rim/hub structure declines in the reverse of the buildup of stress. For a while, it looked as though the hub-biased isotropic design might provide the answer to improved stress distribution and greater specific energy. Isotropic flywheels utilize a geometry in which the mass is concentrated at the hub and decreases toward the rim in proportion to the radial loads. Materials are thereby evenly stressed. Unfortunately, the capacity to store energy is more limited with the isotropic architecture.

More recently, the emphasis has returned to the anisotropic flywheel, where research has concentrated on designs that employ fiber composite rims connected to the hub by spokes or a web. As the rotor expands at high speeds, however, maintaining even stress distribution and dynamic stability between the hub and rim has remained a problem.

The energy stored in a flywheel is calculated as $K = (1/2)IW^2$, where K is kinetic energy, I is the moment of inertia of the rotor, and W is the rotor's angular velocity. Energy management during catastrophic failure of the rotor is one of the most challenging problems. Catastrophic rotor failure suddenly releases the stored kinetic energy. The energy released on failure of a 1-kWh rotor, for example, is enough to lift a midsize automobile 30.5 meters (100 ft.) into the air. Table 5.9 shows the power released when a 1-kWh flywheel fails over various time periods.

In addition, flywheels must be mounted in gimbals and isolated from road shocks. Gyroscopic forces of the rotor are great enough to cause a vehicle to roll over in a turn.

TABLE 5.9 Power Released on Failure of a 1-kWh Flywheel

Energy Dissipated over Time (seconds)	Power (kW)
0.001	3,600,000
0.01	360,000
0.1	36,000
1.0	3600
5.0	720

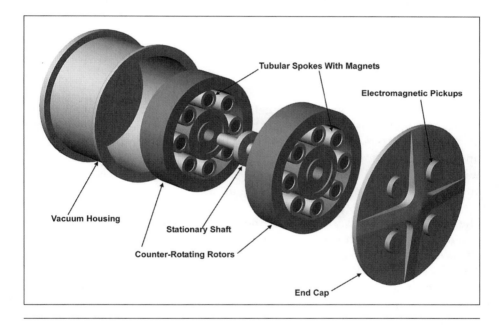

FIGURE 5.21 General Layout of AFS Flywheel Module

American Flywheel Systems (AFS), Bellview, Washington, developed a flywheel design that uses axially positioned tubes spaced evenly around the hub, between the hub and the rim. As the hub and rim expand at different rates, the tubes compensate by distorting, thereby maintaining rim/hub stability and avoiding stress differentials between the two components (Figure 5.21).

TABLE 5.10 Comparison of GM's EV-1 with AFS Kinetic Battery versus Lead-Acid Battery

	GM's EV-1 with Lead-Acid Batteries	GM's EV-1 with AFS Flywheel (estimation)
Total Energy	16.2 kWh (propulsion pack)*	43.6 kWh
Number of Modules	26 (propulsion pack)*	20
Weight of Storage Devices	480 kg (propulsion pack)*	225–270 kg
Specific Energy	30 Wh/kg*	193–226 Wh/kg
System Life	40,000 km	400,000 km
Vehicle Range	80–200 km	480–970 km

Source: American Flywheel Systems, Inc., www.afstrinity.com
Note: Spin-speeds on the order of 200,000 rpm have yet to be demonstrated.
*EV-1 figures updated by author.

AFS planned to package their "kinetic battery" in modules approximately the size of a conventional lead-acid traction battery. Inside each evacuated module, two carbon-fiber counter-rotating rotors are said to spin at speeds of up to 200,000 rpm. Speeds in this range result in specific energy of 225–270 Wh/kg. As with other advanced flywheel designs, specific power is limited only by the capacity of the electrical system.

The ability of materials to withstand the forces of super-high rotor speeds, as well as high manufacturing costs, remain problems for flywheel storage systems. According to AFS estimates, however, lifetime cost may be lower than that of a conventional electrochemical battery system. Life of the kinetic battery is estimated at 4,000 cycles, or about 400,000 km (250,000 mi.). Although manufacturing costs have not been established, AFS believes that cost per vehicle-kilometer could be as little as 10 to 13 percent as much as a comparable lead-acid battery.

In August 2000, AFS announced the purchase of 58 percent of Trinity Flywheel Power. The two companies have combined to become AFS Trinity Power Corporation.

BATTERY-ELECTRIC VEHICLES AND VEHICLE DOWNSIZING

A conflict of technical attributes exists between the idea of vehicle downsizing to minimize energy consumption and the larger battery packs required for greater onboard energy stores. Energy stores increase in proportion to the size and mass of the battery; however, the need for energy decreases when the size and mass of the vehicle are reduced. One strategy for making peace between these two opposing characteristics is to allocate a greater portion of the vehicle's mass to batteries. When more of the vehicle is allocated to the battery pack, vehicle range and performance tend to increase. Nearly 40 percent of the mass of GM's high-performance, two-seater EV-1 consists of batteries. Although this approach can be applied to both smaller and larger vehicles, the most cost-effective approach is to reduce vehicle mass and size in order to reduce energy requirements and thereby allow for a smaller, less costly battery pack.

With the possible exception of GM's EV-1, designers in the United States have tended to favor larger vehicles and larger battery packs with greater energy stores. In Europe, the trend is just the reverse—toward smaller vehicles with more modest appetites for energy. Smaller BEVs require less battery mass for the same range and performance. They also imply a greater emphasis on space-efficient packaging and batteries of greater volumetric energy density. By limiting payload capacity, the size and mass of the vehicle can also be reduced. Two-occupant vehicles are adequate for the 90th percentile trip in the United States. For two-thirds of U.S. drivers, a vehicle with a range of 85–150 km (53–93 mi.) would meet the range requirements of the 95th percentile trip.[19]

When BEVs are no longer envisioned as a replacement for conventional multipurpose cars, range and performance metrics will naturally assume different values. If 10 percent or more of vehicle trips were left to larger vehicles of greater specific power and energy, BEVs could be significantly downsized, battery packs could be smaller, and costs could be reduced. This implies a shift toward mission-specific vehicle design, with larger vehicles for long-distance family mobility and smaller vehicles for personal urban mobility.

PURPOSE-BUILT BATTERY-ELECTRIC CARS

Over the past decade or so, hundreds of battery-electric cars have been developed, both within the automotive industry and by nonautomotive companies, institutions, and individual experimenters. The following selection of purpose-built BEVs includes just a few outstanding examples from within the industry. The selection of vehicles, however, is by no means inclusive, nor does it represent the most successful BEVs in the marketplace. In North America, for example, Ford is the major OEM leader in BEV sales with their electric Ranger pickup. Toyota's highly successful RAV4-EV, Honda's EV PLUS, Nissan's Altra EV, DaimlerChrysler's EPIC EV van, and Chevrolet's S10 Electric pickup are all superbly engineered and noteworthy products.

General Motors EV-1

GM's EV-1 is arguably one of the most advanced BEV designs to emerge from the automobile industry (Figures 5.22 and 5.23). It was introduced with specially designed lead-acid batteries developed by the company's Delco division. GM engineers approached the problem of limited energy stores by designing a highly efficient car with low energy demand. The vehicle's low-friction drivetrain, 345 kPa (50 psi), low rolling resistance tires, and an extremely low 0.19 drag coefficient are responsible for its diminutive energy appetite. On lead-acid batteries, the EV-1 has a highway range of 145 km (90 mi.). It accelerates from 0 to 96 km/h (0 to 60 mph) in 8 seconds. The EV-1

FIGURE 5.22 EV-1 High-Performance Battery-Electric Car. The EV-1 uses regenerative braking, reduced roadload, and an advanced ac tractive system to keep energy consumption to a minimum. Highway range is 145 km (90 mi.) on conventional lead-acid batteries and about twice that on NiMH batteries.

Courtesy: General Motors Corp., www.gm.com

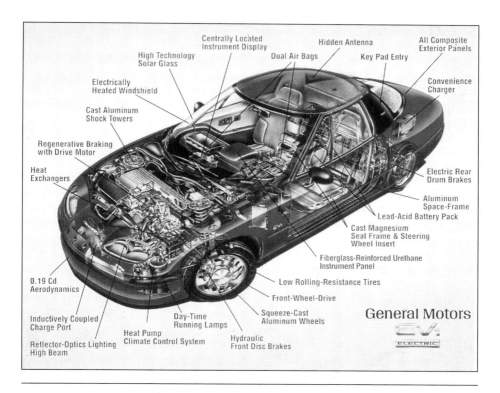

Labels on figure:

Centrally Located Instrument Display
High Technology Solar Glass
Hidden Antenna
All Composite Exterior Panels
Dual Air Bags
Key Pad Entry
Electrically Heated Windshield
Convenience Charger
Cast Aluminum Shock Towers
Regenerative Braking with Drive Motor
Heat Exchangers
Electric Rear Drum Brakes
Aluminum Space-Frame
Lead-Acid Battery Pack
Cast Magnesium Seat Frame & Steering Wheel Insert
Fiberglass-Reinforced Urethane Instrument Panel
0.19 Cd Aerodynamics
Low Rolling-Resistance Tires
Front-Wheel-Drive
Inductively Coupled Charge Port
Day-Time Running Lamps
Squeeze-Cast Aluminum Wheels
Heat Pump Climate Control System
Reflector-Optics Lighting High Beam
Hydraulic Front Disc Brakes
General Motors
EV. ELECTRIC

FIGURE 5.23 Detailed Features of the EV-1.
Motor: Alternating-current, 137 HP; Weight 150 lb.
Batteries: 26, 12-volt lead-acid batteries (1,175 lb.)
Range: 70 miles city, 90 miles highway
Brakes: Rear, by wire; Front hydraulic
Aluminum Space Frame: Weight 290 pounds vs. 600 pounds for conventional steel
Tires: 50psi. Designed by Michelin for low resistance
Body Panels: Plastic. Available in three colors: red, silver-blue, and green
Length: 169.7 in.; Width: 69.5 in.; Height: 50.5 in.; Weight: 2,970 pounds
Cargo Volume: 9.7 cubic feet

Courtesy: General Motors Corp., www.gm.com

was manufactured in two batches of 500 cars each. Manufacturing began in 1996 and was put on hold near the end of 1999. A total of 727 cars were leased through Saturn dealers in select U.S. markets.

Following introduction of the EV-1, GM engineers went on to design two hybrid-electric cars using the highly efficient EV-1 as a basis. One vehicle was a parallel hybrid and the other was a series hybrid.

Ford TH!NK City

Ford's TH!NK City was designed from the beginning as an electric car by Norwegian-based carmaker PIVCO. The City seats two and is powered by a liquid-cooled, three-phase ac induction motor. The electric motor is supplied by 19 nickel-cadmium water-cooled batteries. It has already been introduced in select European markets, and a slightly modified version was planned for introduction into the United States in year 2002 (Figure 5.24). The existing Siemens-supplied powertrain will be replaced in the U.S. model by a new electric motor and controller from EcoStar, a joint venture between Ford and DaimlerChrysler. Performance with the new power system will be similar to the European model: 80–95 km (50–60 miles) range and about 90 km/h (55 mph) top speed.

FIGURE 5.24 Ford Th!nk City. The battery-electric City, which meets European safety standards, is "not trying to compete with a 'normal' car," according to PIVCO founder Jan Otto Ringdal. The first prototype was introduced at the Olympic Games at Lillehammer, Norway, in 1994. Ten cars were available in the no-car zone downtown. Ford became interested and ultimately bought 51 percent of PIVCO.

Courtesy: Ford Motor Co., www.ford.com

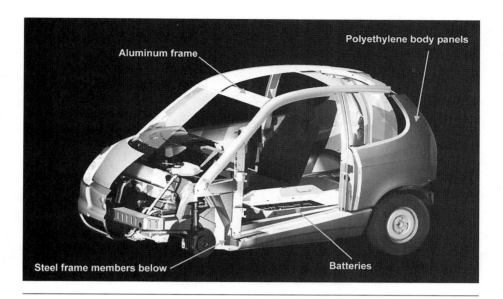

FIGURE 5.25 City Space Frame. The City's body is made of an aluminum and steel space frame, which is covered with polyethylene body panels.

Courtesy: Ford Motor Co., www.ford.com

The City is only about 3 meters long (10 feet), which makes it easy to park. It is still too long, however, to park nose-in to the curb like DaimlerChrysler's Smart Car. The body is made of thermoplastic panels over a steel and aluminum frame (Figure 5.25). This construction makes production costs lower than conventional steel cars because the body is molded in fewer panels. It also results in a very lightweight car. Curb weight is just over 900 kg (2,000 lb.).

Nissan Hypermini

Nissan's Hypermini is an award-winning urban commuter car that has been on sale in Japan since February 2000 (Figure 5.26). It features a lightweight aluminum space frame, a neodymium magnet synchronous motor, and advanced lithium-ion batteries. Batteries can be recharged in four hours using a 200-volt outlet. This ultracompact two-seater BEV delivers a range of 115 km (over 70 miles) and a top speed of 100 km/h (62 mph). According to Nissan, the car's deformable front and rear body zones, driver and passenger airbags, and ABS provide safety equal to that of a conventional small car.

FIGURE 5.26 Nissan's Hypermini Urban Commuter Car. Nissan's award-winning Hypermini is a good example of the extreme small size possible with a BEV. Minimum roadload translates into minimum battery requirements, which then results in the lowest possible cost for the battery pack and other power system components.

Courtesy: Nissan Motor Co., www.nissandriven.com

HYBRID-ELECTRIC VEHICLES

The idea of a hybrid-electric vehicle naturally evolves from the limited specific energy of the secondary battery. Hybrid vehicles began with the idea of augmenting the battery's limited energy stores with energy from an onboard electrical generating system. Range was thereby increased over that available on battery stores alone. Today's hybrid is no longer envisioned simply as a battery-electric vehicle assisted by an auxiliary power unit (APU). Instead, it has evolved into a sophisticated vehicle with an integrated, self-adapting power system and batteries and/or ultracapacitors configured for peak demands and load leveling, rather than as a primary source of energy.

The hybrid power system is intrinsically more complex than a conventional power system, which some may see as a disadvantage. But contrary to the popular adage about keeping it simple, the simple solution is not necessarily the best one. In fact, mechanical and electronic systems tend to perform better as they become more complex. The inherent complexity of the hybrid power system provides abundant design options and many opportunities to reduce energy consumption and emissions. For example, a prime mover can operate more efficiently at a steady-state output when peak demand needs are supplied by the energy storage medium. So a heat engine powering an APU can run much of the time in its region of minimum bsfc, which translates into maximum fuel economy. In addition, emissions are easier to manage when the output band is relatively narrow. Because an energy storage medium is part of the hybrid power system, regenerative braking is a natural option. Regeneration can reclaim as much as half the energy used for acceleration. With a hybrid power system, designers can combine subsystems and manage energy flow in ways that emphasize positive attributes and deemphasize negative ones.

Much of today's research in systems design is oriented toward developing the most effective control strategy, the best biasing between subsystems, and the correct combination of subsystems needed to obtain maximum efficiency from a minimum of hardware. The original concept of an APU-augmented BEV is not necessarily incorrect, but today's shift to a systems approach has opened new opportunities for greater efficiency and lower emissions. Based on the source fuels used to generate electrical power in much of the world today, hybrids may actually be cleaner and consume less primary energy than battery-electric vehicles.

Hybrid Configurations

Hybrids have traditionally been classified as either a series or a parallel design. With the series hybrid, electrical energy from the APU either charges the battery or supplies supplemental power to the propulsion circuit and thereby reduces demand on the battery. A parallel hybrid utilizes two propulsion systems in parallel. Both systems are configured to deliver power to the drive wheels. Depending on the design, the two systems may be utilized independently to propel the vehicle, or both systems may be utilized simultaneously for maximum power. A subcategory of the parallel system is called the *split hybrid*. In the split hybrid, one system powers one pair of wheels and the other powers the other pair. Propulsion power is supplied by two mechanically independent systems. Many other variations are possible within the series and parallel categories.

A second distinction between hybrids is whether the power system is a charge-depleting or charge-sustaining design. With a charge-depleting hybrid, the APU operates mainly as a range extender. Normally, the system is configured so batteries reach 80 percent depletion at the end of the mission. Hybrids in this category are usually designed to run on battery power alone, and the APU is used to extend the vehicle's range. But ultimately, batteries become depleted and have to be recharged, which means that range, although greater than that of a pure battery-electric vehicle, is still limited. Most early hybrids fall within the category of charge-depleting designs.

Modern efforts have focused increasingly on the charge-sustaining hybrid. A charge-sustaining hybrid provides range that is limited only by the amount of fuel on board. With a series charge-sustaining hybrid, batteries provide propulsion power and the APU runs continuously or intermittently, as needed to keep batteries at a high state of charge. Peak demand power is taken from the batteries (or ultracapacitor), and is replenished by the APU during periods of less demand. A parallel charge-sustaining hybrid may be equipped with a downsized IC engine as the primary means of propulsion. Shortfalls in torque are then supplied by an electric assist using energy from the electrical storage medium. Today, this type of hybrid is often called a "mild" hybrid. In other designs, biasing is reversed with the electrical power system functioning as the primary means of propulsion. A charge-sustaining system implies a larger APU or prime mover and a significantly smaller battery pack, in comparison to a charge-depleting hybrid. Batteries are normally configured for maximum specific power rather than maximum specific energy.

Series Hybrid

Series hybrids normally utilize a heat engine to power an onboard generator. A fuel cell, however, may be used instead of the heat-engine/generator combination. The APU is normally configured as a range extender, which means that the vehicle operates on battery power for most of the kilometers traveled. When trip length is not within the capabilities of the battery, the APU is engaged to provide additional range. During longer trips, the APU might be automatically turned on when the battery reaches a predetermined DoD. Early designs, which utilized generators in the 5–10 kW range, could have a significant effect on range only at relatively low urban speeds. To provide greater range at higher demand levels, a significantly more powerful APU is necessary. Improved technology has resulted in greater output from smaller, lighter units, and modern series hybrids can now match the performance of parallel hybrids.

Quincy-Lynn's Town Car is an early series design developed in 1980 (Figure 5.27). At the time, the Town Car was envisioned primarily as an electric urban car with an onboard generator assist. The focus was to minimize the size of the generator and thereby minimize the additional mass and the degree to which the vehicle relied on IC engine power. The relatively light-weight system employed a 3-kW IC engine to drive an onboard dc generator. When the generator was energized it started the engine, which was set to run continuously at approximately 80 percent of maximum output. It pro-duced an output of slightly more than 2 kW, which was delivered either to the batteries or to the propulsion system, depending on the load. Range at a steady-state 56 km/h (35 mph) was 169 km (105 mi.), and fuel consumption at the same speed was 40 km/L (93 mpg). Using electric power alone, range at a steady-state speed of 56 km/h (35 mph) was 105 km (65 mi.). Assist from the APU was conservative, even for the period in which it was designed.

Today, the trend in series hybrids is toward greater range, even equaling the range of multipurpose conventional cars. This implies much larger generat-ing systems. A state-of-the-art generating system is size rated according to

FIGURE 5.27 Quincy-Lynn Town Car. The Town Car, a series hybrid developed at Quincy-Lynn Enterprises in 1980, was a charge-depleting design. Today's development efforts are focused on charge-sustaining hybrids.

the predetermined trip-average demand and vehicle range requirements. Consequently, the maximum-range driving cycle must be defined before the system can be designed. The battery is sized according to peak power demands, and the APU is sized to allow for approximately 80 percent DoD at maximum vehicle range. During hybrid operation at higher speeds, energy may be drawn partially from the battery. Power for acceleration and passing comes largely from the battery. At reduced cruising speeds, excess generating capacity is used to recharge the battery. Although generating system efficiency is high because the heat engine spends much of the time loaded into its region of minimum bsfc, vehicle efficiency is normally highest when wall-plug electricity is used for at least 80 percent of the vehicle kilometers traveled.[20] Figure 5.28 shows a typical series layout.

The catalytic converter in a series system can be preheated to minimize emissions. Once the APU is energized, it then runs continuously at a prede-

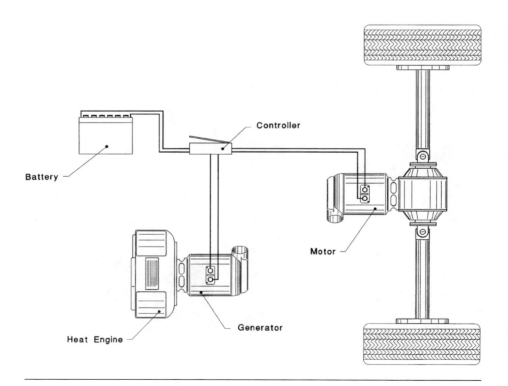

FIGURE 5.28 Series Hybrid Layout

TABLE 5.11 Series Hybrid Performance Characteristics

Vehicle Type	Battery-Electric Mode				Series Hybrid Mode				Acceleration Times (sec)
	FUDS		FHWC		FUDS		FHWC		
	Wh/km	Range* (km)	Wh/km	Range (km)	mpg**	Eff.*** (%)	mpg	Eff. (%)	
Minivan	185	93	188	86	26.1	0.85	26.4	0.88	4.7 12.1
Microvan	136	96	132	93	35.6	0.85	37.5	0.88	4.7 12.5
Compact Car	116	99	103	107	41.8	0.86	47.8	0.87	4.3 11.0

Source: SAE Paper No. 920447, www.sae.org
*Usable range to 80% DoD
**Gasoline fuel and min bsfc = 300 gm/kWh
***Average efficiency from engine output to inverter input

termined output, which minimizes fuel consumption and further simplifies emissions control. Fuel economy for a state-of-the-art series hybrid vehicle over the Federal Urban Driving Schedule (FUDS) and Federal Highway cycles (FWHC) is generally higher than that of a conventional IC vehicle (Table 5.11).[21]

Traditionally, the advantage of the series hybrid has been its mechanically simple design. But new ideas in parallel design have left the presumed advantage of the series hybrid open to debate.

Parallel Hybrid

Parallel hybrids offer a wide variety of power system options. With a parallel hybrid, power from the heat engine is delivered directly to the drivetrain, rather than to a generator. So a parallel hybrid avoids the losses of converting mechanical power into electrical power before putting it to the task of propelling the vehicle. But unlike a series system, which can recharge batteries when the vehicle is at rest, a parallel hybrid system normally works only when the vehicle is under way. The IC engine may be capable of propelling the vehicle by itself, or it might provide only a portion of the necessary propulsive power. Designers are free to install greater IC engine power for better performance and range, or to design for less installed power and thereby reduce fuel consumption and emissions. A typical parallel layout is shown in Figure 5.29. Figure 5.30 shows a split hybrid layout.

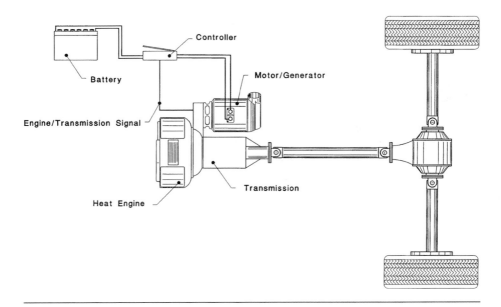

FIGURE 5.29 Parallel Hybrid Layout

Normally, the power system is biased toward either the IC power system or the electric power system. So on a conceptual level a hybrid might be viewed as either an IC vehicle with electric assist or as a battery-electric vehicle with IC assist. This conceptual orientation tends to guide the general approach to design solutions. Superior design, however, comes more from a systems approach than from a predetermined distinction of subsystem dominance. Ideally, each of the two subsystems will be integrated in ways that take greatest advantage of their individual operating characteristics.

With modern hybrids, range and performance are determined mainly by the output of the heat engine. Details of the control scheme determine the vehicle's fuel consumption and emissions. With a battery-dominant design, the vehicle may be configured to operate primarily on electric power at neighborhood speeds or in stop-and-start traffic, then become IC dominant at highway speeds and during periods of rapid acceleration. With an IC-dominant system, the electric power system may be configured to provide extra power during periods of peak demand. This allows for a smaller IC engine than would normally be required, which naturally loads the engine into a region of lower bsfc.

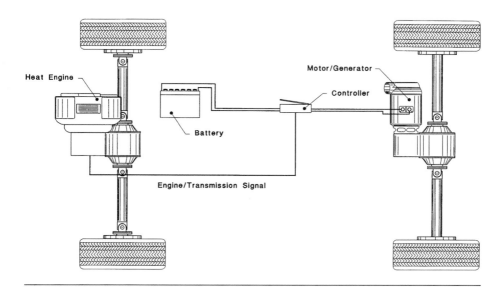

FIGURE 5.30 Split Parallel Hybrid Layout

Energy Storage Systems for HEVs

Energy storage systems for hybrids differ from those of battery-electric vehicles, particularly in the case of IC-dominant and charge-sustaining designs. Because a hybrid generates usable energy onboard by consuming some type of fuel, the storage system may be configured to supply only part of the energy necessary for the vehicle's mission. In a charge-depleting hybrid, the battery pack may closely resemble that of a BEV, but in a charge-sustaining hybrid, the energy storage system assumes more of a load-leveling role. In this case, specific power is more important than specific energy, and the energy storage system can be significantly downsized. Details of the storage system's capacity and performance depend on the details of the energy management strategy.

Ultracapacitors have great promise in HEV applications. Although batteries can be configured for greater specific power, regeneration power levels can easily exceed a battery's ability to accept energy. Peak deceleration loads would then be taken by the vehicle's braking system, which simply converts excess kinetic energy to heat. Ultracapacitors can receive and deliver much greater power levels than batteries, so they can level out peak loads and avoid overloading the battery, or they could ultimately replace the battery storage medium.

An ultracapacitor is somewhat of a cross between a battery and a capacitor. Batteries can store energy for extended periods, but they have comparatively limited specific power, they cannot handle high-level burst loads, and they rapidly break down under repeated charge/discharge cycles because of the phase changes that take place on an electrochemical level. Ordinary capacitors can deliver high power levels with repeated charge/discharge cycles, but they can deliver power only for brief periods, normally microseconds at a time. Ultracapacitors can store many times more energy than ordinary capacitors, and they can accept large input loads and deliver power in more extended bursts. In addition, they are significantly smaller and lighter than batteries of similar specific power, and they have many times greater cycle life. Although the life of a battery may be limited to a thousand cycles or so, an ultracapacitor may be cycled hundreds of thousands of times. Advantages come from the fact that ultracapacitors move the electrical charge between conducting materials, rather than relying on chemical reactions, as do batteries. SRI has been working with Sanyo to develop ultracapacitor technology for several products, including one device with specific energy of 12Wh/kg at 1,000 W/kg specific power.

A perfected flywheel would be an ideal energy storage medium for a hybrid vehicle because of its ability to rapidly receive and deliver high-level power. Unlike a battery, peak power is independent of total energy stores. Power output is limited only by the capacity of the flywheel's electrical interface. In a hybrid, the flywheel will be rated for peak demand, rather than total trip energy needs. In other words, it would assume more of a load-leveling role.

Prime Movers for HEVs

A heat engine or fuel cell is suitable as the primer mover in a hybrid vehicle. The prime mover may be "throttled" for variable output, or it may be set to run predominantly in its region of minimum fuel consumption per unit of output for maximum efficiency. The decision on whether to operate the prime mover at a variable output, to cycle it on and off, or to run it mainly at a continuous steady-state output depends on the type of hybrid and the overall energy management strategy.

A hybrid's load-leveling capability makes it possible to use a downrated prime mover operating closer to its region of greatest fuel economy, and emissions are more easily controlled when a heat engine operates over a narrower output band. A hybrid power system allows designers to design to the natural characteristics of the prime mover for the greatest overall efficiency and lowest possible levels of harmful emissions.

Both diesel and spark-ignition engines are good candidates for hybrid vehicles. The homogeneous charge compression ignition (HCCI) engine is also an excellent, but little explored option. In a hybrid application, the HCCI engine (also known as active thermo-atmosphere combustion [ATAC]) can operate over a variable output band while still remaining in controlled auto-ignition where fuel economy excels. With steady-state operation, the control problem that now challenges HCCI researchers becomes a nonissue. Additional performance gains, both with the engine and with the hybrid power system, are possible with this fundamentally cleaner and more fuel-efficient engine design.

A turbine engine is also more suited to hybrid applications. Turbines tend to respond more slowly to variable demands, and efficiency suffers to a greater extent at part-load operation, in comparison to reciprocating engines. But these characteristic problems in automotive applications become less important in the hybrid power system. Moreover, the turbine's main assets—its high power density, low maintenance, and adaptability to a variety of alternative fuels—make it more attractive in hybrid applications than in conventional vehicles. But much development still remains in order to produce a practical, low-cost turbine for hybrids. After experiments with the turbine-powered Patriot, Chrysler engineers concluded that turbines still present challenges that need to be solved before turbines can find application in hybrid vehicles (Figure 5.31).

Advancements in automotive fuel cells appear to foretell the demise of the IC engine. In the meantime, however, this workhorse of the 20th century keeps becoming increasingly cleaner and more fuel efficient as refinement continues. In contrast, fuel cells tend to become dirtier and less fuel efficient when they are designed to run on conventional fuels. During the window of opportunity for hybrid power systems, before fuel cells reach production, the IC engine could become virtually emissions-free in hybrid applications.

Fuel Cells

Fuel cells are widely hailed as the Holy Grail of automotive power systems. The operating principle of a fuel cell was first developed in 1839 by William Grove. By 1900, engineers were already predicting that fuel cells would be commonplace in a few years for powering vehicles and generating electricity.

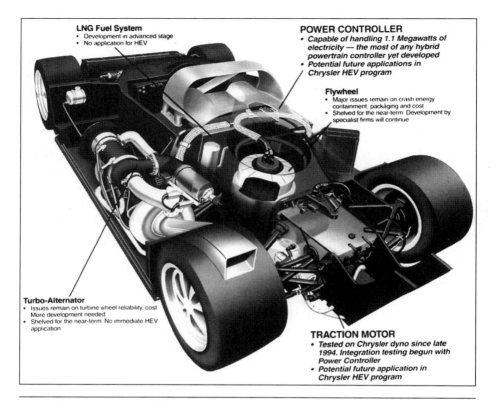

LNG Fuel System
- Development in advanced stage
- No application for HEV

POWER CONTROLLER
- *Capable of handling 1.1 Megawatts of electricity — the most of any hybrid powertrain controller yet developed*
- *Potential future applications in Chrysler HEV program*

Flywheel
- Major issues remain on crash energy containment, packaging and cost
- Shelved for the near-term. Development by specialist firms will continue

Turbo-Alternator
- Issues remain on turbine wheel reliability, cost. More development needed
- Shelved for the near-term. No immediate HEV application

TRACTION MOTOR
- *Tested on Chrysler dyno since late 1994. Integration testing begun with Power Controller*
- *Potential future application in Chrysler HEV program*

FIGURE 5.31 Chrysler's Patriot Turbine-Powered Hybrid. Built in the 1990s, Patriot's power system combined several unconventional subsystems. The APU consisted of an LNG gas turbine engine equipped with an integral alternator that ran at turbine-shaft speed. A high-speed flywheel served as the energy storage medium.

Courtesy: DaimlerChrysler, www.daimlerchrysler.com

Today, 100 years later than predicted, it appears as though this quiet, efficient, and nonpolluting power source may finally become a practical reality.

Fuel cells rely on chemical reactions, instead of combustion, to convert liquid fuel directly into electrical energy. A fuel cell operates in the reverse of the electrolysis of water, in which electricity is used to split water into gaseous hydrogen and oxygen. A fuel cell catalytically recombines hydrogen and oxygen into water, and in the process creates a flow of electrons from the anode to the cathode. Essentially, the electrical energy needed to split water into hydrogen and oxygen is given up when the two are recombined into water.

In terrestrial applications, oxygen is normally taken from the air. Hydrogen can be carried on board in gaseous or liquid form, or it can be extracted on board from a variety of hydrocarbon fuels. Depending on the type of fuel cell, conversion efficiency ranges from 40 to 60 percent, or about 2.5 times greater than an IC engine at the upper extreme. When pure hydrogen is used, efficiency is greatest and emissions are in the form of water vapor. When alternative fuels are substituted for pure hydrogen, efficiency drops and harmful emissions of various species are produced, depending on the particular fuel.

A fuel cell consists of catalytically activated electrodes (anode), the oxidant (cathode), and the electrolyte. Alkaline fuel cells (AFCs), which utilize an aqueous solution of potassium hydroxide in combination with hydrogen and oxygen as fuel and oxidant, have powered spacecrafts since the mid-1960s. Alkaline fuel cells suffer from slow startup, low specific power, and slow response to variable demand. Consequently, they are not good candidates for automotive application. The fuel cells singled out for automotive application include the proton exchange membrane (PEM) and the solid oxide fuel cells (SOFCs). Depending on the type of fuel cell, a separate reformer may be needed to extract hydrogen from ambient-temperature liquid fuels, but newer direct-methanol fuel cells (DMFCs) do not require reformers. Solid oxide fuel cells, which operate in the range of 1,000°C, exhibit higher fuel efficiency, greater power density, and do not require separate reformers.

Several problems remain to be solved before fuel cells can make large inroads in automotive application. High manufacturing cost is still a major factor, but costs are rapidly dropping. In 1990, for example, fuel cells were roughly 1,000 times too costly. Ten years later they were only about 10 times too costly. As of year 2000, the estimated cost of a mass-produced fuel cell was about $200 per kW, whereas a conventional powertrain can be manufactured for less than $30 per kW.

Another major problem is that fuel cells run on hydrogen, and no consumer-oriented hydrogen supply infrastructure exists. Hydrogen can, however, be reformed onboard vehicles from several hydrocarbon fuels. Methanol, ethanol, and gasoline are all suitable fuels, but conversion efficiency drops and "tailpipe" emissions are produced when hydrogen is reformed from a hydrocarbon fuel. Direct methanol fuel cells, for example, produce significant CO_2 emissions, and conversion efficiency drops to about 35 to 40 per-

TABLE 5.12 Fuel Cell CO$_2$ Emissions by Hydrogen Source

Engine Type	Water Vapor (kg/km, lb/mile)	Carbon Dioxide (kg/km, lb/mile)
Gasoline combustion	0.11, 0.39	0.24, 0.85
Hydrogen reformed from gasoline	0.09, 0.32	0.20, 0.70
Hydrogen reformed from methane	0.07, 0.25	0.04, 0.15
Renewable hydrogen	0.07, 0.25	0.00, 0.00

Source: Desert Research Institute, www.dri.edu

cent. Gasoline fuel cells are only somewhat more efficient than IC engines, but they produce almost as much CO$_2$ emissions, considering emissions over the entire fuel chain. Even using pure hydrogen in the fuel cell itself, fuel chain emissions may still be present, depending on how the hydrogen is produced. Today, hydrogen reformed from methane is the most inexpensive and cleanest source. When produced by electrolysis using electricity from nuclear, solar, wind, hydroelectric, or biofuels, however, there is a zero net gain of atmospheric CO$_2$. A comparison of IC engine and fuel-cell water vapor and CO$_2$ emissions by source is shown in Table 5.12.

The amount of CO$_2$ emissions produced by an engine is a factor of the chemistry of the fuel and the quantity of fuel consumed. It does not come from the conversion technology. Hydrocarbon fuels consist mainly of hydrogen and carbon. So when hydrogen is reformed from a hydrocarbon fuel, or reacted inside a fuel cell, carbon is released in the form of carbon dioxide. But when the fuel is made from a renewable biomass, such as bioethanol or biomethanol, the result is a zero net gain of CO$_2$ in the environment because the carbon in biofuels is absorbed from the environment when the biomass is grown. So the carbon is cycled in a closed loop: it migrates from the environment to the biomass during plant growth, then from the biomass to the fuel during production, and then to the environment again in the form of CO$_2$ emissions when the fuel is consumed. Overall fuel-chain CO$_2$ emissions are therefore zero. But even without bio-fuels, FCVs running on hydrocarbon fuels are still cleaner than IC engines (gasoline fuel cells are only marginally cleaner) and could cut CO$_2$ emission levels per distance traveled in half because of improved fuel economy alone.

Fuel cells have an efficiency curve that is the inverse of that of an IC engine. Efficiency goes up during part-load operation. In the abstract, "throttling" a fuel cell for the low outputs typical of urban driving would not be as detrimental to fuel economy as it is with a conventional power system.

Production and Prototype Hybrids

At this writing, hybrid vehicles are just beginning to appear on the showroom floors of some of the major OEMs. These early production hybrids represent just the beginning in terms of potential fuel savings and reduced emissions. The general consensus within automotive circles is that combustion hybrids are transition vehicles on the way to fuel-cell vehicles. Combustion hybrids appear to have a window of opportunity through about 2010 or so, at which point the emphasis in production vehicles may switch to fuel cells.* But much depends on future progress with fuel-cell development and the establishment of a refueling infrastructure for hydrogen or a hydrogen carrier fuel, as well as on the actual fuel economy and emissions reduction achievable with refined combustion hybrid power systems.

Honda Insight

Honda's production Insight is an excellent example of an IC-dominant parallel hybrid in its most simplified form (a mild hybrid) (Figure 5.32). Honda uses a system called Integrated Motor Assist (IMA), wherein the electric motor directly assists the IC engine. The Insight is equipped with a 0.995-liter, three-cylinder, 50-kW (68-hp) lean-burn gasoline engine weighing only 56 kg (124 lb.) fully dressed. A 10-kW pancake motor/generator, about the diameter of the flywheel and 64 mm thick (2.4 in.), is located between the flywheel and the five-speed manual-shift transmission. The permanent magnet rotor turns with the engine crankshaft, and the stator is fixed to the adapter housing. So the length of the three-cylinder inline engine/motor assembly is about equal to that of a conventional four-cylinder inline engine.

Honda's IMA power system provides several benefits. First, IMA allows for a downrated IC engine, so the engine runs closer to its minimum bsfc through much of the operating schedule. During periods of peak torque demand, the motor provides extra torque to assist the downrated engine. The motor is capable of supplying up to 50 percent of the torque available

---

* Although DaimlerChrystler is committed to introducing fuel-cell vehicles in 2004, these will be limited-production vehicles. A large-scale switch to fuel cells appears unlikely until about 2010, and perhaps even later.

FIGURE 5.32 Honda Insight. Honda's Insight hybrid-electric vehicle was introduced in 1999 and has won several awards for its high fuel economy and ultra-low emissions.

Courtesy: Honda Motor Company, www.honda.com

from the engine alone (engine torque × 1.5 = engine/motor torque). Second, it provides regenerative braking. Because the motor turns in unison with the engine crankshaft, regeneration is available on deceleration.

Another benefit is the motor's ability to smooth out the engine's power pulses. Immediately following ignition on a power stroke, the motor momentarily switches to the regeneration mode, which reduces the sharp increase in crankshaft speed. Microseconds later the motor switches to the assist mode to provide momentary torque input, which slows the decline in crankshaft speed that occurs between power strokes. So a third benefit of IMA is that it works to smooth out the power-stroke vibrations typical of a three-cylinder engine.

Idle shut-off is another fuel-saving benefit. Because the 144-volt, 10-kW motor is fixed to the engine's crankshaft, it can rapidly and reliably restart the engine without the noise of a flywheel pinion. The engine is shut off when the clutch is depressed below a predetermined speed, or when the transmission is shifted into neutral at a stop. It is immediately restarted when needed, according to normal operator/driving inputs. The restart system is smooth, quiet, and reliable. Except for the silence at a stoplight, the driver is hardly aware that anything unusual is happening. (When the air conditioning is on, the engine stays on at a stop.)

A final benefit is the impact of the IMA system on the necessary electrical energy stores. The operating schedule keeps electrical energy demand to a modest level, which allows for a significantly smaller battery. The Insight's 144-volt, NiMH battery pack consists of 120 D-size cells providing 1.2 volts each. The battery pack weighs only 22 kg (48 lb.) and has a storage capacity of 0.936 kWh. No external charging is necessary. The Insight's IMA control system keeps the battery pack at a high state of charge throughout normal driving.

Extensive use of aluminum and superior aerodynamic design combine with Honda's innovative hybrid power system to produce a vehicle that gets 3.4 L/100 km (EUDC), or 61 mpg city/70 mpg highway (EPA). Curb weight is a mere 834 kg (1,847 lb.), or only slightly more than the original VW Beetle. Aerodynamic drag coefficient is 0.25, the lowest of any mass-produced car available.

Toyota Prius

Toyota's Prius hybrid-electric vehicle is a compact four-seat sedan weighing 1,255 kg (2,765 lb.) (Figure 5.33). Toyota engineers designed the power system around the 1.5-liter, double overhead cam, four-cylinder engine from the subcompact Echo. It has been modified, however, to run as an Atkinson cycle engine, which improves thermal efficiency. The electric motor is a permanent magnet type of 33.0-kW (44-hp) output. The battery pack consists of 38 sealed NiMH modules.

The Prius is somewhat of a cross between a series hybrid and a parallel hybrid. The IC engine powers a generator, and it is connected to the drivetrain by a continuously variable transmission. With this configuration, the engine can deliver power directly to the drivetrain, to both the drivetrain and the generator, or to the generator only, depending on the battery's state of charge and road load demands. The Prius starts off under electric power alone. The IC engine starts when load reaches 10 kW, or when the battery state of charge drops below a predetermined threshold. Power from the engine is automatically managed and split between the generator and drivetrain by the vehicle's

FIGURE 5.33 Toyota Prius Hybrid-Electric Car. An inpanel display shows a real-time graphic representation of energy flow in the power system (inset). Other than by viewing the display, however, the driver is virtually unaware of transitions between subsystems. As of November 2000, more than 45,000 of these efficient hybrid cars were already on the road in Japan.

Courtesy: Toyota Motor Corporation, www.toyota.com

control module. Transitions are smooth, and the driver is unaware that anything out of the ordinary is happening in the powertrain.

More than 35,000 of these highly efficient hybrids are already on the road in Japan (year 2001), and the vehicle was recently introduced into the United States. Base price in the United States is $20,450, which is a Toyota-subsidized price that results in a loss on every car sold. The true base price is estimated at about $40,000, assuming Toyota would make a profit on sales.

Ford Prodigy

The Prodigy is a family-size parallel hybrid that combines a small, energy-efficient diesel engine and a three-phase ac electric drive system, along with an automatic-shift manual (ASM) transmission (Figure 5.34). Like Honda's IMA, the motor is packaged between the engine and transmission. It provides up to 35 kW (46 hp) in supplemental power on demand. The starter/alternator functions as an electric motor to assist the IC engine when extra power is needed, and it also works as a starter. On deceleration, it switches to alternator mode to reclaim the vehicle's kinetic energy. The engine is a 1.2-liter, four-cylinder, direct-injection aluminum diesel (Figure 5.35). It delivers 54 kW (72 hp) at 4,100 rpm. Combined, the motor and engine have a total output of 89 kW (119 hp), which gives the lightweight Prodigy performance comparable to today's Ford Mondeo. The Prodigy also features idle shut-off, which eliminates fuel wasted at idle. When shut off, the engine restarts in less than 0.2 seconds at the touch of the throttle.

FIGURE 5.34 Ford Prodigy. Ford's Prodigy HEV is a full-size family sedan that gets up to 34 km/L (80 mpg) fuel economy.

Courtesy: Ford Motor Company, www.ford.com

FIGURE 5.35 Prodigy's Diesel Engine. The Prodigy's lightweight direct-injection diesel engine is assisted by a three-phase ac motor, which also serves as an alternator and a starter.

Courtesy: Ford Motor Company, www.ford.com

The Prodigy's variable ride height, grille shutters, and fairings underneath the car work together with its slippery aerodynamic shape to give it a drag coefficient of only 0.199. Because of extensive use of lightweight materials, curb weight is only 1,085 kg (2,390 lb.), which is about 450 kg (1,000 lb.) less than the typical full-size family sedan. The Prodigy sips fuel at the rate of less than 3 liters per 100 km (80 mpg).

General Motors Precept

General Motors' Precept, a five-passenger midsize sedan developed under PNGV, was built in two versions: a fuel-cell vehicle and a diesel-powered hybrid (Figures 5.36 and 5.37). In terms of gasoline-equivalent fuel economy, the Precept's fuel-cell version consumes just slightly over 2 liters per 100 km (108 mpg) in urban driving. It accelerates from 0 to 100 km/h (0 to 60 mph) in 9 seconds and reaches a maximum speed in nearly 200 km/h (120 mph).

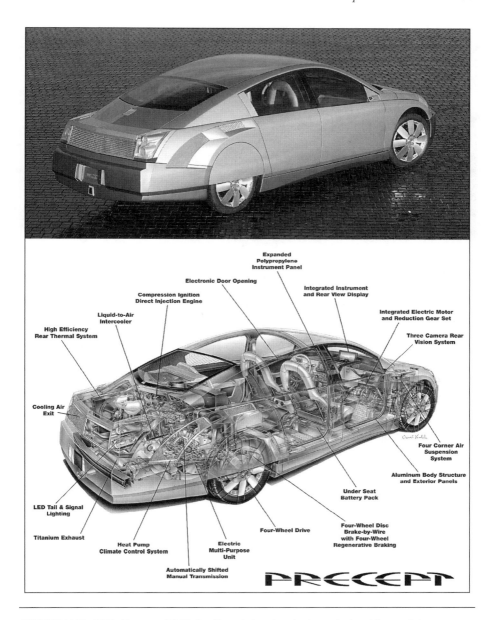

Expanded Polypropylene Instrument Panel

Electronic Door Opening

Integrated Instrument and Rear View Display

Compression Ignition Direct Injection Engine

Liquid-to-Air Intercooler

Integrated Electric Motor and Reduction Gear Set

High Efficiency Rear Thermal System

Three Camera Rear Vision System

Cooling Air Exit

Four Corner Air Suspension System

Aluminum Body Structure and Exterior Panels

Under Seat Battery Pack

LED Tail & Signal Lighting

Four-Wheel Disc Brake-by-Wire with Four-Wheel Regenerative Braking

Four-Wheel Drive

Titanium Exhaust

Heat Pump Climate Control System

Electric Multi-Purpose Unit

Automatically Shifted Manual Transmission

PRECEPT

FIGURE 5.36 GM's Precept HEV. Cooling air is taken in through the sides and then exhausted at the rear into the car's wake. This airflow scheme helps control turbulence along the sides. Dumping air into the wake makes the body think that it's tapered like the EV-1. Rearward-looking cameras, instead of conventional rearview mirrors, reduce drag by an additional 17 percent.

Courtesy: General Motors Corporation, www.gm.com

FIGURE 5.37 *Precept's Interior.* This smooth and simple high-tech interior previews the direction in interior design.

Courtesy: General Motors Corporation, www.gm.com

Power comes from a 400-cell, 100-kW PEM fuel-cell stack operating at 260 to 340 volts. The Precept's fuel cell receives pure hydrogen from a newly developed chemical hydride.

The HEV version is powered by an Isuzu three-cylinder, lean-burn compression-ignition, direct-injection (CIDI) diesel engine, which is located in the rear. This location helps improve airflow around the body and contributes to the vehicle's extremely low 0.16 drag coefficient. An electric-assist motor is built into the engine and acts as a combination starter/generator. A separate 35-kW (47-hp) three-phase electric motor independently powers the front wheels. Both motors work to assist the 40-kW (54-hp) engine during acceleration and to reclaim energy during braking. A low-energy-storage NiMH battery pack is located under the seats. Fuel economy is 3 liters per 100 km (80 mpg), in terms of gasoline equivalency. Actual consumption of diesel fuel is 2.6 liters per 100 km (90.4 mpg).

DaimlerChrysler Dodge ESX3

DaimlerChrysler's hybrid-electric ESX3 (Figure 5.38) builds on the knowledge gained from its predecessors, the Dodge Intrepid ESX (1996) and the ESX2 (1998). A major focus in designing the ESX3 was to produce a super-efficient vehicle at an affordable price. Because of the cost of the technology, if the first-generation ESX had been manufactured, it would have cost $60,000 more than a conventional car. With the ESX2, the cost penalty was reduced to $15,000. The ESX3 brings the cost differential down to about $7,500. Concurrently, fuel economy has been steadily improved with each successive version.

Progress in fuel economy and cost comes from several technology improvements: The vehicle's mild hybrid (or "mybrid") powertrain combines a three-cylinder, 1.5-liter DICI diesel engine with a 15-kW motor and a low-energy-storage lithium-ion battery pack to achieve 3.3 liters per 100 km (72 mpg) fuel economy (gasoline equivalent). That is 0.09 liters per 100 km (2 mpg) better than the ESX2, and close to the PNGV goal of 2.9 liters per 100 km (80 mpg).

FIGURE 5.38 Dodge ESX3 Hybrid Concept Car. DaimlerChrysler's EXS3 hybrid concept car gets 3.3 liters per 100 km (72 mpg) fuel economy. Production costs have been continually reduced over three prototypes to bring the cost to within $7,500 of that of a conventional car. The ESX3 has a lightweight, high-strength body made of a new proprietary mix of thermoplastic, aluminum, and foam.

Courtesy: DaimlerChrysler, www.daimlerchrysler.com

The injection-molded thermoplastic body made of only 12 pieces is one of the key factors in keeping the vehicle's weight and cost down. The ESX3 weighs only 1,020 kg (2,250 lb.), and the body will cost less to manufacture than a conventional steel body. A patent is pending on DaimlerChrysler's proprietary mix of thermoplastic, aluminum, and structural foam from which the car is built. Despite its featherweight body, computer-simulated crash tests show that it will stand up to federal crash-test requirements. In addition, the entire vehicle is more than 80 percent recyclable.

DaimlerChrysler NECAR 5 Fuel-Cell Vehicle

DaimlerChrysler's NECAR 5 (New Electric Car) is the first vehicle in which the entire fuel-cell system is small enough to package within the underbody of a Mercedes A-Class car (Figures 5.39 and 5.40). By using methanol as a carrier fuel, the hardware and space needed for cryogenic or gaseous hydrogen storage is eliminated. The NECAR 5 is equipped with a simple fuel tank

FIGURE 5.39 DaimlerChrysler NECAR 5 Methanol Fuel Cell Car. The NECAR 5 runs on methanol. Emissions are water vapor and carbon. Methanol produced from biomass, however, would result in a zero net gain of CO_2 because carbon released from methanol reforming will have first been absorbed from the environment during growth of the biomass.

Courtesy: DaimlerChrysler, www.daimlerchrysler.com

FIGURE 5.40 Entire Fuel-Cell System Mounts to the Car's Underside. Technicians prepare to fit the fuel cell to NECAR 5's underbody.

Courtesy: DaimlerChrysler, www.daimlerchrysler.com

for its ambient-temperate liquid fuel, methanol. Energy efficiency of the system is nearly double that of an equivalent IC-powered car. DaimlerChrysler expects to begin selling fuel-cell cars in limited quantities by 2004. Large market penetration will not happen until the technology has been proven out and a refueling infrastructure is in place.

REFERENCES

1. Quanlu Wang and Daniel Sperling, "Energy Impacts of Using Electric Vehicles in Southern California," Institute of Transportation Studies, UCD-ITS-RR-92-13, May 1992.

2. Quanlu Wang and Mark A. DeLuchi, "Impacts of Electric Vehicles on Primary Energy Consumption and Petroleum Displacement," Institute of Transportation Studies, July 1991.

3. *Ibid.*

4. *See* note 1.

5. Stacy C. Davis, "Transportation Energy Data Book: Edition 17," Oak Ridge National Laboratory, September 1997. (Because of reporting inconsistencies, fuel economy figures may not be directly comparable between countries.)

6. GM uses a figure of 170 Wh/mi to calculate Impact's city range. The author has used 180 Wh/mi in the preceding table, which is based on GM's published specifications of 26 propulsion modules supplying 16.2 kW/hr energy, and 72 miles city range at 80 percent DOD. GM's internal figures would result in a slightly greater savings of primary energy. Conversion to SI by the author.

7. *See* note 2.

8. "Electric Vehicles: The Beginning of a Bright Future," Promotional Brochure, Electric Power Research Institute, Inc., Pleasant Hill, CA, 1991.

9. Jeffrey Skeer, "A Brief Survey of EIA-Sponsored Activity on Technology for Transport," Published in OEDC Document, *The Urban Electric Vehicle: Policy Options, Technology Trends, and Market Prospects*, ISBN 92-64-13752-1.

10. Quanlu Wang and Danilo L. Santini, "Magnitude and Value of Electric Vehicle Emissions Reductions for Six Driving Cycles in Four U.S. Cities with Varying Air Quality Problems," Paper presented at the 72nd Annual Meeting of Transportation Research Board, January 10–14, 1993, Washington, D.C.

11. Mark A. DeLuchi, "Greenhouse-Gas Emissions from the Use of New Fuels for Transportation and Electricity," *Transportation Research*, Vol. 27A, No. 3, pp. 187–191, 1993.

12. P. J. McConachie, "Electric Vehicle Performance Requirements for Traffic Compatibility," Motor Vehicle Fuel Conservation Workshop, Australia Dept. of National Development and Energy, 1981.

13. Donald W. Kurtz et al., "Performance Testing and System Evaluation of the DOE ETV-1 Electric Vehicle," SAE Paper No. 810418.

14. J. P. Zagrodnik et al., "Development of Advanced Battery Systems for Vehicle Applications," SAE Paper No. 890783.

15. "Battery and Electric Vehicle Update," *Automotive Engineering*, September 1992, p. 20.

16. S. R. Ovshinsky et al., "Recent Progress in the Development of Ovonic Nickel Metal Hydride Batteries for Electric Vehicles," SAE Paper No. 921571.

17. D. Reisner and M. Eisenberg, "A New High-Energy Stabilized Nickel-Zinc Rechargable Battery System for SLI and EV Applications," SAE Paper No. 890786.

18. M. Altmejd and E. Spek, "Capabilities of Na/S Batteries for Vehicle Propulsion," SAE Paper No. 890784.

19. William Hamilton, "Basic Requirements for Urban Cars," General Motors Research Corp., SAE Paper No. 780219.

20. A. F. Burke, "Hybrid/Electric Vehicle Design Options and Evaluations," SAE Paper No. 920447.

21. *Ibid.*

CHAPTER SIX

THREE-WHEEL CARS

with Tilting Three-Wheel Vehicle Information by Tony Foale

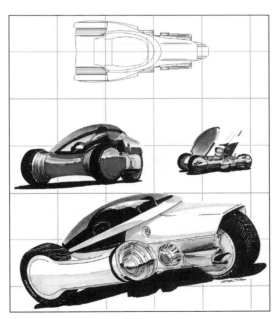

XR-3 tilting three-wheel vehicle concept by Robert Q. Riley. Rendering by Bill Ortis.

What, sir, would you make a ship sail against the wind and currents by lighting a bonfire under her deck? I pray you excuse me. I have no time to listen to such nonsense.

—Napoleon to Robert Fulton

Automotive product planners have generally disregarded the three-wheel platform as a legitimate layout for anything other than low-speed special-purpose vehicles. A proposal for a high-performance consumer product designed around three, instead of four, wheels is unlikely to receive much support from automotive decision makers. Like Robert Fulton's idea to "make a ship sail against the wind and currents by lighting a bonfire under her deck," the notion of three-wheel cars goes against conventional wisdom. In truth, however, high-performance three-wheel vehicles with excellent performance and stability are technically feasible. Bold new concepts in personal mobility products—products that can capture the imagination of consumers and carve out new territories of their own—may actually be more powerful in the marketplace when rendered in three-wheel instead of four-wheel designs. Moreover, tilting three-wheelers, vehicles that lean into turns like motorcycles, point the way toward a new category of products of unparalleled performance and cornering power, and nearly infinite design possibilities. Today, the three-wheel platform offers a unique opportunity to define cutting-edge personal mobility products with significantly lower mass, much greater fuel economy, and even superior performance.

Chapter Two includes a brief discussion of the three-wheel layout as a marketing tool. This chapter centers on the technical aspects of the layout. Criticism of the platform has centered mainly on its presumed overturn instability, but stability problems of particular production three-wheelers have been largely the result of poor design, rather than an inherent shortfall of the layout. An hour at the controls of a well-designed three-wheel car can light new fires of enthusiasm under people who are tired of routine driving experiences.

Three-wheel overturn resistance is affected by the same relationship of center of gravity (cg) height to effective half-tread that determines the resistance to overturn of four-wheel vehicles. A margin of safety against rollover equal to that of four-wheelers can be attained through traditional design practices. But three-wheelers are more sensitive to center of gravity displacement, both longitudinally and vertically, and payload variations can have a significant effect on vehicle center of gravity. In addition, dynamic conditions introduce a variety of forces that can affect a vehicle's rollover stability. Under dynamic conditions, a reduction in the margin of safety against rollover results whenever cornering and braking or acceleration forces combine to cause a result-ant that projects toward the single-wheel end of the vehicle. But these char-acteristics need not result in a vehicle with low resistance to overturn. Designers need only to provide an adequate margin of safety, and vehicle cg variations caused by variable payloads can be appropriately controlled by cor-rectly locating storage and occupant zones.

Mechanical advantages of the three-wheel layout result primarily from the elimination of one wheel. By eliminating the redundant fourth wheel, the chassis becomes inherently less costly to manufacture. A three-wheel chassis is also inherently lighter, and in many cases may be comparatively stronger. Torsional loads are reduced, but still remain because of dynamic inputs and the location of the cg above the groundline. A chassis with only three wheels will have less rolling resistance than a similar chassis with four wheels, mainly because of its lighter weight and reduced bearing and brake drag. Aerodynamic drag is also reduced by the simple act of eliminating one wheel. Given three- and four-wheel vehicles of equal coefficient of rolling resistance and equal mass, however, both will have roughly the same rolling resistance because each of the wheels is more heavily loaded in a three-wheel vehicle of equal mass. So the reduction in road load typical of three-wheel vehicles comes from reduced aerodynamic drag, reduced bearing and brake drag, and the platform's inherently lower mass.

As a legal entity, three-wheel cars do not exist in the United States. The U.S. Code of Federal Regulations classifies any vehicle with "not more than three wheels in contact with the ground" as a motorcycle. This releases any three-wheel vehicle from the need to comply with stringent passenger car safety regulations. An entrepreneurial company considering the market for an alternative personal mobility product in the United States could see this legal anomaly as a powerful argument in favor of the three-wheel configuration, but today's consumers are very safety-conscious. So a manufacturer produc-ing such a vehicle for the consumer market will be wise to consider passen-ger car safety provisions, regardless of legal requirements. Despite their legal classification, three-wheel full-bodied passenger vehicles will be referred to as three-wheel cars in this chapter.

Mechanically Simple Design

The three-wheel layout naturally results in a simplified chassis of fewer parts and reduced weight, which translates into reduced manufacturing costs. The savings in cost and weight come mainly from the simple elimination of one wheel. This results in a 25 percent reduction in the combined weight and cost of tires, wheels, brakes, and suspension components, when compared to a similar four-wheel design. A three-wheeler with the single wheel at the rear also makes it possible to eliminate the differential, both drive axles, and some of the supportive structure. A single-front-wheel design simplifies front suspension and steering.

Less obvious is the reduced need for torsional rigidity in the three-wheel chassis. With one wheel at each corner, the four-wheel vehicle is subjected to greater torsional loads resulting from uneven road surfaces and uneven bounce inputs at the four corners. These torsional loads require greater torsional rigidity in the four-wheel chassis. This does not imply that a three-wheel chassis can dispense with torsional rigidity. Because the center of gravity is located some distance above the groundline, torsional loads caused by inertia are transferred to the chassis.

In general, a three-wheel chassis is mechanically simplified and lighter, which allows equal performance with less installed power. Reduced power places less demand on the drivetrain, which allows for lighter and less costly components. In a variety of ways, the simple act of eliminating one wheel has a cumulative effect throughout the design of the vehicle. Figure 6.1 provides a comparison of three-wheel and four-wheel layouts.

Sports Car Handling

Designing to the three-wheeler's inherent characteristics can produce a high-performance machine that will outcorner many four-wheelers. A well-designed three-wheeler will likely be one of the most responsive machines one will ever experience over a winding road. It is difficult to place a thrill value on the feeling of cornering response. In a study of three-wheeler stability for the U.S. Department of Energy, Dr. Paul Van Valkenburg documented very fast yaw response times that were far superior to four-wheel vehicles.

Yaw response time is the time it takes for a vehicle to reach steady-state cornering after a quick steering input. A softly sprung four-wheeler will have a yaw response time of about 0.30 seconds. A four-wheel sports car will respond in about half that time. The most poorly designed three-wheelers tested by Van Valkenburg achieved steady-state cornering approximately 0.20 seconds after steering input. The best three-wheelers, however, reached steady-state cornering in as little as 0.10 seconds, which is approximately 33 percent quicker than the best four-wheel car tested. Three-wheel cornering agility is a quality that must be experienced to be fully appreciated.

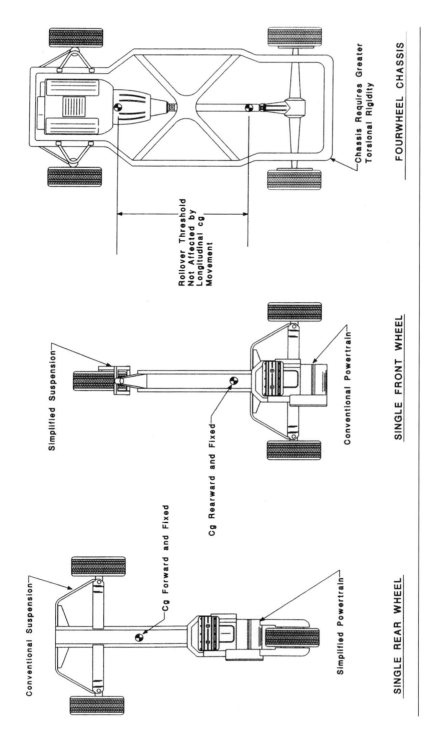

FIGURE 6.1 Three-Wheeler and Four-Wheeler Layout Comparison

311

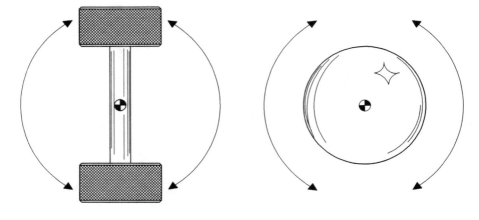

10-Lb. Dumbbell
High Polar Moment of Inertia

10-Lb. Ball
Low Polar Moment of Inertia

FIGURE 6.2 *Polar Moment of Inertia*

The attributes responsible for the quick response to steering inputs have nothing to do with the number of wheels or how they are configured. Instead, they are byproducts of the reduced mass and low polar moment that results when a three-wheel car is correctly designed (Figure 6.2). The typical three-wheeler has approximately 30 percent less polar moment than a comparable four-wheel design. Most passenger cars are designed with a relatively high polar moment of inertia. The engine is located over the front or rear axle, and the fuel and luggage are located at the opposite end. The center of the vehicle is devoted primarily to the occupants.

A low polar moment of inertia results in a vehicle with more responsive handling, but it also produces a choppy ride. A vehicle with high polar mass is less nimble, but it rides more smoothly. The relationship between polar mass and ride was discovered back in the 1920s when it was found that cars rode better when the engine was placed far ahead over the front wheels. Sports cars tend to have a low polar moment of inertia for nimble handling, and they also tend to ride more roughly than passenger cars. Normally, a good balance between ride and handling can be achieved, regardless of the number of wheels or the car's performance orientation. One does not have to choose between a rough ride and poor handling.

TABLE 6.1 Rigid-Body Rollover Threshold Comparison

Vehicle Type	CG Height inch (mm)*	Tread inch (mm)*	Rollover Threshold (lateral g-load)*
Sports Car	18–20 (457–508)	50–60 (1270–1524)	1.2–1.7
Compact Car	20–23 (508–584)	50–60 (1270–1524)	1.1–1.5
Luxury Car	20–24 (508–610)	60–65 (1524–1651)	1.2–1.6
Pickup Truck	30–35 (762–889)	65–70 (1651–1778)	0.9–1.1
Passenger Van	30–40 (762–1016)	65–70 (1651–1778)	0.8–1.1
Medium Truck	45–55 (1143–1397)	65–75 (1651–1905)	0.6–0.8
Heavy Truck	60–85 (1524–2159)	70–72 (1778–1829)	0.4–0.6

Source: Fundamentals of Vehicle Dynamics, www.sae.org

*SI equivalents inserted by the author

ROLLOVER THRESHOLD

A vehicle's rigid-body rollover threshold is established by the simple relationship between the cg height and the maximum lateral forces capable of being transferred by the tires. Modern passenger car tires have a friction coefficient on the order of 0.8, which means that the vehicle can negotiate turns that produce lateral forces equal to 80 percent of its weight (0.8 g) before the tires lose adhesion.* The cg height in relation to the effective half-tread of the vehicle determines the length-to-height (L/H) ratio, which establishes the lateral force required to overturn the vehicle. As long as the side-force capability of the tires is less than the side-force required for overturn, the vehicle will slide before it overturns. This analysis provides a useful figure for comparing the rollover threshold of various vehicles, as shown in Table 6.1.[1] Under dynamic conditions, a vehicle's rollover threshold is a more complex issue.

During dynamic conditions, overshooting the roll angle or tripping against an obstacle can cause the vehicle to overturn, even though it has a high rigid-body margin of safety against rollover. Rapid-onset turns impart roll acceleration to the body of a magnitude that can cause the body to overshoot the steady-state roll angle. This happens with sudden steering inputs, and it also occurs when a skidding vehicle suddenly regains traction and begins to

---

* Suspension design has a significant effect on cornering capabilities. The discussion is necessarily simplified.

turn again. A hard turn in one direction followed by an equally hard turn in the opposite direction (a slalom turn) can also cause the vehicle to overshoot the roll angle. Roll moment depends on the vertical displacement of the center of gravity above the vehicle's roll center. The degree of roll overshoot depends on the balance between the roll moment of inertia and the roll-damping characteristics of the suspension. An automobile with 50 percent (of critical) damping has a rollover threshold that is nearly one-third greater than the same vehicle with zero damping.

Overshooting the steady-state roll angle can lift the inside wheel(s) off the ground. Once lift-off occurs, the vehicle's resistance to rollover rapidly diminishes, which results in a condition that quickly becomes irretrievable. The roll moment of inertia reaches much greater values during slalom turns, wherein the forces of suspension rebound and the opposing turn combine to "throw" the body laterally through its roll limits from one extreme to the other. Inertia forces that result from overshooting the steady-state roll angle can exceed the forces produced by the turn rate itself.

Tripping is another cause of rollover in an otherwise rollover-resistant vehicle. Tripping occurs when a vehicle skids against an obstacle such as a curb (Figure 6.3). In this case, the lateral speed of the vehicle is suddenly arrested, and high momentary loads are imposed across the vehicle's center of gravity. If the load spike exceeds the vehicle's rollover threshold, rollover will occur.

Dynamic conditions and the resulting forces are difficult to predict in random real-world events. Vehicles that appear to have an adequate margin of safety against rollover may nevertheless overturn under the right conditions. Consequently, the best design for rollover protection will include adequate roll damping and the greatest possible margin of safety against rollover. Roll damping in a three-wheeler can come only from the two side-by-side wheels at one end of the vehicle.

SINGLE FRONT OR SINGLE REAR WHEEL

The debate over the preferred choice between the single-front-wheel (1F2R) or single-rear-wheel (2F1R) layout centers on which configuration will result in a more stable vehicle. Both configurations can be designed for an adequate rigid-body margin of safety against rollover. Under dynamic conditions, the two designs have different characteristics. With the single wheel in front, the vehicle's margin of safety against rollover is diminished during braking turns and increased during accelerating turns. If the single wheel is at the

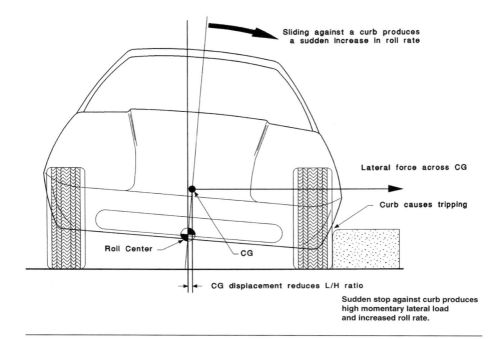

Sliding against a curb produces a sudden increase in roll rate

Lateral force across CG

Curb causes tripping

Roll Center

CG

CG displacement reduces L/H ratio

Sudden stop against curb produces high momentary lateral load and increased roll rate.

FIGURE 6.3 Rollover Caused by Tripping

rear, the margin of safety against rollover is degraded during accelerating turns and enhanced during braking turns. Because braking forces can exceed acceleration forces, the single-rear-wheel configuration normally exhibits the best dynamic margin of safety against rollover for a given L/H ratio.

The two configurations exhibit opposite oversteer/understeer characteristics as well. Vehicles with the single wheel in front inherently oversteer, and single-rear-wheel designs inherently understeer. Although suspension tuning and the selection of tires and wheels can compensate for the layout's natural tendency, these variables normally cannot make a single-front-wheel design understeer, nor a single-rear-wheel design oversteer at the limit of adhesion. Regardless of the inherent characteristics, neither configuration need result in a vehicle with unsafe handling characteristics. Oversteer with VW's Beetle was greater than with the single-front-wheel Trimuter built at Quincy-Lynn Enterprises. Quincy-Lynn's Tri-Magnum (Figure 6.14) understeered about like an average, softly sprung American sedan. As for sheer performance, the Tri-Magnum will outcorner the majority of four-wheelers on the road today.

Single Front Wheel (1F2R)

The single-front-wheel layout is an inherently sleek design. The narrow front and wide rear of the body blends well with the natural shape of two occupants sitting side by side with legs extending forward. It is naturally wide across the occupants' shoulders where extra room is needed and narrower near the occupant's feet where less width is required. This basic shape tends to be attractive and aerodynamically clean.

Basic handling characteristics and the margin of safety against rollover result primarily from the relationship between the cg location, the effective half-tread, and the wheelbase of the vehicle. Roll forces in a three-wheeler are taken exclusively by the two side-by-side wheels. Moving the cg toward the side-by-side wheels increases the effective half-tread, which causes the vehicle to behave as though the side-by-side wheels had been moved farther apart. The relationship of the effective half-tread to the cg height and the friction coefficient of the tires determines whether the vehicle will slide or overturn at high lateral g-loads.

With a three-wheel platform, the L/H ratio is normally modeled as a three-dimensional cone to account for braking and acceleration forces and the relationship between longitudinal cg displacement and the triangular wheel layout. As long as the base cone projects to the ground-plane inside the effective half-tread, the vehicle will slide before it overturns (Figure 6.4). If the base cone projects outside the effective half-tread, or moves outside because of dynamic forces, the vehicle will overturn before it slides. The margin of safety against rollover can be increased by increasing the effective half-tread and by reducing the height of the cg. The effective half-tread increases when the two side-by-side wheels are moved outboard (increasing the actual tread), to a lesser degree when the wheelbase is increased, and when the cg is moved closer to the side-by-side wheels.

With the single-front-wheel platform, the attributes that produce a high resistance to rollover also create a rear-heavy vehicle, which leads to oversteer. Oversteer can be reduced by placing larger, wider tires at the rear and by appropriate suspension design, but the basic tendency cannot be eliminated.* An oversteering vehicle is considered more difficult to control for the

---

* Tilting three-wheel designs, discussed later in this chapter, promise to eliminate oversteer in the single-front-wheel layout.

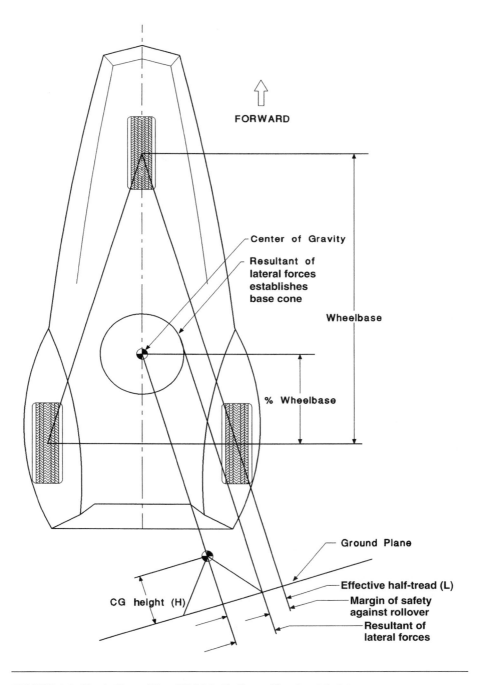

FORWARD

Center of Gravity

Resultant of
lateral forces
establishes
base cone

Wheelbase

% Wheelbase

Ground Plane

Effective half-tread (L)

Margin of safety
against rollover

Resultant of
lateral forces

CG height (H)

FIGURE 6.4 Single-Front-Wheel Vehicle Rollover Margin of Safety

average driver. Additional disadvantages of the single-front-wheel configuration include an increased sensitivity to crosswinds, aerodynamic lift at the light (front) end of the vehicle, and the unfavorable effects of having the resultant of forces project forward and to the outside during braking turns.

The single-front-wheel platform tends to yaw more readily when subjected to crosswinds because the aerodynamic center of pressure of the body is normally ahead of the center of gravity. Also, at high speeds aerodynamic lift can cause the already light front end of the vehicle to become even lighter. Quincy-Lynn's Trimuter began to feel nose-light at approximately 130 km/h (80 mph). Despite these attributes, however, the single-front-wheel configuration is still a viable layout for high-performance passenger vehicles. The single-front-wheel layout can be designed for high dynamic stability, and aerodynamic lift need not create a problem as long as the body is correctly designed.

Single Rear Wheel (2F1R)

When the single wheel is located at the rear and the cg is moved forward toward the side-by-side front wheels, handling characteristics change. The resulting front-heavy vehicle tends to understeer, just like four-wheel cars with which consumers are familiar. Crosswind characteristics also improve when the cg is located nearer to the center of aerodynamic pressure (usually about 25 to 30 percent of body length). The relationships that determine the margin of safety against rollover are identical to those of the single-front-wheel layout, except they are reversed (Figure 6.5).

Braking turns no longer present a problem for the three-wheeler with the side-by-side wheels at the front. The forward projection of forces caused by a braking turn increases the rollover threshold, which tends to offset the negative effect of the forces acting toward the outside of the turn. The net effect is that resistance to rollover is not necessarily degraded. Conversely, an accelerating turn produces an unfavorable resultant toward the outside rear; however, acceleration normally produces dynamic forces of less magnitude than braking. Consequently, the unfavorable projection of forces is not as great with the single-rear-wheel configuration.

The disadvantages of the single-rear-wheel configuration include the difficulty of entering or leaving the vehicle over or behind the outboard wheels, reduced interior space resulting from the steering angle of the front wheels, and the limited ability to transfer power through a single powered rear

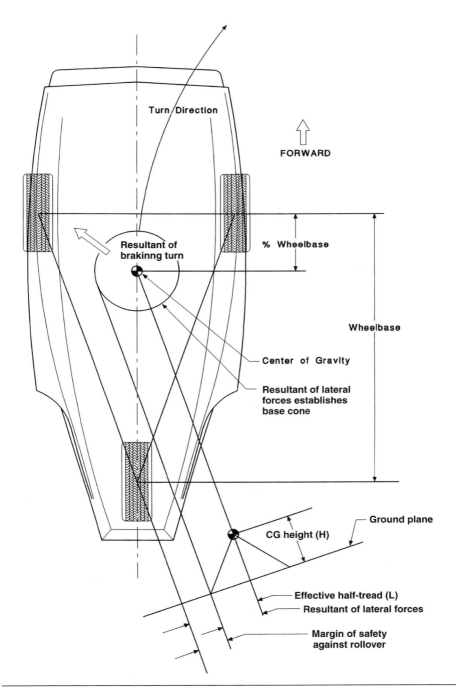

FIGURE 6.5 Single-Rear-Wheel Rollover Margin of Safety

wheel. In an effort to appropriately locate the cg, occupants may be placed between the two side-by-side wheels. This location has the advantage of offering side intrusion protection; however, convenience of ingress and egress is sacrificed, along with substantial interior room that must be dedicated to the wheelhouse. The engine is often located behind the occupants, where it powers the single rear wheel. This configuration was successfully used in several designs, including Ron Will's TurboPhantom, Walter Korff's Duo-Delta, and Quincy-Lynn's Tri-Magnum. Both the TurboPhantom and Tri-Magnum are high-performance sports cars capable of speeds in excess of 160 km/h (100 mph).

A three-wheel design incorporating front-wheel-drive and a transverse-mounted engine ahead of the wheels offers the ability to transfer more power to the road than is possible with a single powered rear wheel. The powertrain also creates a natural weight bias toward the two side-by-side wheels, and thereby allows the passenger zone to move aft for easier ingress and egress behind the front wheels. The tradeoff with this layout is a drivetrain of increased complexity and cost, when compared to the powered single-rear-wheel design. The VW Scooter and the Trihawk are two excellent examples of front-wheel-drive three-wheel cars.

Tandem or Side-by-Side Seating

With side-by-side seating, frontal area is not appreciably affected by reducing vehicle capacity from four to two occupants. Placing one occupant behind the other in a tandem arrangement reminiscent of the German Messerschmitt can reduce the frontal area of a design by as much as 40 percent. The effect on aerodynamic drag can be significant. At 80 km/h (50 mph), aerodynamic drag of a four-wheeler typically accounts for approximately half of the total road load. In the case of a low-rolling-resistance three-wheel configuration, aerodynamic drag would represent an even greater portion of the vehicle's total road load. Using Trihawk's relationship between aerodynamic drag and rolling resistance, a 40 percent reduction in frontal area would theoretically translate into a 28 percent reduction in road load at 80 km/h (50 mph).

There are few areas in which a simple design change can result in such large reduction in road load, but the tandem seating arrangement does have drawbacks. As a general rule, tandem seating tends to extend the vehicle's wheelbase and normally results in a greater cg displacement with different loads (one or two occupants). A tandem-seater also tends to reduce the social feeling, which is enhanced by side-by-side seating.

Examples of Nontilting Three-Wheel Vehicles

Unfortunately, many three-wheelers of the traditional nontilting type, both experimental and production vehicles, have not been well designed. Even among production vehicles, an inadequate margin of safety against rollover is a common attribute. The Bon Bug, which was manufactured for years in the United Kingdom and enjoyed limited success in the marketplace, would lift an inside wheel in hard turns. When Paul Van Valkenburg performed stability tests on three-wheelers in 1980, some of the test vehicles had to be equipped with outriggers to keep them from overturning during cornering tests. Several well-designed but less known experimental three-wheelers have actually performed better.

Turbo Phantom

The TurboPhantom is one of the vehicles originally tested by Van Valkenburg when he studied three-wheeler stability (Figures 6.6 and 6.7). It is not one of the vehicles that needed an outrigger to stay upright in turns. The

FIGURE 6.6 *Ron Will's TurboPhantom.* The dry lakebed at Lake El Mirage, California, gives the perfect backdrop for one of the most exotic three-wheelers ever built.

Courtesy: Ronald J. Will

FIGURE 6.7 A View from the Rear. The steering wheel lifts up with the canopy to provide wide-open access to the TurboPhantom's interior.

Courtesy: Ronald J. Will

TurboPhantom is a high-performance sports car that will maintain a 0.82 g corner with only two degrees of body lean (Figure 6.8). The vehicle's 60 percent margin of safety against rollover gives it greater resistance to overturning than most four-wheel vehicles.

The TurboPhantom was designed around a Honda GL-1000 motorcycle powertrain. An 82-kW (110-hp) turbocharged version would reach a top speed of 209 km/h (130 mph). Curb weight was 680 kg (1,500 lb.), which put the TurboPhantom in the weight category of the early VW Beetle. Ronald J. Will, the car's designer, actually put the vehicle into production and marketed it as a "three-wheel luxury sports car." In 1981 the price started at $20,000. Today, Ron is with Subaru in New Jersey.

VW Scooter

The VW's Scooter is a mid-1980s prototype built by the VW Research and Development Division in Germany (Figures 6.9 and 6.10). One of the most

FIGURE 6.8 High-Speed Cornering on Three Wheels. The TurboPhantom cornering at 0.82 g during handling tests at Edwards Air Force Base for Van Valkenburg's study on three-wheeler stability.

Courtesy: Ronald J. Will

impressive features of the Scooter is that the vehicle is said to meet European and North American safety standards. The vehicle is designed to pass the 50 km/h frontal collision test, which is the European equivalent of the U.S. 30 mph barrier test.

With the engine located ahead of the front drive axles, Scooter has exceptionally good handling characteristics. The 545-kg (1,200-lb.) three-wheeler can reach a top speed of 120 km/h (75 mph) and achieve fuel economy of 25 km/L (60 mpg) at 88 km/h (55 mph). Urban cycle fuel economy is on the order of 15 km/L (35 mpg). It is maneuverable in traffic, easy to park, and has excellent cornering characteristics. Interior room across the shoulders is about the same as a Japanese Kei car. Gull-wing doors and the rear window are detachable, which gives the Scooter a sporty, open-air feeling. Aerodynamic drag with the car closed up is 0.25. The concept behind the vehicle was to mix as much as possible of the flavor of the motorcycle with the safety and convenience of the automobile.

FIGURE 6.9 VW Scooter Concept Car. A state-of-the-art three-wheel prototype built by Volkswagen. The Scooter features front-wheel drive, removable gull-wing doors, and safety features that meet European and North American safety standards.

Courtesy: Volkswagen AG, www.vw.com

FIGURE 6.10 VW Scooter on Skid Pad. VW Scooter undergoes skid tests in the rain.

Courtesy: Volkswagen AG, www.vw.com

Trihawk

The Trihawk is an excellent example of the potential of the front-wheel-drive powertrain in a three-wheel vehicle (Figures 6.11, 6.12, and 6.13). Vehicle engineering was done by Bob McKee of CanAm and Indy fame. David Stollery, a former GM stylist who also established Calty, Toyota's U.S. design studio, did the styling. The open-bodied roadster has a curb weight of 613 kg (1,350 lb.) and can accelerate from 0 to 96 km/h (0 to 60 mph) in slightly more than 10 seconds. Maximum speed is nearly 160 km/h (100 mph). At 96 km/h (60 mph), the Trihawk can be brought to a standstill in 45 m (148 ft.), and in about 76 m (250 ft.) at 129 km/h (80 mph).

The Trihawk is noted for its exceptionally high turn rate. The center of gravity is just 305 mm (12 in.) above the ground. *Road and Track* magazine recorded 0.83 g turns, *Car and Driver* recorded 0.87 g turns, and in independent tests the Trihawk is said to have developed turn rates as high as 0.91 g before predictably understeering to the outside. For a while in the early 1980s, the Trihawk was manufactured under the direction of David Stollery at facilities in

FIGURE 6.11 Trihawk Three-Wheel Sports Car. The gleaming Trihawk is a high-performance front-wheel-drive sports car that will reach 160 km/h (100 mph).

Courtesy: Harley-Davidson, www.harley-davidson.com

FIGURE 6.12 Trihawk View from the Front. A wide 1,676-mm (66-in.) stance and low center of gravity give the Trihawk 0.87 g cornering capability before predictably understeering to the outside.

Courtesy: Harley-Davidson, www.harley-davidson.com

FIGURE 6.13 Trihawk Interior. Trihawk's businesslike interior is reminiscent of early European sports cars.

Harley-Davidson, www.harley-davidson.com

Dana Point, California. Price was in the $14,000 range. Ultimately, the project was purchased by Harley-Davidson and then subsequently shelved. According to officials at Harley-Davidson, the potential for liability problems was the deciding factor.

Tri-Magnum

This motorcycle-based three-wheeler can outperform most production sports cars and still delivers fuel economy roughly equivalent to that of the original motorcycle. The key constant is the power-to-weight ratio. With 60 kW (80 hp) powering Tri-Magnum's 568-kg (1,200-lb.) curb weight, power-to-weight ratio is about 9.5 kg/kW (15 lbs./hp).

Tri-Magnum's styling is both functional and in character with its aggressive performance (Figures 6.14 and 6.15). Slippery aerodynamics, engine cooling requirements, accessibility to the cockpit and engine compartment, and safety considerations are all integrated into the design. The impact-absorbing foam-filled front bumper, which ties into the frame with large steel members, is designed to spill air onto the body. Body lines flow smoothly from

FIGURE 6.14 Tri-Magnum. The Tri-Magnum is a high-performance sports car capable of speeds in excess of 160 km/hr (100 mph). More information on Tri-Magnum is available at www.rqriley.com.

front to rear, where they break sharply around taillight nacelles to create a clean separation point. The rear-facing duct on top and the two shark-gill louvers on the sides draw hot air from the engine room while cool air is ducted in from underneath. A small electric fan, located just ahead of the engine, keeps it cool at idle.

The lift-up canopy, although exotic, is simple, functional, and strong. It leaves the main body area integral for maximum strength. When the canopy is open, it presents an entirely open cockpit, so one doesn't have to duck under a low roofline when getting in and out. The canopy is counterbalanced with two air cylinders, so it is easy to lift and stays up by itself. A steel framework is laminated inside the canopy for extra strength and rigidity. Seating is side-by-side, and the seat, which is contoured like a space capsule from head to knees, is molded into the car's body (Figure 6.16). Passengers sit between the two front wheels.

FIGURE 6.15 Tri-Magnum Side Rear View. Taillight treatment is used to blend the wide body at the front into the narrow aft section. The engine room cover opens to reveal an entire motorcycle, sans the front wheel and fork.

FIGURE 6.16 *Tri-Magnum Pistol-grip Shifter.* The pistol-grip shifter holds the motorcycle clutch lever and handlebar switch modules. The front axle beam passes under the occupants' knees. The steering wheel tilts up for ingress and egress.

Sparrow

Corbin Motors in Hollister, California, manufactures an electric single-seat three-wheeler called Sparrow (Figure 6.17). Power comes from thirteen 12-volt lead-acid batteries, which deliver current to an 18.6-kW (25-hp) series motor (37.2 kW peak power). Top speed is 110 kmh (70 mph), and range is 65–97 km (30–60 miles). The Sparrow, which the company calls a "Personal Transit Module," retails for $14,900 (year 2001).

Gizmo

The Gizmo is a single-occupant electric three-wheeler manufactured by Neighborhood Electric Vehicle Company in Eugene, Oregon (Figure 6.18). The vehicle has no steering wheel. Instead, it is steered by side-mounted levers, which also carry the accelerator, brakes, and horn. According to the company, the control system is logical, intuitive, and easy to use.

FIGURE 6.17 Sparrow by Corbin Motors. As of February 2001, 183 Sparrows had been shipped to customers. The Sparrow is equipped with a single door on the right side.

Courtesy: Corbin Motors, www.corbinmotors.com

FIGURE 6.18 Gizmo Electric Urban Car. The Gizmo is about as minimal a vehicle as you'll find. Top speed is 65 km/h (40 mph), and range is 40 km (25 miles).

Courtesy: Neighborhood Electric Vehicle Company, www.nevco.com

TILTING THREE-WHEELERS

As previously discussed, the variables with which designers achieve high rollover resistance in a three-wheel vehicle include reducing the height of the cg, moving the cg closer to the two-wheel end of the vehicle, and providing a greater separation between the single pair of side-by-side wheels. Another option, which is only recently being well explored, is the idea of a suspension system that lets the vehicle lean into turns. The application of the tilting suspension system to the three-wheel vehicle opens new horizons in potential cornering power and stability, and provides designers with much greater latitude in cg placement and separation between the side-by-side wheels. With a tilting suspension, designers of three-wheel vehicles no longer have to rely on a wide, low vehicle for high rollover stability, and cornering power can actually exceed that of conventional, nontilting four-wheel vehicles.

To appreciate the benefits of tilting, consider that single-track vehicles, such as bicycles and motorcycles, do not overturn when going around corners. This is because they lean into turns. Such vehicles do not need opposing wheels to stay upright because their lean angle keeps them in balance with the turn forces that are acting on them, as shown in Figure 6.19. Rollover, in

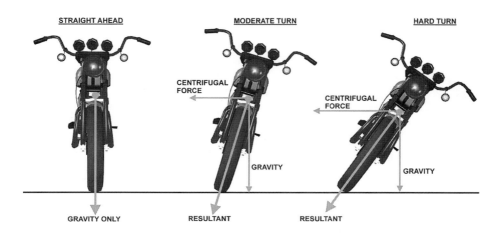

FIGURE 6.19 *Motorcycle Turn Behavior.* A motorcycle does not roll over because it leans into turns in order to remain in line with the resultant of turn and gravity forces. Height of the cg is unimportant as long as the motorcycle remains in a balanced turn, but the cg height does have a small effect on lean angle and can also affect the roll rate of establishing a turn.

the traditional sense, is no longer possible as long as the vehicle leans appropriately and turn forces do not exceed the ability of the tires to grip the road. In addition, occupants experience cornering loads as an increase in their apparent weight, rather than as side loads forcing them sideways across the seat. So occupants and packages stay put, and open drinks do not spill.

When a multitrack vehicle leans inward, emulating the turn of a motorcycle, a narrower track can still provide good rollover stability (Figure 6.20). In a

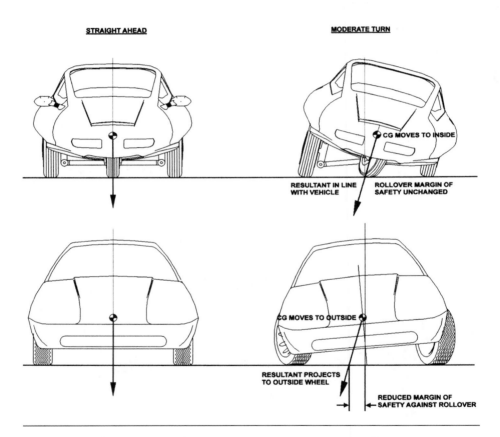

FIGURE 6.20 Comparison of Resultant in Tilting and Nontilting Vehicles. In a balanced turn, a tilting multitrack vehicle behaves like a single-track vehicle. By leaning into turns, the resultant meets the groundline at or near the centerline of the vehicle, depending on the degree of lean. In contrast, a nontilting vehicle rolls to the outside of the turn, so the resultant projects to the groundline progressively closer to the contact patch of the outside tire as the turn rate increases. Once the resultant projects past the contact patch centerline, the vehicle will overturn.

balanced turn, no increased loading is transferred to the outside tires. In contrast, a conventional nontilting vehicle transfers an increasingly greater load to the outside tires as the turn rate increases. So conventional vehicles have to rely on a wide separation between opposing wheels in order to provide an adequate margin of safety against rollover.

Although vehicles with either three or four wheels can be made to lean into turns, most of today's activity in tilting vehicle development centers on three-wheelers. In comparison to a four-wheel platform, a three-wheeler naturally lends itself to greater lean angles with fewer mechanical challenges because of the single wheel at one end of the platform.

TTW Classification

Development efforts with tilting three-wheelers (TTWs) have resulted in a wide diversity of mechanical layouts and control schemes. Generally, TTWs can be classified according to the following four main design attributes: wheel layout, lean control, lean limit, and number of tilting wheels.

Wheel Layout

As with conventional three-wheelers, TTWs may be classified according to platform layout, which was described earlier in the chapter (1F2R and 2F1R).

Lean Control

Lean control may be either free or active, or a scheme for switching between the two modes may be used in a hybrid form of control. The simplest lean control is that used by motorcycles. This "natural" or "free-leaning" system relies on rider input to balance the machine under dynamic conditions. At rest, however, a free-leaning, fully enclosed three-wheeler must be equipped with some form of lean lock or brake to hold the leaning section upright. Unlike a motorcycle, the rider cannot simply put a foot down.

Another form of lean control is the active lean mechanism. With active lean control, either hydraulic or electromechanical actuators are used to establish the vehicle's lean angle. The appropriate angle of lean is normally determined by processing signals from various sensors, which monitor variables such as steering angle, vehicle yaw and lean, and lateral acceleration. Although electronic control is normally used, various schemes of exclusively hydraulic control mechanisms have also been tried.

A dual mode of control allows for a switch between free and active lean control. Such systems normally require a method of determining when changeover is appropriate, and some sort of damping is usually necessary to "soften" the actual transition between modes of operation.

Lean Limit

Motorcycles typically have a physical lean limit of 45 to 50 degrees before grounding the lowest parts of the machine. Lean angles in this range represent a lateral cornering acceleration of 1.0 g to 1.2 g. High lean angles are necessary only under extreme cornering conditions, and beyond the upper range, cornering forces are likely to exceed the friction limit of tires. So machines with a lean limit approaching 50 degrees, in effect, have no practical lean limit. An extremely high limit of lean is necessary with a single-track machine because it must be capable of assuming a lean angle that will balance cornering forces under every conceivable condition. If a motorcycle cannot lean to the angle necessary to balance cornering forces, it will overturn. Because a completely free-leaning three-wheel vehicle behaves just like a motorcycle, it must be capable of equally high lean angles in order to accommodate the maximum expected lateral acceleration.

Due to mechanical interference and geometry problems, however, some three-wheel designs do not lend themselves to the high lean angles necessary to remain in balance with turn forces at the extremes. But a three-wheel vehicle does not necessarily have to assume a lean angle that places it precisely in balance with turn forces. Instead, a portion of the cornering force can be transferred to the outside wheel. The wider the track, the less critical the angle of lean will be. Some type of active lean mechanism will, however, be necessary in order to allow the machine to operate at a lean angle different from that needed to balance turn forces.

Number of Tilting Wheels

Conceivably, TTWs could be designed around configurations with one, two, or three leaning wheels. Some have been designed with a leaning body and no leaning wheels. It is unnecessary to lean all of the wheels, or even any of the wheels, in order to gain at least some of the dynamic benefits of a tilting vehicle.

NO TILTING WHEELS. This system uses a leaning passenger compartment in combination with a nonleaning chassis equipped with a traditional suspension system. It is equally suited for either the 1F2R or 2F1R layout. The system was suggested for a tilting tricycle in the late 1980s by Cliff Ingram and is used today on tilting trains. It is important to keep the roll axis of the leaning compartment as low as possible in order to provide the maximum margin of safety against rollover.

ONE TILTING WHEEL. This is the system used for the final version of GM's Lean Machine—a highly successful free-leaning design under development from the late 1970s through the early 1980s. The Lean Machine is a 1F2R layout wherein the passenger compartment and front wheel assembly lean in relation to a nonleaning rear module containing the powertrain, rear suspension, and wheels. Depending on the details of the design, this layout can have a very high maximum lean angle, typically on par with that of a motorcycle. No examples of the 2F1R layout in this category are known.

TWO TILTING WHEELS. It is difficult to imagine a reason for using this scheme, and no examples are known.

THREE TILTING WHEELS. A system using three tilting wheels is appropriate for either the 2F1R or the 1F2R layout. This is the only system capable of providing a high degree of stability in vehicles with a very narrow track. But as vehicle track is increased, the amount of wheel movement required for an equivalent lean angle also increases and can become excessive at large lean angles (Figure 6.21). For this reason, wide-track designs such as the Project 32 Slalom (see Figure 6.30, page 351) typically limit the maximum angle of lean. A restricted angle of lean normally implies active lean control throughout the range of operation or a limited range of free-leaning operation.

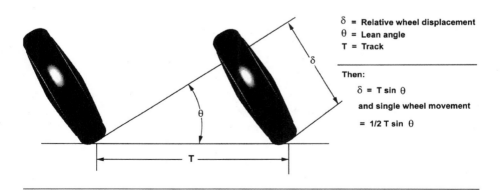

δ = Relative wheel displacement
θ = Lean angle
T = Track

Then:

$\delta = T \sin \theta$

and single wheel movement

$= 1/2\ T \sin \theta$

FIGURE 6.21 Effect of Vehicle Track on Total Wheel Travel

TABLE 6.2 Total Wheel Movement with 76-mm (3-inch) Bounce
and 51-mm (2-inch) Rebound

Track (mm/inch)	Total Movement for 45-Degree Lean Angle (mm/inch)	Total Movement for 20-Degree Lean Angle (mm/inch)
318/12.5	445/17.5	242/9.5
635/25	762/30.0	358/14.1
1270/50	1397/55.0	589/23.2

Another reason for limiting maximum lean angle is to stay within the angular limitations of the flexible couplings used on powered half-shafts.

Table 6.2 shows the total wheel movement necessary for a tilting suspension with 76 mm of bounce travel (upward from center) and 51 mm of rebound travel (downward from center) at 45 and 20 degrees of lean. Notice that a high lean angle in combination with a wide track results in an unrealistically large wheel movement.

It is also possible to use three tilting wheels in combination with a nontilting or minimally tilting body. The author's XR-3, shown in the rendering at the beginning of this chapter, uses a body that rolls at a lesser rate and assumes a smaller angle than the tilt of the wheels. Systems like this one produce an inboard shift of the cg during cornering, which improves rollover margin. But lateral loads on occupants are not relieved in designs using a nontilting body and are only partially relieved in designs using a minimally tilting body (depending on the tilt angle).

Dynamic Behavior of Tilting Three-Wheelers

In order to understand the behavior of a TTW, it will be helpful to review the dynamic behavior of single-track vehicles such as motorcycles and bicycles. A free-leaning three-wheeler behaves the same as a single-track vehicle in many aspects. Even with vehicles that have active suspension systems, some of the same dynamic forces are at work.

Although many motorcyclists are unaware of it, in order to initiate a turn to the left, the rider of a single-track vehicle must initially steer slightly to the right. This is called "counter-steering," and for most riders, it is done instinctively on a subconscious level. In motorcycle racing, riders often use deliberate counter-steering to achieve the high roll rates necessary for quick maneuvers. Fundamentally, it is the combination of precessional moments and centripetal force that causes or requires this counter-steering action. Imagine riding a motorcycle and turning the front wheel to the right. There will be four main effects:

1. A steering action to the right will produce a small gyroscopic tendency to lean the machine to the left.
2. This steering action of the front tire will also cause the machine to start turning right, and just as in a car, centripetal force will cause a lean to the left.
3. Gravity will then augment the banking effect during the early part of the lean-in.
4. Gyroscopic reaction from the roll velocity will then produce a steering feedback torque to the rider.

Unlike a car, a motorcycle has no static roll stiffness, and it will continue to lean until all of the forces are in balance. The same dynamics are at work in the case of TTWs that are equipped with all leaning wheels. Because there must be little or no roll stiffness with a free-leaning vehicle, it follows that no roll moment has to be supported by differential loading of the tires.

Although motorcyclists are quite comfortable with this natural form of control, automobile drivers will have a certain learning curve, and in an emergency, may apply inappropriate control responses. For this reason, designers with an eye toward the automotive market have tended to favor active leaning systems. Active lean control also offers several other benefits.

Active Lean Control

Unlike the free leaners, steering inputs with active lean control are similar to those used with a conventional automobile. To turn right, one steers to the right. But the same left-leaning tendencies still exist as with free-leaning systems and must be overcome by the vehicle's lean actuators. In addition to overcoming the counter-leaning tendencies, the lean actuators also must provide the additional force necessary to impart the required roll acceleration. Active lean systems typically use a sensor to monitor errors between the actual lean angle and the lean angle needed for a balanced turn. Signals from the sensor are electronically processed, and then appropriate control commands are sent to the actuators.

Control System Strategies

Although optimal design parameters are still being researched, several possible strategies exist for active lean systems. One approach is to maintain a fully balanced, motorcycle-like turn up to the mechanical limit of lean, and then hold the system at its limit of lean during periods of greater lateral acceleration (sharper turns). The balance between lean angle and turn forces can be monitored by a lateral accelerometer. An electronic control unit (ECU) then works to zero the accelerometer signal. For vehicles with a lean limit, some form of signal ramping on approach to lean limit may be needed to avoid hard bottoming against the limit stop and to reduce the unsettling effects on the machine's balance and the occupants' comfort level.

An alternative algorithm aims to reach mechanical lean limit at maximum lateral acceleration, while allowing a differential to exist between actual lean angle and that required for a balanced turn. With this approach, the differential between lateral acceleration and lean angle may be gradually increased as lateral acceleration increases. This results in a lean angle that is always less than that required for a balanced turn, and occupants feel gradually increasing lateral force as the turn rate increases. Tests have shown that some sense of lateral acceleration is desirable in order to provide cornering feedback to the driver. When a vehicle leans like a motorcycle, there is less feedback through the steering system, so lateral force on the driver is an important means of indicating the rate of cornering.

In another control strategy, the required lean angle is calculated on the basis of vehicle speed and steering angle. A shortfall of this approach is that the steering angle required to produce a particular lateral acceleration varies according to roadway composition, surface conditions, and gradient.

Another important factor is the amount of lateral acceleration perceived by the driver during transitions in lean angle. A driver who feels excessive lateral force during a rapid roll, for example, may have difficulty distinguishing between the force resulting from the roll rate and the force resulting from the actual turn rate, which may lead to inappropriate control inputs.

The effect of roll acceleration on vehicle rollover stability also must be considered. The torque required to put the vehicle into a lean is necessarily transferred to the tires at the two-wheel end of the vehicle. This roll torque results in an increased loading on the outer tire and a decreased loading on the inner tire. It is possible to completely unload the inner tire and produce a rollover at high roll rates, depending on vehicle track. This potential exists during rapid-onset rolls following quick steering inputs. The resulting roll torque can actually lift the inside wheel free of the ground. Moreover, the magnitude of roll torque necessary to lift the inside wheel becomes less at greater lean angles. Control algorithms must therefore account for these input limitations and dynamic variables. Increasing the vehicle's track increases the maximum tolerable roll torque. This important safety aspect makes forced lean systems unsuitable for very narrow track vehicles, such as the Calleja described later.

The effect of roll acceleration becomes more critical under hard cornering with a configuration having only a single front wheel that leans. In a steady-state turn, this system naturally transfers greater load to the outside rear tire, which results in a certain degree of oversteer. If the vehicle were steered deeper into the turn, then extra transient roll torque will further load the outside rear tire, exacerbating the oversteer condition and reducing the margin of safety against rollover. In extreme cases, this could lead to loss of adhesion at the rear, even though lateral acceleration may be well within the steady-state turn capabilities of the vehicle.

When the all-leaning-wheels layout is operating in a balanced, steady-state turn, equal load will be placed on the outside and inside tires, and no roll torque will be required of the lean actuators to maintain the lean. During transient conditions, however, roll torque will place an increased load on either the outside or inside tire (depending on roll direction), as described previously.

In addition to the control inputs and dynamic forces associated with normal turns, it is also important to consider lean requirements and dynamic forces that may be experienced under abnormal conditions. For example, if a tilting vehicle were to spin out while appropriately leaned to the inside of a turn, after having rotated 180 degrees in a skid, it will then be leaned inappropriately to the outside of the turn. So a lean angle that enhances stability under normal conditions could make the vehicle more susceptible to overturning during loss of control. Presumably, control algorithms can be developed to account for atypical and loss-of-control scenarios.

Tires

Leaning vehicles present new considerations for tire designers. Motorcycles generate much of their cornering force through camber angle (camber thrust), whereas automobile tires rely mainly on slip angle to produce cornering force. Nonleaning tires on a tilting vehicle are subjected to loads similar to those experienced with a conventional automobile, so passenger car tires will presumably have the correct characteristics. Leaning tires operating in a coordinated turn will experience loading akin to that in a motorcycle application. Provided that the load ratings are not exceeded, these tires will be well suited for this application.

The choice of tire, however, is less straightforward in the case of leaning wheels on a vehicle with a lean limit below that required for balanced cornering. Consider, for example, a turn of 0.8 g with a machine that has a 20-degree lean limit. A 20-degree lean equates to a lateral acceleration of 0.36 g, which leaves an excess lateral acceleration of 0.44 g to be supported as a side load mainly on the outside tire. Side loading of this magnitude is significantly greater than that normally expected of a motorcycle tire, and the implied slip angle also exceeds motorcycle standards. But passenger car tires are also unsuitable because they are not designed to operate at a camber angle of 20 degrees. Consequently, new tires may have to be developed if tilting vehicles are to achieve their full design potential.

The profile and dimensions of rims also deserve review. Motorcycle rim profiles are not designed to handle large side forces and slip angles, and vehicles with all leaning wheels present another consideration. Tilting the side-by-side wheels changes the track at the contact patch, and hence introduces lateral tire scrubbing as the vehicle leans into curves and straightens up coming out.

Suspension Loading

When a vehicle leans into a turn, cornering forces translate into an increased load on the suspension system of the tilting wheels. This will be greatest at high cornering speeds in machines without a lean limit. In some cases this may require stiffer springing than might otherwise be selected for optimum ride comfort (Figure 6.22).

General Characteristics

For purposes of analysis, it will be useful to regard the TTW operating within its lean limit (in a balanced turn) as a virtual motorcycle. The longitudinal plane of symmetry of the leaning portion of the TTW can be considered as equivalent to the similar plane of a motorcycle. At the end of the vehicle with the side-by-side wheels, one can then, on a conceptual level, replace the two opposing wheels with one "virtual" motorcycle wheel centrally located between them. By envisioning the TTW as a virtual motorcycle,

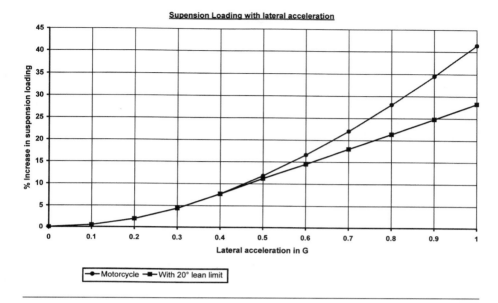

FIGURE 6.22 Suspension Loading with Lateral Acceleration

the various dynamic factors and geometric relationships become more obvious. Notice in Figure 6.23 that rollover stability is achieved when the contact patch of the virtual wheel remains inboard of the two actual wheels during lean maneuvers. In a balanced turn, the virtual wheel in the center will always be inline with the resultant, as will be the case with a motorcycle in an equivalent turn.

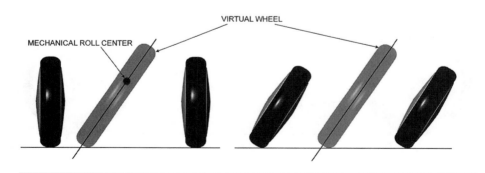

FIGURE 6.23 Virtual Wheel

1F2R with Nonleaning Rear Wheels

With this layout, the leaning front section typically houses the passengers, controls, front suspension, and wheel. A nonleaning rear section contains the powertrain, suspension, and rear wheels, which are normally carried by a conventional suspension system. These two sections of the vehicle are typically connected by a mechanical pivot, which defines the roll axis (Figure 6.24). The mechanical roll axis, however, is not the same as the dynamic roll center of the vehicle as a whole, but the location and orientation of the mechanical roll axis affects the vehicle's steering and dynamic behavior, as well as its margin of safety against rollover. Although all known machines of this type use a fixed roll axis, conceivably a multilink system could be designed to provide a virtual roll axis that would move about according to the vehicle's roll angle.

This layout offers a wide range of lean angles and can achieve a lean equivalent to that of motorcycle. It is equally well suited for either free or active lean control, provided the track is wide enough. The suspension of the leaning front wheel is subjected to compression resulting from cornering loads, but no such loads are imposed on the nonleaning rear wheels.

Rollover Threshold

With vehicles in this category, rollover threshold depends on the separate contributions of the leaning and nonleaning sections. The nonleaning section behaves like a conventional nonleaning three-wheeler. Its rollover threshold is determined mainly by the L/H ratio, as described earlier in this chapter.

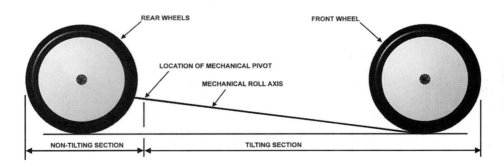

FIGURE 6.24 Front Passenger Module Pivots around Roll Axis

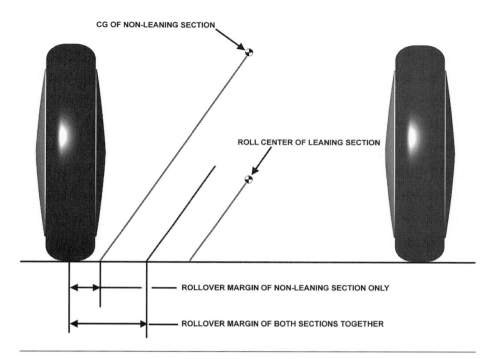

CG OF NON-LEANING SECTION

ROLL CENTER OF LEANING SECTION

ROLLOVER MARGIN OF NON-LEANING SECTION ONLY

ROLLOVER MARGIN OF BOTH SECTIONS TOGETHER

FIGURE 6.25 Roll Axis Height at Rear Axle Affects Rollover Margin

In a balanced turn, the height of the mechanical roll axis at the point at which it intersects the axle centerline of the nonleaning wheels determines the effect of the tilting section on the vehicle's rollover threshold. If the roll axis is lower than the cg of the nonleaning section (at the point of intersection), then the influence of the leaning section will be to increase the vehicle's resistance to rollover. If the roll axis is higher than the cg of the nonleaning section, resistance to rollover will be decreased in a turn, relative to that of the nonleaning section alone. So it is important to locate the mechanical roll center as low as possible. Figure 6.25 illustrates the effect of roll axis height on rollover threshold.

As long as the turn remains balanced, the cg height of the leaning section has no effect on the vehicle's static rollover threshold. But the height of the cg does influence rollover threshold in vehicles with a residual roll moment, as when operating at a lean angle less than that required for a balanced turn. In unbalanced turns, the cg height of the leaning section has a significant influence on the vehicle's rollover threshold, and should therefore be as low as possible in order to ensure the highest degree of stability. In addition to affecting a vehicle's rollover threshold, the inboard migration of the cg during a leaning turn reduces lateral load transfer, and thereby mitigates the inherent oversteer tendency of 1F2R three-wheelers.

Roll/Yaw Coupling

Unless the mechanical roll axis projects to the ground at the front axle (Figure 6.26), the vehicle will have a roll/yaw coupling between front and rear wheels. In other words, the longitudinal inclination of the mechanical roll axis can cause the rear section to steer, relative to the front, emulating an effect commonly referred to as rear-wheel steering. If the roll axis projects to a point above ground level at the front wheel, then as the leaning section rolls into a turn, the roll axis will move to the inside and steer the nonleaning rear section into the turn (Figure 6.27). To eliminate this roll/yaw coupling, the roll axis must project to the ground directly under the front axle.

One possible utilization of this roll/yaw coupling centers on the difference between the way in which the leaning and nonleaning wheels generate cornering forces. The leaning front wheel generates much of its cornering force through camber angle. In contrast, the nonleaning rear wheels develop cornering force by means of a slip angle. In the absence of rear-wheel steering, the slip angle must be provided by the yaw attitude of the virtual motorcycle. Therefore, the yaw/steering coupling may be useful to quickly set up the required yaw attitude of the rear wheels. But despite potential steering benefits, a potential downside also exists: Motorcycles can exhibit various wobble/weave instabilities, and analysis has shown that TTWs can be subject to the same phenomenon. This type of roll/yaw coupling could exacerbate wobble/weave type instabilities.

Pitch/Lean Coupling

Another implication of a longitudinal roll center that does not project to ground level at the front wheel is that of the resulting pitch/lean coupling (Figure 6.28). When the roll center projects to a point above ground level, the vehicle will rotate forward around the rear axle as it leans. With this type of coupling, changes in lean angle translate into pitch movements, and pitching moments translate into lean inputs. A rearward pitch caused by acceleration, for example, will unload the front and tend to reduce the lean angle. Consequently, accelerating while in a turn will result in a moment that tends to

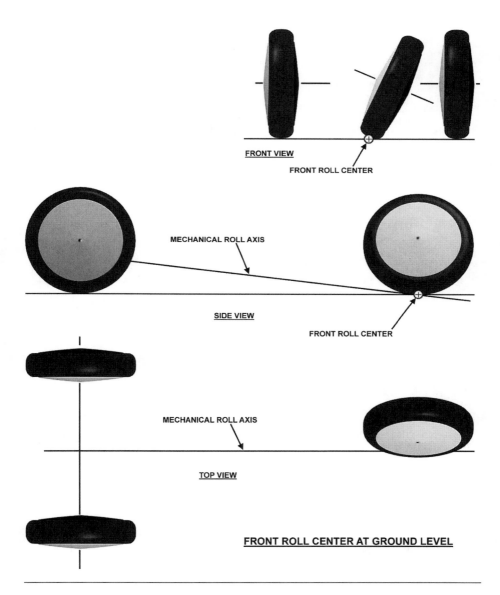

FRONT VIEW

FRONT ROLL CENTER

MECHANICAL ROLL AXIS

SIDE VIEW

FRONT ROLL CENTER

MECHANICAL ROLL AXIS

TOP VIEW

FRONT ROLL CENTER AT GROUND LEVEL

FIGURE 6.26 No Roll/Yaw Coupling When Roll Axis Projects to Ground at Front Axle

bring the vehicle out of the lean. Conversely, braking in a turn will tend to increase the angle of lean. This pitch/lean coupling is a more important consideration in free-leaning designs. With vehicles that have active lean control, the lean angle will be determined by the control system.

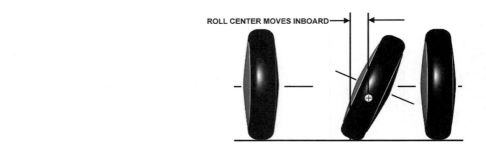

ROLL CENTER MOVES INBOARD

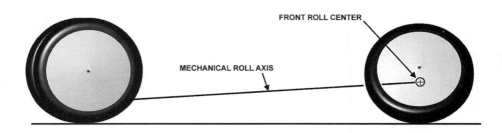

FRONT ROLL CENTER

MECHANICAL ROLL AXIS

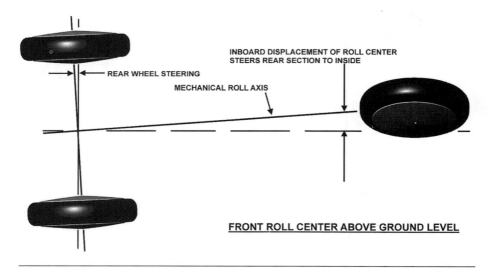

INBOARD DISPLACEMENT OF ROLL CENTER
STEERS REAR SECTION TO INSIDE

REAR WHEEL STEERING

MECHANICAL ROLL AXIS

FRONT ROLL CENTER ABOVE GROUND LEVEL

FIGURE 6.27 Roll/Yaw Coupling Caused by Roll Axis Height at Front Axle

PITCH LEAN COUPLING

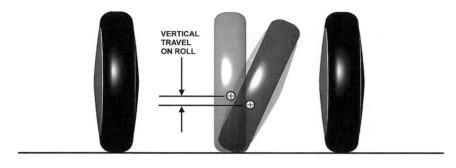

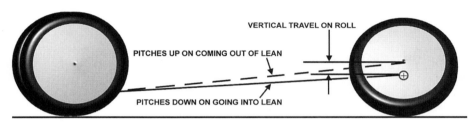

FIGURE 6.28 Lean/Pitch Coupling Caused by Projection of Roll Axis

TTWs with All Leaning Wheels

Designs in this category include vehicles with a very narrow track, almost emulating a single-track vehicle, and ranging upward in width to those having a relatively wide track approaching that typical of conventional automobiles. Maximum achievable lean angle is largely determined by the track width. The narrower the track, the greater the maximum angle of lean because a wider track translates into greater wheel movements at given angles of lean (see Figure 6.21, Table 6.2, Figure 6.29). Consequently, vehicles of relatively wide track typically have a limit on maximum lean angle. When lean is limited to an angle less than that required for a balanced turn at maximum lateral acceleration, the vehicle must be equipped with active lean control, or the lean angle must be locked at lean limit during unbalanced turns. Vehicles with a very narrow track are unsuited for active lean control because they cannot support large roll moments.

In many respects, the rollover threshold of vehicles with all leaning wheels is simpler to analyze than vehicles in the foregoing category. With this type of vehicle, the virtual motorcycle wheel will be located midway between the two

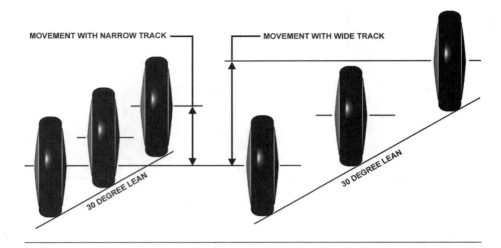

FIGURE 6.29 Effect of Track on Total Wheel Movement in Lean

actual wheels. In a balanced turn, no residual roll moment will exist, and consequently, no rollover tendency will be present. Narrow-track machines in this category emulate the cornering characteristics of motorcycles. But wide-track vehicles, when operating at the limit of lean or otherwise held in an unbalanced turn by the control system, will have a residual roll moment. As a result, calculation of the margin of safety against rollover becomes relevant. Rollover threshold is easily calculated using the base cone method. Unlike nontilting three-wheelers, however, a rollover analysis of a TTW must include the geometric effects of suspension articulation caused by lean angle and the dynamic effects of the inward shift of the cg in a turn.

Designs of the 1F2R configuration are typically equipped with a steerable front wheel and two powered rear wheels. In fully coordinated turns, the lack of lateral load transfer will eliminate the oversteer characteristics typical of this layout. But when the lean angle is less than that required for a balanced turn, some lateral load transfer is experienced, which may translate into oversteer, depending on the magnitude of the load transfer. Oversteer can be reduced or eliminated by proper tire and wheel selection, adjusting the inflation pressure, and other traditional methods.

With vehicles of the 2F1R layout, the front wheels are typically used solely for steering, and the single rear wheel is powered. This layout introduces no special problems for powertrain designers. The characteristic understeer of this layout, however, is reduced because of the reduced front lateral load transfer resulting from the lean angle.

TTWs with No Leaning Wheels

This is the layout used on tilting trains. Typically, all three wheels, along with their suspension systems, are mounted to a nonleaning chassis, which also carries the powertrain. This layout may be less suited to road vehicles because of the mechanical difficulties of locating the roll axis of the tilting body low enough. Generally, the special requirements of the nonleaning chassis tend to force the roll axis into a higher than desirable location. As a result, the vehicle's margin of safety against rollover may not be much improved over that of a nontilting machine. Reduction or elimination of lateral turn forces experienced by occupants, however, is still an important benefit.

EXAMPLES OF TILTING THREE-WHEELERS

Although the application possibilities of TTWs are still being explored, the following examples provide an overview of how the various design features have been implemented in actual vehicles (Table 6.3). At the time of this writing, most but not all of these vehicles have been proof-of-concept machines, with only a few either in current or anticipated production.

TABLE 6.3 TTW Primary Features

Vehicle	Wheel Layout	No. of Tilting Wheels	(A)ctive or (F)ree Lean	Lean Limit (in degrees)	Special Features
Transit Innovations Project 32 Slalom	1F2R	3	A	Approx. 17	Two side-by-side seats
Mercedes Lifejet F300	2F1R	3	A	30	Two tandem seats
Millennium Tracer	2F1R	3	F	37	Motorcycle based, one seat
Calleja	1F2R	3	F	45	Very narrow track
GM Lean Machine	1F2R	1	F (+ pedals)	50	By major car manufacturer
Jephcott Micro	1F2R	1	A	30	Early all-active control
Vanderbrink Carver	1F2R	1	A	45	All-hydraulic control
Ingram Zero G	2F1R	0	A	unknown	Proposal only

Transit Innovations Project 32 Slalom
(U.S. Patents Pending, 1997–2001)

The Project 32 Slalom is a 1F2R design developed by Transit Innovations. Its suspension system is the brainchild of company president/owner Larry K. Edwards. Edwards first became known for his role in leading the Lockheed team that developed the world's first underwater-launched nuclear delivery system, the Polaris Missile. Later, he served as Director of Engineering at NASA on the space shuttle program. His land transportation contributions include the Gravity-Vacuum Transit System, a 500-mph superefficient intercity tube train, and the highly acclaimed System 21 monobeam urban transit system. The Slalom is Edwards's most recent venture and represents the third leg of his "total solution to the world's land transportation challenges." The author joined the Project 32 development effort to provide input on vehicle dynamics and to complete the preliminary layout, packaging, and styling.

One of the most noteworthy features of Edwards's Project 32 Slalom is its conventional automotive-style body with fully enclosed tilting wheels and side-by-side seating for two (Figures 6.30 and 6.31). Normally, TTWs must seat occupants in tandem because high lean angles make it necessary to limit the width of the body. And tilting wheels normally must be exposed because they swing though large arcs during tilt movements. With the Slalom, tilt angle is limited, which makes it possible to seat occupants side by side and enclose the wheels. Notice that fenders are carried forward across much of the occupant zone, where they provide extra room for the occupants.

The Slalom is equipped with an active lean control system and has a lean limit of approximately 17 degrees. A 17-degree lean angle allows for balanced turns of up to 0.28 g, which accounts for most maneuvers in ordinary driving. At greater turn rates, a portion of the turn force is transferred to the outside rear wheel, and occupants feel a lateral force.

Motive power for the electric version is provided by two 18.6-kW (25-hp) electric motors, one powering each rear wheel. Conventional IC engine versions are also possible.

To avoid problems caused by excessive drive shaft angularity at high lean angles, each motor is mounted in a yoke attached to the lower suspension arm. This adds only slightly to the unsprung mass. During bounce, the motors move at a lesser rate than the wheel assemblies. Leaning is accomplished by means of a hydraulic cylinder. Active control is provided by an

FIGURE 6.30 Project 32 Slalom. Project 32 Slalom is the only tilting three-wheeler known to seat occupants side by side. Dimensions: Wheelbase, 2,490 mm; Track, 1,295 mm; Target Weight, 455 kg. More information on Slalom is available at www.rqriley.com.

Courtesy: Transit Innovations

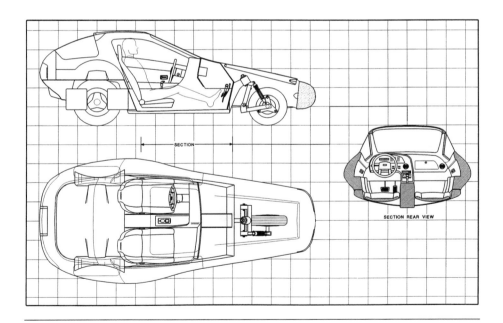

FIGURE 6.31 Project 32 Slalom Section Views

Courtesy: Transit Innovations

integrated ECU. The vehicle's comparatively wide 1,295-mm (51-in.) track was selected in order to reduce lateral load transfer and the resulting over-steer tendency. Moreover, studies indicate that the Slalom may be the first 1F2R design to entirely eliminate the characteristic oversteer of the single-front-wheel layout.

A parallelogram linkage system controls the location and leaning of the rear wheels. One of the most innovative features of the suspension design is the use of a wide transverse leaf spring as the upper suspension link.

Mercedes F300 Life Jet (1997–unknown)

Built by Mercedes, a company known for producing high-quality luxury cars, the F300 is designed mainly as a fun vehicle. Its midmounted IC engine drives the single leaning rear wheel, and the two front wheels lean and steer (Figures 6.32 and 6.33). Two occupants are seated in tandem. The vehicle is

FIGURE 6.32 *Mercedes F300 Life Jet.* Dimensions: Length, 3,954 mm; Width, 1,730 mm; Height, 1,527 mm; Weight, 800 kg. The Life Jet is a high-performance concept car built by Mercedes.

Courtesy: DaimlerChrysler, www.mercedes-benz.com

FIGURE 6.33 F300 Life Jet Front Suspension. Front cowling is removed to show the Life Jet's tilting suspension system. An electronic control unit drives a hydraulic actuator, which can put the vehicle into any lean angle up to 30 degrees.

Courtesy: DaimlerChrysler, www.mercedes-benz.com

equipped with an active lean control system. Lean is limited to 30 degrees, which allows for coordinated turns of up to 0.58 g. To provide the driver with a sense of cornering rate, the control algorithm allows some cornering force to be transferred to the occupants. Driver feedback is achieved by leaning less than the angle required for balanced turns. The F300's adaptive control system adjusts the system's lean response rate according to the style of driving. Mercedes reports 0.9 g cornering capability.

Motorcycle tires were used on the prototype, but a test rig has been commissioned to investigate tire performance and requirements at large camber angles and loads.

A 1.6-liter engine from the A-class Mercedes develops 75 kW (100 hp), which gives the machine performance similar to a sports car. It can accelerate from 0 to 60 mph in 7.5 seconds. Maximum speed is 130 mph.

Millennium Tracer (1996–ongoing)

This Australian machine was originally conceived as an all-weather motorcycle (Figure 6.34). Although it shares the same configuration as the F300, the Tracer is a free-leaning machine with a lean limit of 37 degrees. This lean limit allows for balanced turns up to the maximum cornering acceleration of 0.75 g, which is great enough for all but the most spirited driving.

Unlike the motorcycle that inspired the design, it is not possible to put a foot on the ground to hold the machine upright while stationary, so an electromechanical, operator-controlled lean lock has been incorporated (Figure 6.35). To date, one vehicle has been built and tested in daily use. A second

FIGURE 6.34 Millennium Tracer. Dimensions: Length, 3,150 mm; Track, 860 mm; Wheelbase, 1,830 mm; Weight, 280 kg. Aluminum-bodied Tracer leans hard into a turn.

Courtesy: Millennium Motorcycles, members.iinet.net.au/~jsulman/tracer/

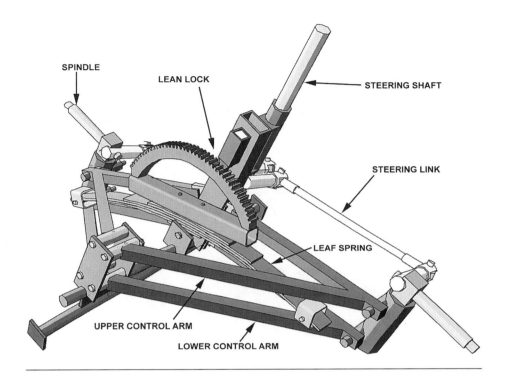

FIGURE 6.35 General Layout of Tracer's Lean Mechanism

Courtesy: Millennium Motorcycles, members.iinet.net.au/~jsulman/tracer/

machine is slated for an automatic lean-locking system and active control system that will assume control in the lower speed ranges. According to the designers, the new control system will make the machine more user friendly.

Calleja (U.S. Patent No. 5,611,555: 1997–ongoing)

The Calleja is an all-wheels-leaning 1F2R vehicle designed around a motorcycle (Figures 6.36 and 6.37). Its two rear wheels are located close together for a very narrow track. This Spanish-designed proof-of-concept vehicle has been built to very high standards. Ultimately, the lean system will be applied to narrow cars as well. Because of its narrow track, it is a free-leaning vehicle with no active control. The Calleja has the feel and handling characteristics of a motorcycle. Lean limit is approximately 45 degrees, which provides enough lean for 1.0-g turns.

FIGURE 6.36 *Calleja Three-Wheel Motorcycle.* Invented by Carlos Calleja, this three-wheel motorcycle still retains the balanced lean and fast cornering capability of a conventional motorcycle. In addition, the side-by-side rear wheels are no wider than an ordinary motorcycle with the rider onboard.

Courtesy: Tony Foale, www.tonyfoale.com

FIGURE 6.37 *Calleja in a Lean.* The Calleja can lean to an angle of about 45 degrees, which is enough to handle a 1-g turn.

Courtesy: Tony Foale, www.tonyfoale.com

The most innovative feature of the design is the rocking yoke actuator (balancer), which permits free leaning in combination with normal suspension movements (Figure 6.38). A manually controlled electrically actuated drum brake, located in the center of the yoke, holds the vehicle upright at rest and can account for lateral inclines.

A differential is located on the pivot axis of the trailing arms, and power is delivered individually to the two rear wheels by chains. Because the differential is located on the trailing arm axis, the center distance between drive and driven sprockets remains consistent during suspension bounce movements. The Calleja has the narrowest track of the vehicles reviewed, and its designers believe it provides a viable basis for a small one- or two-place city car. Its narrow track, however, makes the design unsuitable for active lean control, which means that prospective drivers would need motorcycle-type riding skills.

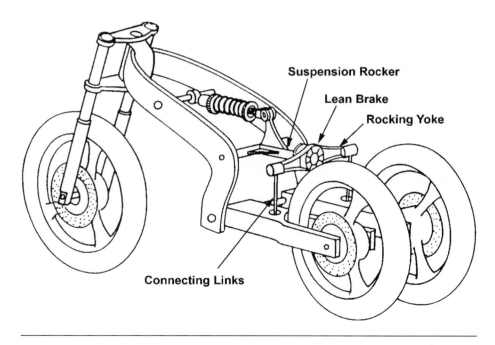

FIGURE 6.38 Calleja Lean Mechanism. Rocker yoke has a brake in the center to hold the machine upright when stopped.

General Motors Lean Machine (late 1970s–early 1980s)

General Motors Lean Machine, a 1F2R design with nonleaning rear wheels, is a good example of what happens when a group of engineers are given free rein to develop an idea. The version shown here was the last in a series of 15 prototypes developed over a seven-year period (Figures 6.39 and 6.40). The Lean Machine borrows much from the world of the motorcycle. A bubble-shaped driver's pod leans into turns, while the narrow-tracked power pod follows along in standard, flat-cornering fashion. Lean angle is controlled by floor-mounted pedals and steering input. Direct-link handlebars steer the single front wheel. With practice, the driver can balance the control pressures for effortless, gliding turns that emulate the leaning, zero-g turns of a motorcycle.

FIGURE 6.39 GM's Lean Machine. The Lean Machine leans into turns for high-speed cornering.

Courtesy: Road & Track *magazine, www.roadandtrack.com*

A cd of 0.35 and a curb weight of 159 kg (350 lb.) contributed to the Lean Machine's ability to deliver high performance on a modest power budget. The 11-kW (15-hp) Honda ATV engine and five-speed transmission located in the power pod pushed the Lean Machine to 129 km/h (80 mph) and delivered fuel economy of 50 km/L (120 mpg) at a steady 64 km/h (40 mph). A 50-degree lean limit would allow for balanced turns of 1.2 g, provided the tires could maintain adhesion. The rear tread was only 711 mm (28 in.).

The goal was to combine the economy of a motorcycle with the lean-into-the-corner thrill in a vehicle that resembled a full-bodied automobile. Officially, GM never had plans to produce the Lean Machine. The vehicle's learning curve and potential product liability problems were the two main drawbacks;

FIGURE 6.40 Lean Machine on Test Track. Dimensions: Track, 710 mm; Length, 2,615 mm; Width, 915 mm; Height, 1,220 mm; Weight, 181 kg. A narrow 710-mm (28-in.) rear track would normally lead to poor cornering capabilities, but with its leaning passenger pod, the Lean Machine can negotiate 1.2-g turns, according to *Road & Track* magazine.

Courtesy: Road & Track *magazine, www.roadandtrack.com*

FIGURE 6.41 Lean Machine's Aircraft-Style Canopy. Project Engineer Jerry Williams lowers the canopy for a test drive.

Courtesy: Road & Track *magazine, www.roadandtrack.com*

however, facilitators who were present during Lean Machine product clinics conducted in the 1980s report that consumers were exceptionally enthusiastic about the design. Market potential appeared to be about twice the level that had been expected. According to project engineer Jerry Williams (retired) (Figure 6.41), by the time GM shut down the Lean Machine development effort, they had designs on the drawing boards that were far more advanced, both stylistically and mechanically, than any of the prototypes.

Jephcott's Micro (U.S. Patents allowed to lapse, early 1980s)

Two versions of the Micro were built by Dr. Jephcott in England in the early 1980s. The first was a proof-of-concept scooter-based machine, which used foot pedals for balancing, in much the same fashion as GM's Lean Machine

FIGURE 6.42 Jephcott's Micro. Dimensions: Length, 2,590 mm; Width, 915 mm; Height, 1,420 mm. Photo shows an early engineering testbed on the left and a finished prototype on the right.

(left photo in Figure 6.42). The second prototype was built to higher standards and included tandem automotive-type seating in an attractive fully enclosed fiberglass body designed by Richard Oakes (right margin in Figure 6.42). It was equipped with an innovative all-hydraulic active lean control system, although an electronic control system was acknowledged in patent documents. A pendulum device mounted on the leaning pod acted directly on a hydraulic valve to keep lean in balance with turn forces. When the front pod was not in a balanced turn, the pendulum would not be lined up, and the valve would direct hydraulic fluid to an actuating cylinder to increase or decrease the lean angle and restore balance. As implemented, the lean response time was very slow, but lean performance would probably have improved with a more sophisticated control system and increased hydraulic power.

Vandenbrink Carver

Winner of an inventor's award, the Carver is a 1F2R design with a 45-degree lean capability. The vehicle was conceived at a Dutch company, Brinks Dynamics, and is now in limited production. It is equipped with a sophisticated all-hydraulic lean control system called Dynamic Vehicle Control (DVC)

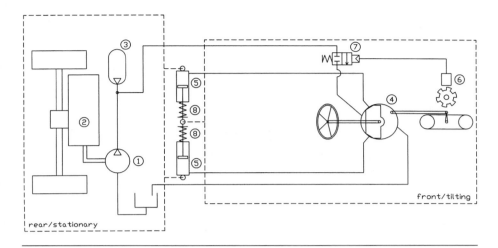

FIGURE 6.43 General Layout of Carver's Lean Control. 1. Hydraulic pump; 2. Engine; 3. Accumulator, keeps pressure at 120 bar; 4. Steering valve; 5. Actuating cylinders; 6. Speed sensor; 7. Shut-off valve, in combination with the speed sensor. This prevents lean actuation under 10 km/hr or in reverse; 8. Springs to keep the machine upright, in event of hydraulic failure or parking.

Courtesy: Vandenbrink b.v., www.carver.nl

(Figure 6.43). The control system is based on the premise that steering torque will be zero when executing a balanced, steady-state turn.* A valve, similar to the torque-sensitive valves in automotive power steering systems, controls hydraulic output to the roll actuators, so the vehicle's roll rate is proportional to applied steering torque. Additional controls hold the roll rate within predetermined values, and internal damping softens transitions in rate. The system also provides "artificial" lean-dependent steering feedback to the driver for a sense of cornering rate. Little or no delay exists between a steering/ leaning request and system response, which results in a very responsive machine.

Many TTWs have a mechanical roll axis that uses a structurally poor cantilevered bearing at the front of the nonleaning module. In contrast, the Carver features an up-and-over addition to the front structure, which connects to the pivot axis at both the front and rear of the nonleaning section. The production vehicle is shown in Figure 6.44.

---

* Author's note: This is not strictly true, but in general, the torque value will be low and will depend mainly on tire width.

FIGURE 6.44 Vandenbrink Carver. Vandenbrink's Carver is a top-quality, low-production tilting three-wheeler made in the Netherlands. Dimensions: Wheelbase, 2,600 mm; Length, 3,400 mm; Width, 1,200 mm; Weight empty, 550 kg.

Courtesy: Vandenbrink b.v., www.carver.nl

LIMITATIONS OF THE THREE-WHEEL CONFIGURATION

Although the ability to lean into turns enhances three-wheel vehicle rollover stability and handling characteristics, the layout does have inherent limitations. The nontilting three-wheel platform does not work well for high-profile vehicles or vehicles in which cg location might be significantly displaced by placing cargo at one end of the platform. Under certain conditions, it may not be desirable to rely on a single wheel for all of the traction at one end of the vehicle.

In terms of the traditional nontilting platform, cg location has a greater effect on the rollover stability of a three-wheel vehicle. The designer must therefore design for a stable and predictable cg location in order to ensure an adequate margin of safety against rollover. In contrast, a four-wheel vehicle is more tolerant of cg displacement. Consequently, if a large shift in cg can be expected, such as when placing heavy cargo at one end of the platform, the four-wheel layout is the preferred design. In addition, the option of moving the cg location closer to the side-by-side wheels is inherently limited. Hardware and occupants cannot all occupy the same space at one end of the vehicle. Load must be distributed along the wheelbase to some degree, if for no other reason than to provide room for components and occupants. Also, the single-wheel end of the vehicle cannot be too lightly loaded or the vehicle will become unstable because of inadequate traction.

Maneuvering at high speed over rough surfaces can destabilize a three-wheeler to a greater degree than a four-wheeler. A four-wheel vehicle relies on the traction of two wheels at each end to maintain directional stability. Irregularities in the road can cause transient losses of traction at the wheel encountering the disturbance. A critical loss of traction may occur more often at the single-wheel end of a three-wheel vehicle.

Limitations on suspension tuning are another consideration. One of the most effective tools for tuning a chassis is the ability to adjust the ratio of roll stiffness between the front and rear suspension. But the roll stiffness of a three-wheel layout can only come from the end with the side-by-side wheels. This limits the options for tuning the suspension primarily to that of varying the size and inflation pressure between the front and rear tires, and a large difference in tire size introduces the problem of selecting a spare tire that will work at either end of the vehicle.

Finally, blending with the existing infrastructure can be one of the most challenging problems of any unconventional product. If new solutions do not blend well, better technology often must give way to more marginal solutions simply because the lesser technology is more amenable to the infrastructure. Automobile manufacturing, transportation, and service facilities are designed for four-wheel vehicles. As a result, trailers, hoists and pits, alignment racks, and inspection stations often will not accommodate a vehicle with a wheel located in the center. Although fixtures for adapting to a single, centrally located wheel are easily designed, a vehicle with a wheel at each corner would need no such adapting. But designers must account for the inherent characteristics of any vehicle layout, regardless of the number of wheels. A correctly designed three-wheel vehicle can result in an exceptionally lightweight, highly fuel-efficient machine with superior handling characteristics and excellent rollover stability.

SPECIAL ACKNOWLEDGMENT

The author would like to extend a special note of gratitude to Tony Foale for contributing the material on tilting three-wheelers. He was educated in

Australia and graduated with a degree in electrical engineering, followed by a Master's in Engineering Science in nuclear engineering. He is the author of *Motorcycle Chassis Design: The Theory and Practice* and is well known in Europe for his theoretical work in developing mathematical models of motorcycle stability and dynamic behavior. Tony Foale resides in Spain where he consults on various computing and vehicular projects, specializing in chassis and suspension design. Tony also writes technical articles and is technical editor of the U.S. based "Motorcycle Consumer News." His website is located at: http://www.tonyfoale.com/.

REFERENCE

1. Thomas D. Gillespie, *Fundamentals of Vehicle Dynamics* (Warrendale, PA: SAE, 1992).

CHAPTER SEVEN

SAFETY AND
LOW-MASS VEHICLES

Courtesy: TRW Safety Systems, Inc., www.trw.com

Fooling around with alternating current is just a waste of time. Nobody will use it, ever. It's too dangerous—it could kill a man as quick as a bolt of lightning. Direct current is safe.

—Thomas Edison

The idea of introducing ultra-low-mass (ULM) cars into traffic with conventional cars brings up significant questions about small-car safety. In a two-car crash, the vehicle with the lowest mass receives the greatest blow, and matters get worse as mass differential increases. On a statistical basis, smaller cars also overturn more often as a result of an accident. In addition, the implications of low-mass vehicles go beyond the issues of vehicle crashworthiness alone. Concerns about product liability exist whenever unconventional vehicle designs are considered. Engineering and legal precedents have been established with conventional cars. No such precedents exist with unconventional vehicle types. As a result, personal injury involving a lightweight three-wheel car, for example, could offer rich new opportunities for exploitation both in and out of the courtroom.

Motor Vehicle Safety Standards that were developed for high-mass cars may be inappropriate for low-mass vehicles, especially with vehicle types that operate primarily in the urban environment. New vehicle types may require different standards. New standards specifically designed for ULM vehicles could provide engineering guidelines and help define accepted practices. Finally, designs for increasing low-mass vehicle safety must ultimately rely more on crash-avoidance countermeasures. Given equal technology, a low-mass car will likely remain the less crashworthy design. Crash-avoidance technology will put all cars on a more equal footing.

Vehicle Safety in Perspective

Automobile safety is an issue that is highly charged with emotion and complicated by the random and variable nature of the crash event. Crash tests measure the physical capabilities of the automobile structure to withstand a predefined event and protect the occupants inside. Accident statistics introduce the effects of the operating environment in combination with human behavior. Several factors associated with operating environment and human behavior determine the probability of an accident taking place, as well as the most likely magnitude and direction of forces, if an event occurs. Once an event does occur, vehicle crashworthiness and crash survival systems determine the likelihood and severity of injury. Change at any level can affect the overall safety of the transportation system.

Accident statistics are often used to extrapolate the effects of operating environment and human behavior, as well as to evaluate the effects on safety of particular vehicle design attributes; however, evaluating statistical relationships for specific meanings is a difficult and demanding task. A simple review of data often leads to inconclusive results. The true nature of the risk of driving in traffic with the varied mix of drivers, potential encounters, and vehicle designs is clouded by many interwoven factors that are difficult to separate from the whole. Concurrent events are often causally unrelated or related in ways that are not obvious. Accurate evaluations of risk and the factors influencing risk can easily become obscured by voluminous and often contradictory data.

Automobile safety is normally measured according to fatalities and injuries per distance traveled. In the United States, transportation safety statistics are compiled as injuries and fatalities per 100 million vehicle-miles traveled (vmt). Figures based on vmt (1 mi. = 1.609 km) provide a measure of how safely the vehicle moves about. It can be converted into passenger-miles traveled (pmt) by multiplying the average vehicle occupancy rate (1.6 persons in the U.S.) by vmt. Regardless of how it is measured, all forms of transportation are becoming safer. Statistically, driving today is approximately 10.5 times safer than in 1930 and 3.5 times safer than in 1970. Cars built today are nearly five times safer than vehicles built in 1969, and they are approximately 10 percent smaller and 20 percent lighter.*

Figure 7.2 shows the U.S. fatality rate of a few everyday activities, measured on the basis of fatalities per 100,000 population. On the basis of population only, the statistical relationship shows that bicycling is 7.5 times safer than walking, and motorcycles are nearly four times safer than walking. But many more people walk than ride a bicycle or a motorcycle, so fatality rates based on population do not actually reflect the relative safety of walking, bicycling, and motorcycling. Moreover, most pedestrian, bicycle, and motorcycle fatalities are, in fact, automobile accident fatalities. For example, 5,362 pedestrian fatalities in 1998 were caused by motor vehicle traffic accidents (including encounters with motorcycles), and only 957 were caused by other events. In regard to falls, nearly 70 percent of fall-related fatalities occur with those

---

* The apparent discrepancy between the safety level cited here and the one in the previous sentence has to do with the mix of older vehicles reflected in current accident statistics, as opposed to the independent safety level of new vehicles.

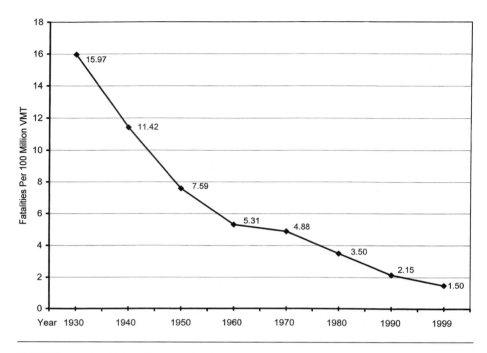

FIGURE 7.1 Traffic Fatalities 1930–1999

older than 70 years of age. Although children younger than 12 years old fall far more often than adults older than 70, falls are far more hazardous to older people.

So age turns out to be a significant factor in the hazards associated with falling. If these factors were not taken into account, any attempt to reduce the fatality rates of the associated activities would likely be ineffective. Most causal relationships in accident statistics are significantly more obscure. The relationship between cause and effect may lie deep within statistical data. Once accident data are deciphered, the next step would be to define an appropriate and realistic relationship among system costs, consumer convenience, and absolute system safety.

Several years ago I met a fellow who had purchased a homebuilt gyrocopter that wouldn't fly. The whole idea of such devices is to be able to soar with the eagles on a flea-market budget. For that benefit, one is obliged to accept a certain level of risk, which this prospective pilot was all too eager to do. He had tried several times, but the machine simply would not leave the ground.

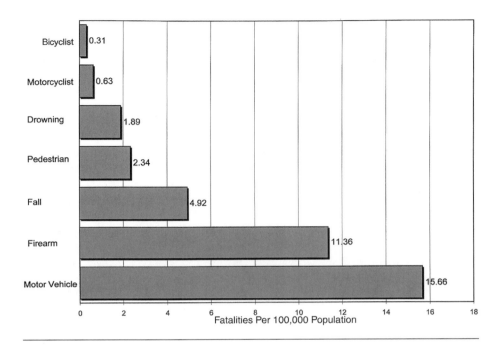

FIGURE 7.2 1998 U.S. Fatality Rate for Select Activities Per 100,000 Population

Credit: National Center for Injury Prevention and Control, www.cdc.gov

Over a period of perhaps an hour, he recounted a string of stories about ill-fated attempts to fly that all ended with the gyrocopter on its side in the nearest ditch. For me, it quickly became clear that here was a man who surely leaped before he looked. As it turned out, he had installed the rotor upside-down and overtightened the teeter-pin so the rotor could not pivot (the blade assembly could not flap). This guaranteed that the gyrocopter would remain forever earthbound. I remember thinking that his was probably the safest gyrocopter around, and perhaps he ought to leave the rotor as it was and keep his safety record intact.

The problems experienced by the gyrocopter pilot resulted first from a desire to fly, and second from a decision to circumvent the system and fly on a shoestring. Appropriate training and reliable equipment would have been expensive, and he was unwilling to pay the price or to give up the idea of flying. In reality he is not too different from the rest of us. An appropriate balance between cost and convenience, and the freedom to pursue activities, are relevant to all of our decisions, including those involving personal safety. There is always a level of risk associated with activities that may otherwise add to the quality of life. The fellow with the gyrocopter simply functioned closer to the outer boundaries of the envelope.

Improved technology has already significantly reduced automobile accident injury and fatality rates. With advanced crash survival systems and crash-avoidance technologies, it may soon be possible to eliminate most occupant fatalities resulting from automobile accidents, regardless of the size of the vehicle. Because of the differences in operating environments, occupant protection in low-mass urban cars may require different priorities and techniques than might be appropriate for larger highway cars. The per-passenger cost of safety equipment may grow as vehicle size shrinks. Ultra-low-mass cars may require a greater emphasis on safety engineering, as well as on technologies that can prevent vehicle crashes in the first place. As a result, a greater portion of first costs may be allocated to safety features in significantly smaller vehicles.

Given equal technology, the smaller car will probably continue to be the less crashworthy vehicle. It is not necessary, however, that low-mass cars actually equal the crashworthiness of larger cars in order to provide equally safe transportation. Operating environment has a large effect on the magnitude of forces that are typically experienced in an accident involvement. Vehicles that operate primarily in the urban environment are typically subjected to far less force when a collision occurs.

ACCIDENT STATISTICS AND AUTOMOBILE SIZE

In the United States, as well as in most of the world (except for Japan), automobile fatalities are inversely related to car size. That is, the smaller the vehicle, the greater the occupant fatality rate. The Insurance Institute for Highway Safety provides the following comparison of size versus occupant fatalities (Figure 7.3).[1]

In the early 1980s, A. C. Malliaris reviewed the statistical association of fatalities and car size and developed the following table to correlate risk increase in relation to reduced wheelbase and reduced vehicle mass.[2]

Figure 7.3 and Table 7.1 are based on simple associations between vehicle size and weight, and fatality risk. Focusing on different statistical associations reveals a more comprehensive, but perhaps more inconclusive, picture.

U.S. PASSENGER CAR FATALITIES PER 10,000 CARS

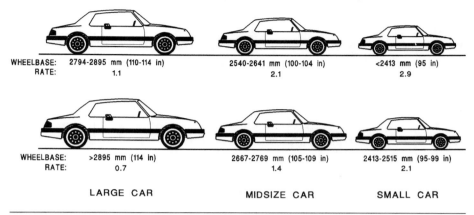

WHEELBASE:	2794-2895 mm (110-114 in)	2540-2641 mm (100-104 in)	<2413 mm (95 in)
RATE:	1.1	2.1	2.9

WHEELBASE:	>2895 mm (114 in)	2667-2769 mm (105-109 in)	2413-2515 mm (95-99 in)
RATE:	0.7	1.4	2.1

LARGE CAR	MIDSIZE CAR	SMALL CAR

FIGURE 7.3 Car Size Effect on Traffic Fatalities

TABLE 7.1 Occupant Fatality Risk Variation by Impact Type as a Function of Vehicle Weight and Wheelbase

Impact Type	Percent Risk Increase	
	Per 500-lb. (227-kg) Reduction in Curb Weight	**Per 5-in. (127-mm) Reduction in Wheelbase**
Frontal	16.6 ± 1.1	14.8 ± 0.9
Side	17.4 ± 1.1	14.9 ± 0.9
Rear	24.6 ± 2.0	21.8 ± 1.7
Other	32.3 ± 2.1	30.1 ± 1.6
All Types	18.6 ± 1.0	16.4 ± 0.8
	Per 500-lb. (227-kg) Reduction in Curb Weight at Fixed Wheelbase	**Per 5-in. (127-mm) Reduction in Wheelbase at Fixed Weight**
Frontal	−0.4 ± 3.2	15.1 ± 2.7
Side	5.3 ± 3.4	10.7 ± 2.8
Rear	1.6 ± 6.1	20.5 ± 5.2
Other	−14.6 ± 5.9	41.8 ± 5.0
All Types	0.9 ± 3.0	15.7 ± 2.5

Source: A. C. Milliaris, NHTSA Office of Vehicle Research, www.nhtsa.dot.gov

ACCIDENT STATISTICS AND OTHER VARIABLES

Automobile safety is statistically associated with a variety of conditions that are independent of vehicle type and design. For example, if data are controlled for driver age, large cars are often overrepresented in accidents.[3] It turns out that drivers younger than 20 years of age are "at fault" in fatal crashes at twice the rate per license holder of drivers aged 20 to 34, and five times the rate of drivers aged 35 to 64. These more youthful drivers are also overrepresented as the drivers of the smallest cars. Moreover, the percentage of teenage drivers increases in proportion to the decrease in car size.[4] Drivers age 16 and 17 represent only about 2 percent of licensed drivers but are involved in 11 percent of crashes. Drivers who are under the influence of alcohol are another major contributing factor in automobile accidents and fatalities. Alcohol is involved in 38 percent of all traffic fatalities. Many alcohol-related and driver-age–related fatalities may be incorrectly attributed to vehicle size.

Studies have suggested that drivers may adjust driving habits according to vehicle size. Wasielewski researched driver aggressiveness in 1983 and dis-

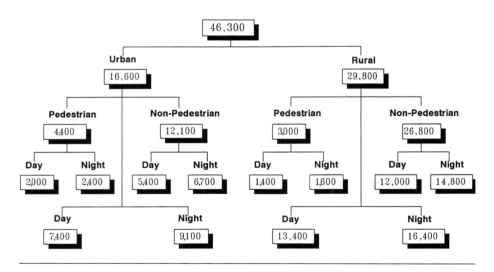

FIGURE 7.4 Total U.S. Traffic Fatalities. In 1990, rural and urban fatalities stood at 64 and 36 percent, respectively. By 1999, total fatalities had declined to 41,611, but the percentage relationship had varied only slightly to 61.5 and 38.5 percent, respectively, between rural and urban areas.

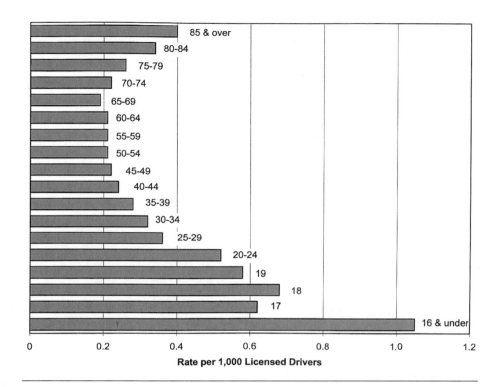

FIGURE 7.5 *Drivers Involved in Fatal Crashes by Age*

Credit: National Highway Traffic and Safety Administration 1996, www.nhtsa.dot.gov

covered that drivers of small cars take fewer risks in everyday driving.[5] Driver risk taking was measured by close following in heavy traffic, greater vehicle speeds on a two-lane road, and the number of drivers who did not wear seat belts. Wasielewski found that driver risk taking increased in proportion to vehicle size, with the drivers of the largest cars driving most aggressively. This finding appears to be supported by statistics on pedestrian traffic involvements, wherein large cars are responsible for the greatest number of fatalities. In the United States the smallest cars are involved in pedestrian deaths at about half the rate of the largest cars.[6]

Another consequence of increased large-car driver aggressiveness was pointed out by Sparrow in 1984. He noted that small cars are overrepresented as the struck vehicle in tow-away rear-end collisions and offered the data in Table 7.2.[7]

TABLE 7.2 NCSS Data on Rear-End Collisions

	Struck Car Size				
	Med/Large	Compact	Subcompact	Super Sport	Total
No. of rear collisions (5–7 o'clock position)	225	50	69	1	345
Total collisions	11,745	2,484	2,941	108	17,278
% Rear total	1.91%	2.01%	2.35%	—	1.99%

Source: Automotive Transportation Center, Purdue University, www. purdue.edu

Rural driving is far more hazardous to motorists. Given that an accident occurs, statistics indicate that occupants are approximately five times more likely to be killed on a rural road than on an urban road, even though most accidents occur in the city. Urban areas account for roughly three-quarters of all accidents but only about one-third of traffic-related fatalities. Rural driving is much more lethal at 26.4 percent of the accidents and 65 percent of

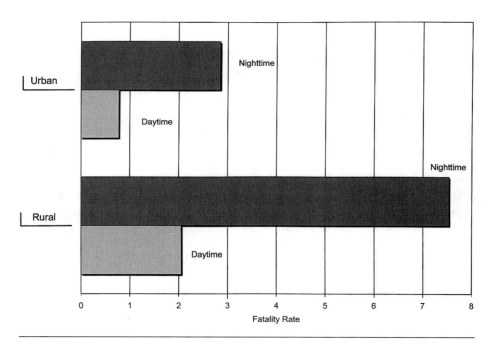

FIGURE 7.6 Urban and Rural Fatality Rate

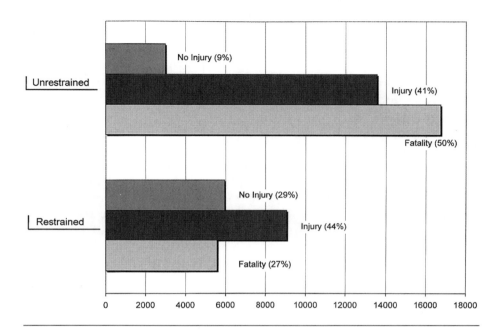

FIGURE 7.7 Effect of Three-Point Restraints on Injuries and Fatalities

the total fatalities.[8] Another interesting observation is the difference between day and nighttime driving. On the basis of vehicle miles traveled in the United States, urban commuting during daylight hours is 10 times safer than nighttime driving on a rural road, and urban commuting during the day is nearly four times safer than the same trip at night. This variable alone has a greater impact on occupant safety than the statistical association between vehicle size and fatality rates.

Where, when, and how a vehicle is operated has an enormous effect on the safety and well-being of the occupants. All cars today are equipped with seat belts, but many occupants do not wear them. Fatalities are approximately double for unrestrained occupants in a fatal automobile crash.

Many factors other than vehicle size and mass influence the level of driving hazard. Using the data from Table 7.1, a 450-kg (1,000-lb.) ULM car with a 2,032-mm (80-in.) wheelbase would be assigned nearly three times the risk of a car of 2,794-mm (110-in.) wheelbase and 1,589-kg (3,500-lb.) curb weight. Based on the foregoing statistics, daytime commuting in the 450-kg Malliaris-rated vehicle would therefore be safer than nighttime urban driving in a full-size car. By controlling driver age, for example, daytime driving in the hypothetical ULM car with a driver aged 34 years or older would be nearly twice as safe as the same trip in a full-size car with a teenage driver.

JAPANESE EXPERIENCE

In Japan the mix of vehicles by size category is much different than in the United States. The average car in Japan is about the size of a U.S. midsize car. The smallest Japanese cars are also much smaller than the smallest U.S. cars. The Japanese have established a special "Kei" (small) classification that includes vehicles with an engine of up to 660 cm^3 displacement and a curb weight of 590 kg (1,300 lb.) or less. Vehicles in this category are also referred to as the K-car. These microcars are restricted to a top speed of 80 km/h (50 mph). Fuel economy is on the order of 21–25 km/L (50–60 mpg).

The K-car has a safety record in Japan that runs contrary to the U.S. experience with small automobiles. When Sparrow researched K-car safety in the early 1980s, he found that they caused slightly fewer accidents and about 40 percent fewer fatalities per vehicle-kilometer traveled. On the basis of vehicle population, the K-car accident rate today is nearly 15 percent lower than the accident rate for regular-size cars in Japan; however, the accident rate is approximately 25 percent higher than it was when Sparrow researched the issue. The accident rate of regular-size cars in Japan is virtually identical to the 1982 rate.

Another characteristic that Sparrow noted had to do with driver aggressiveness. In general, the drivers of large cars were more aggressive, both with other cars and with pedestrians, and the drivers of small cars were less aggressive. Large-car aggressiveness and risk taking have been observed both in the United States and in Japan, but they are not well understood. Wasielewski documented that small-car drivers in the United States take fewer risks than drivers of larger cars.[9] He found that small-car drivers drive more slowly, wear their seat belts more often, and do not follow as closely as do the drivers of large cars.

The relative aggressiveness of drivers can be measured by several characteristics, but most significant, the more aggressive drivers are responsible for the greatest number of accidents. The magnitude of this characteristic becomes more significant when one considers the fact that younger Japanese drivers are the predominant users of the K-car, and, as in the United States, Japanese teenagers are considerably more accident prone with their automobiles.

TABLE 7.3 Japan Accident Rate and Driver Aggressiveness (K-Car Compared to Regular-Size Car)

	Population		Accidents		Caused by	
	Count	% Total	Events	Rate	Events	% Total
K Car:						
1990	2,060,591	6.3	29,958	0.015	16,951	56.6
1982*	2,065,886	8.4	24,582	0.012	14,678	59.7
Std. Car:						
1990	30,877,222	93.7	556,447	0.018	343,356	61.7
1982*	22,512,638	91.6	392,890	0.017	258,381	65.8

Source: Japan Ministry of Transport, www.mlit.go.jp/english/
*1982 figures taken from Sparrow (1985).

When an accident does occur, the occupant of the K-car in Japan has a slightly better chance of surviving than the occupant of a regular-size car. This finding strongly contradicts the U.S. experience where the occupant of the smallest U.S. vehicle is approximately four times less likely to survive than the occupant of the largest vehicle. Injury rate, however, is higher in K-cars. The increase in K-car injury rate is probably caused by the vehicle's reduced crashworthiness, while the decreased fatality rate may be the result of speed restrictions. Interestingly, Traffic Bureau statistics indicate that in 1982 approximately the same percentage of small- and large-car drivers who were injured in an accident were not wearing their seat belts (96 percent in both cases).

The pedestrian is always the loser in any encounter with an automobile. Although the K-car operates primarily in more congested urban areas where

TABLE 7.4 Japan Accident and Injury Rate by Size of Car

	Vehicles				Percentage of Accidents Resulting in Injury by Level of Seriousness					
	Population	%	Accidents	Rate	Fatal	%	Serious	%	Slight	%
K-Car	2,060,591	6.3	29,958	0.015	147	0.49	1,127	3.76	17,784	59.4
Regular Car	30,877,222	93.7	556,447	0.018	2,831	0.50	15,612	2.80	277,060	49.8

Source: Japan Ministry of Transport, www.mlit.go.jp/english/

TABLE 7.5 Pedestrian/Automobile Accidents versus Fatalities

	Accidents	Rate	Fatalities	% of Accidents
K-Car	2,202	0.0010	53	2.4
Regular Car	41,130	0.0013	1,533	3.7

Source: Japan Ministry of Transport, www.mlit.go.jp/english/

it is more exposed to pedestrian/automobile encounters, such encounters occur nearly 25 percent less often than with larger vehicles. When an accident does occur, the pedestrian has a 1.5 times greater chance of surviving an encounter with a K-car. Therefore, given an equal number of exposures, the larger car is twice as life-threatening to pedestrians.

The Japanese have achieved a remarkable safety record with very lightweight vehicles, and much can be learned from their experience. Legally restricting the speed of all low-mass vehicles, as they do in Japan, may not feasible in North America and Europe, however. For residents of many of the world's metropolitan areas, freeway/expressway driving is a necessary part of commuting. A special-purpose commuter car must therefore be capable of driving the freeways at the legal speed limit. Urban cars by definition operate at lower speeds. A separation of vehicle types, along with distinctions between the most effective strategies for safety engineering, may therefore be appropriate.

The Japanese experience with K-cars indicates that very small cars can be safely merged with larger vehicles, even though the smaller vehicles may be less crashworthy. If small urban cars were designed to protect the occupant during the most likely urban crash events, occupant safety could be optimized. Commuter cars designed for freeway driving obviously require more sophisticated crash survival technology.

SMALLER CARS REQUIRE BETTER SAFETY ENGINEERING

Accident statistics may indicate strategies in addition to improved vehicle crashworthiness that can reduce the casualty rate of smaller cars. Significantly smaller cars will benefit from lower urban speeds, and drivers appear to adopt defensive driving strategies with smaller vehicles. The difference between daytime and nighttime casualties may point to better lighting and lighter colors for smaller vehicles. Alarms or vehicle deactivation in the presence of alcohol might be mandated for significantly smaller vehicles. Special training and licensing for more youthful small-car drivers is another option. In addition, vehicle crashworthiness must be significantly improved.

Even with the reduced speeds in the urban environment, smaller cars will still experience greater deceleration forces in car-to-car collisions, especially during involvements with significantly larger vehicles. Moreover, freeway-traveling commuter cars will have the same exposure as larger cars, and the compounded risk caused by the smaller car's higher delta-v resultants cannot be ignored. A greater emphasis on vehicle safety engineering with strategies oriented toward the operating environment is therefore essential for smaller vehicles. Recent data indicate that improved vehicle design is already closing the gap between large- and small-car accident casualties. For example, statistics involving passenger cars of the 1976–1978 vintage indicate that cars weighing 900 kg (2,000 lb.) or less had a fatality rate that was 3.5 times higher than cars weighing 1,800–2,000 kg (4,000–4500 lb.). For cars of the 1986–1988 model year, the ratio is approximately 2.5 to 1.[10] In a report outlining options for improving automobile fuel economy, the U.S. Office of Technology Assessment (OTA) made the following observation about small car safety: "It seems clear that, were significant downsizing of the fleet to occur, a good portion, and perhaps all, of any resulting loss in safety could be balance by improvements in safety design."[11]

Given the different driver behavior patterns and the operating environments that may be characteristic of large, multipurpose cars in comparison with smaller, special-purpose vehicles, a less crashworthy urban car could have the better overall safety record. The entire system, of which the vehicle is one component, establishes the ultimate level of safety. Nevertheless, smaller cars are destined to crash, and when they do, the consequences are likely to be worse for the occupants unless countermeasures are designed into the vehicles. When an accident occurs, meaningful statistics are reduced to those associated with speed, impact vectors, vehicle crashworthiness, and crash survival technologies. Experience has shown that improved safety engineering produces a significantly greater reduction of harm in smaller cars than it does in larger cars.[12] With appropriate technology, it is possible to design lightweight vehicles that have greater occupant crash protection and greater overall safety than today's large cars. Safe small cars are technically feasible.

AUTOMOBILE CRASH DYNAMICS

Crash kinematics are primarily concerned with the forces of acceleration and the transfer of kinetic energy that take place during a collision. Designing a vehicle for occupant crash survival centers on the idea of managing the kinetic energy in a way that does not cause injury to those inside.

When a vehicle stops or changes direction, the occupant continues on in the same direction and at the same velocity as before the maneuver. Under normal driving conditions this does not cause problems because the resulting forces are minimal. Steering and braking forces normally cannot exceed the friction coefficient of tires, which might be on the order of 0.8. Consequently, the most abrupt maneuvers will rarely result in loads as high as 1 g. Unrestrained occupants can therefore change direction and speed along with the vehicle.

If the vehicle runs into an obstacle, loads imposed on occupants can rise dramatically. In this event, unrestrained occupants can no longer change speed and direction with the vehicle and, instead, continue to move within the interior space at the preimpact speed until colliding with one of the interior surfaces. In the case of a frontal impact, the most likely interior surfaces include the steering wheel, the windshield, and instrument panel structures. Collision with the interior of the automobile is responsible for most occupant injuries and fatalities related to crashes. Methods for preventing casualties involve various techniques for restraining occupants from ballistic movement within the vehicle, controlling occupant deceleration to keep loads within tolerable limits, and cushioning potential contact areas inside the vehicle.

The magnitude of the forces depend on the difference between the preimpact speed and the postimpact speed, and the time elapsed during the event. The forces of acceleration and deceleration are identical. Consequently, impact forces are usually referred to as the forces of acceleration even though the event is one of deceleration. The more rapidly the vehicle changes

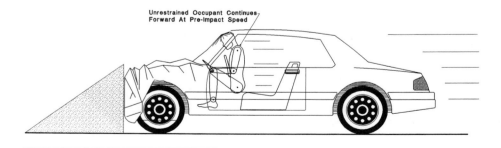

FIGURE 7.8 Occupant Continues to Travel inside the Vehicle

speed, the greater the (acceleration) forces and the more hazardous it is for the occupants. Fortunately, the passenger compartment of an impacting or impacted vehicle stops over a period of time, rather than instantaneously.

Even when a fixed object is impacted, the vehicle does not instantly stop. Instead, it is brought to rest over the distance of the vehicle crush zone. The crush zone provides part of the time/distance required to decelerate the occupants. The remaining ride-down space comes from the free space inside the passenger compartment. If the vehicle did not crush, it would stop much more rapidly, and greater forces would be transferred to the restraint system, requiring more ride-down space inside the vehicle in order to prevent injury. When an impact takes place with another vehicle, the kinetic energy of the striking and the struck vehicles combine to produce a resultant of forces, which are affected by the mass and the combined crush characteristics of both vehicles. Most collisions occur at an angle and with vehicle-to-vehicle contact areas that react differently to impact. Consequently, the dynamics of a two-vehicle crash are complicated by the random contact characteristics and the mass differentials between impacting vehicles.

In the United States, crash protection system performance is evaluated on the basis of the 30-mph (48-km/h) crash into a fixed barrier according to FMVSS 208. Most cars, however, run into another car at an angle and, therefore, experience forces that are much less than those produced by a 30-mph (48-km/h) Barrier Crash Test. Ninety percent of accidents involving fatalities occur at a barrier equivalent velocity less than that of the 30-mph (48-km/h) Barrier Crash Test. Barrier equivalent velocity is determined by the velocity required to produce equal damage if the vehicle were crashed into a fixed barrier, instead of the actual test event.

The most lethal impact—and one that will probably never have a high survival rate—is the direct head-on collision. Head-on collisions in the United States account for roughly 3 percent of all car-to-car involvements and 29 percent of all car-to-car events involving fatalities.[13] Fortunately, the largest percentage of head-on collisions happen at lower, and consequently more survivable, speeds in urban traffic. Rural collisions, which occur at greater speeds, account for 72 percent of fatal head-on, two-car involvements. A head-on collision produces extremely high forces, and a large-car/small-car involvement transfers the largest share of energy to the smaller car. Delta-v expresses the magnitude of the velocity change and serves as a measure of the difference in energy transfer between the two impacting vehicles.

At lower urban speeds, significantly improved survival rates are possible with improved safety engineering, and fatalities need not increase even if

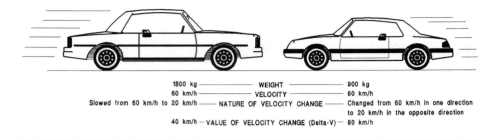

1800 kg	WEIGHT	900 kg
60 km/h	VELOCITY	60 km/h
Slowed from 60 km/h to 20 km/h	NATURE OF VELOCITY CHANGE	Changed from 60 km/h in one direction to 20 km/h in the opposite direction
40 km/h	VALUE OF VELOCITY CHANGE (Delta-V)	80 km/h

FIGURE 7.9 Head-On Collision between Large Car and Small Car

vehicles on the order of 450-kg (990-lb.) curb weight are introduced. According to Donald Friedman of Minicars Inc., with established techniques it is possible to achieve 80-km/h (50-mph) frontal and oblique barrier performance for all cars, and 160-km/h (100-mph) small-car/large-car oblique impacts without injury.[14] Friedman refers to conventional small cars, not to vehicles on the order of 450 kg (990 lbs.). Nevertheless, low-mass vehicles do not preclude safe vehicle designs.

OCCUPANT CRASH PROTECTION

Anyone who has watched an Indy car race has undoubtedly witnessed the results of state-of-the-art crash engineering. Essentially, the driver is strapped inside a rigid enclosure that is designed to remain intact while the rest of the vehicle is sacrificed to the process of dissipating kinetic energy. This technique accounts for the high oblique-impact survival rate at speeds on the order of 320 km/h (200 mph). Passenger car crash engineering is based on the same principle.

Items such as improved safety glass, recessed controls, frangible protrusions, and high-density padding contribute to the passenger car's safety. As a result, the interior of a modern automobile is a relatively safe environment, even for the unrestrained occupant, at speeds up to 25 km/h (15 mph). At greater speeds, other means are required in order to prevent injury.

Traditionally, the restraint system and the vehicle structure work together as a complete system to protect occupants during a crash.[15] A relatively rigid passenger compartment serves a purpose similar to that of the race car enclosure: It is designed for minimal deformation and to prevent intrusion into the occupant zone. The parts of the vehicle that lie outside the passen-

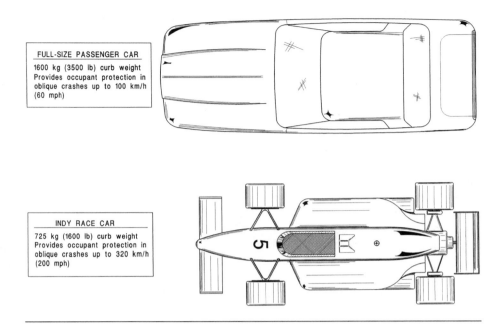

FULL-SIZE PASSENGER CAR

1600 kg (3500 lb) curb weight
Provides occupant protection in
oblique crashes up to 100 km/h
(60 mph)

INDY RACE CAR

725 kg (1600 lb) curb weight
Provides occupant protection in
oblique crashes up to 320 km/h
(200 mph)

FIGURE 7.10 Passenger Car and Race Car Share Crash Protection Strategy

ger compartment are designed to crush at a controlled rate in response to
impact forces. Restraints have an elongation rate (in the case of seat belts),
which functions to decelerate the occupant at a controlled rate while evenly
distributing loads over the body. Restraints also prevent ejection and collision with the interior of the vehicle. Saving occupants from injury therefore
depends on having enough total stroke or ride-down space (crushing of cars
and interior stroke space) to allow the occupants to decelerate over a
time/distance interval that will keep loads from exceeding the maximum tolerable level.

Stated in abbreviated terms of whole-body deceleration limits, the U.S.
National Highway Traffic and Safety Administration (NHTSA) has determined that the human body can tolerate acceleration loads up to a maximum of 60 g without injury, provided that loads are properly distributed.[16]
(A complete description of the load limitations specified for Barrier Crash
Test procedures are contained in the Code of Federal Regulations, Title 49,
Part 571-FMVSS and Part 572-ATD's.) This 60-g limitation establishes the
necessary vehicle crush rate and the performance characteristics of the
restraint system. If the occupant is decelerated over a greater time/distance,
loads will be reduced and the likelihood of injury caused by improper load

distribution or load spikes will also be reduced. If deceleration time/distance is reduced, loads will climb above the 60-g level and injury will become increasingly more severe.

In actual practice, it is difficult to achieve a uniform deceleration rate during an automobile collision. When a vehicle strikes an obstacle, it immediately begins to decelerate, but not at a uniform rate. Figure 7.16 shows the actual deceleration curve of a Ford sedan during a 60-km/h (38-mph) frontal barrier crash. When variations in vehicle deceleration result in occupant deceleration loads in excess of 60 g for longer than 3 milliseconds, the design fails to meet the NHTSA's performance requirements. Historically, vehicle restraint systems have been designed around a 30-g curve in order to account for deceleration variables and ensure barrier test performance.[17] Experience has shown that modifications to the vehicle structure can eliminate the extremes in deceleration and produce a much more uniform crush rate.

Upon impact, the occupant does not immediately begin to decelerate with the vehicle. Instead, the occupant continues on at the preimpact speed until

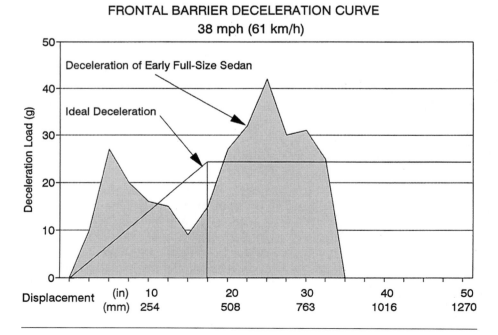

FIGURE 7.11 Load Spikes Cause Injury

the free-flight distance is taken up in the restraint system and the loads can reach the maximum level throughout the system. As a result, the amount of time elapsed between obstacle impact and maximum occupant deceleration varies depending on the onset rate, free-flight distance, restraint system characteristics, and position of the occupant at the time of impact. A smaller car has a smaller crush zone and is generally subjected to higher forces during a car-to-car collision. As a result, demands on the vehicle structure and the restraint system are greater. Smaller cars also have less interior space, which demands a more efficient use of the available time/distance. Optimal efficiency throughout the vehicle structure and restraint system is essential with smaller vehicles.

Figure 7.12 shows a uniform 50-g deceleration schedule. Ride-down schedules on the order of 50 g are technically feasible and tolerable, provided that the vehicle crush zone and the restraint system operate together at optimal efficiency.[18] In general, however, building a uniform crush rate into the vehicle structure, which must fulfill a variety of other requirements, is much more challenging than building the same degree of uniformity into the restraint system.

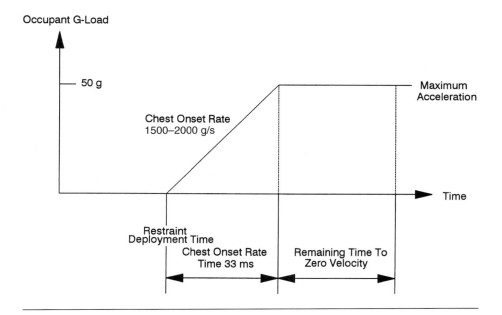

FIGURE 7.12 Load/Time Curve

Restraint Systems

State-of-the-art restraints include the mandatory three-point seat belt, air bags, and the knee bolster. After two decades of experimentation, it is generally agreed that a combination of approaches is necessary to provide the broadest possible protection.

Restraint systems are designed primarily to control loads imposed from the front. Because the vehicle is presumed to be traveling forward, collision with an obstacle will produce loads on the occupants that are predominantly opposite to the direction of travel. If, however, the vehicle is the struck, rather than the striking, vehicle, loads can be imposed from any direction. A review of fatal accidents in relation to the direction of the principal force reveals the degree to which frontal loads are the predominant cause of injury.

Forces from the 5, 6, and 7 o'clock directions are managed primarily by the seats and headrests. These forces are usually the product of relatively low barrier equivalent velocities, and seats should be designed to evenly transfer loads to the occupant and prevent head and neck injuries. Approximately 35 percent of the loads occur from one side or the other. Loads imposed from the 8, 9, 10, and 2, 3, and 4 o'clock directions are also predominantly at low barrier equivalent velocities. They are also much more difficult to protect against because of the minimal stroke available in the sideways direction. According to the National Crash Severity Study, the highest injury rates result from impacts toward the side of the struck vehicle, even though loads are the result of sideswipes and angular impacts, which may not transfer high energy values. Side intrusion members are designed to prevent intrusion into the occupant zone, and seat belts prevent ejection and ballistic movement within the vehicle. At high barrier equivalent velocities, however, protection is largely inadequate in the sideways direction. Impacts that result in loads from the frontal 60 degrees include the highest barrier equivalent velocities and result in the greatest number of fatalities (approximately 60 percent). An adequate vehicle crush zone and a properly designed restraint system can eliminate most fatalities resulting from these collisions, even with significantly smaller vehicles.

Seat Belts

The integrated three-point (lap/torso) safety belt with the emergency locking retractor is standard equipment on today's cars. A self-retracting reel allows relatively free movement while maintaining minimal tension on the belt.

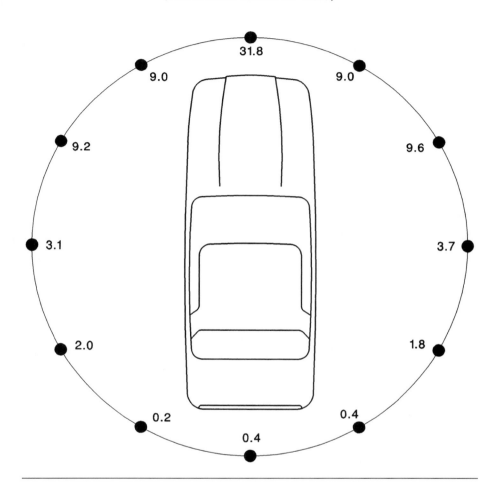

FIGURE 7.13 Fatalities Based on Principal Direction of Force

This lets the belt spool out as the occupant leans forward and then retracts upon return to a normal seating position. Retractors are equipped with inertial locks to prevent spool-out during a collision. Even mild braking forces are adequate to actuate the locking mechanism.

The three-point seat belt is an exceptionally effective device. Statistics show that wearing seat belts alone (without air bags) reduces the likelihood of serious or greater injury by 60 percent and the number of fatalities by 45 percent. Seat belt use also results in a 300 percent reduction in injury rates for those occupants who survive fatal vehicle crashes. After years of encouragement, and mandatory requirements on a state level, occupant seat belt use has risen from 11 percent in 1980 to 68 percent in 1996, but approximately 32 percent of automobile occupants still do not buckle up. Fatalities and injuries could be significantly lowered if seat belts were faithfully used.

Seat belts restrain occupants and prevent ejection or impact with the interior surfaces of the vehicle. A belt's elasticity also softens the transfer of energy to the occupant. A "soft" belt works somewhat like an air bag in that the system—either by belt elongation or by a force-limiting device—has a built-in elasticity that reduces the maximum load on the occupant. Modern seat belts also have characteristic problems that sometimes prevent performance from matching theory. The lap portion of the belt sometimes slips above the top of the pelvis (submarining) during a collision and can thereby produce serious internal injuries. In order to perform correctly, the lap belt must engage the pelvis during the event. Belts that are appropriately angled toward the floor usually perform correctly; however, improper placement or an out-of-position or undersize occupant can result in submarining during a collision.

In general, slack in the belt is the single major contributing factor in the failure of the seat belt to perform at maximum effectiveness. Slack is caused primarily by the portion of the belt that is loosely wound around the retractor reel. Consequently, the belt is essentially ineffective during the first few milliseconds after an impact as the buildup of forces locks the retractor and the forward movement of the occupant pulls the belt tightly around the spool. Forward movement unfortunately eliminates valuable interior stroke space that might otherwise provide room in which to decelerate the occupant. If seat belts were worn as tightly as those of race car drivers, belt slack would cease to be a problem and system efficiency would be much greater. Because a tight-fitting belt is not acceptable to consumers, an alternative approach is to leave the belt loose and tighten it immediately upon impact. Modern pretensioning devices utilize propellants or pretensioned springs that activate upon impact to tighten the take-up spool and eliminate belt slack. Spring-actuated pretensioners can remove 100 mm of slack in about 8 milliseconds.

A final problem has to do with the belt itself. The single narrow strap across the chest is limited in its ability to safely transfer loads to the occupant. At higher loads, the sternum and ribs have been broken by the torso restraint. Wider belts and even inflatable pads built into the belt (often referred to as the air belt) have been tried. An inflatable restraint developed by Simula and TRW appears destined for production. The device called an Inflatable Tubular Torso Restraint looks like a conventional three-point harness. During a crash, however, a flat tube inside the torso belt inflates to hold passengers snugly in place and provide a cushion against torso loads.

Air Bags

Beginning in 1998, all new passenger cars were required to be equipped with driver- and passenger-side air bags. Air bags were introduced some 25 years ago as the panacea of automobile accident injuries, but inflatable restraint systems have limitations. Air bags are designed to protect occupants during single-impact events that result in frontal loads across an arc of approximately 60 degrees. Forces from outside the arc may not activate the system, or the air bag may provide little or no protection if it is activated. Secondary impacts are also largely unprotected by inflatable restraints because of the system's rapid deflation. Actual deployment happens within 1/25th of a second. The entire deployment and deflation of an air bag are over within one second of impact.

Air bags are not stand-alone restraint systems. For the most effective protection, they must be part of a system that includes the three-point harness as the primary restraining device. In a direct frontal crash, air bags alone reduce fatalities by 35 percent. When all frontal crashes are considered (impacts from the 10 through 2 o'clock direction), air bags result in an 18

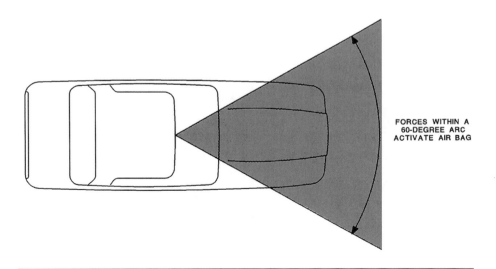

FORCES WITHIN A
60-DEGREE ARC
ACTIVATE AIR BAG

FIGURE 7.14 Activation Zone for Air Bags

percent reduction in fatalities. When all crashes are considered, regardless of impact direction, fatalities are down by about 11 percent when only air bags are used, but overall fatality risk is cut in half when air bags and three-point restraints are used together. On the basis of accidents involving moderate injuries, air bags alone result in an 18 percent reduction, three-point restraints alone result in a 40 percent reduction, and air bags in combination with three-point restraints result in a 60 percent overall reduction.

Correct air bag operation depends on a series of precisely timed events. The timing of these events (measured in milliseconds) is different in different applications. In general, a smaller vehicle with reduced interior space will require a system that works within a more condensed schedule. Consequently, a generic inflatable restraint system cannot be designed to work in a multitude of automobiles. Each system is designed according to the crash characteristics of the vehicle.

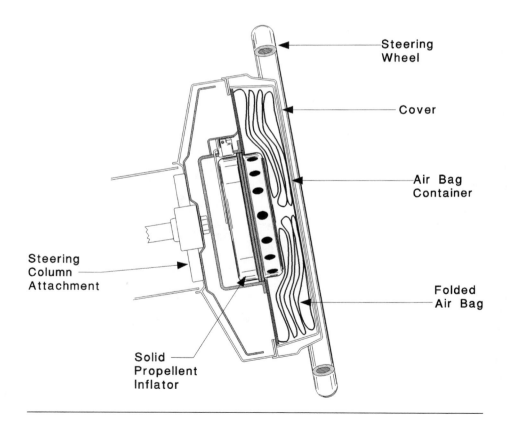

FIGURE 7.15 Driver's-Side Air Bag in Steering Wheel

The sequence of events begins when a sensor detects rapid deceleration and closes contacts to energize the propellant. A solid propellant then ignites and inflates the air bag. Forward occupant excursion is arrested by the inflating bag, which also prevents contact with the supporting structure. The air bag dissipates the energy to avoid undue rebound and then rapidly deflates. In a

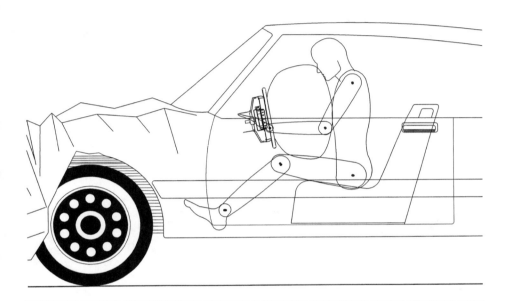

FIGURE 7.16 Vehicle Crush Zone and Air Bag Work as a System

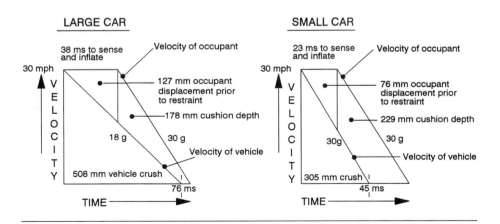

FIGURE 7.17 Large-Car/Small-Car Ride-Down Schedule

typical 30-mph (48-km/h) Barrier Crash Test, the entire sequence of events takes place within a window of approximately 150 milliseconds. Air bag deployment through the beginning of deflation occurs in as little as 30 milliseconds.

The smaller the vehicle, the less time is available in which to complete the operations because of the reduced crush distance and the more limited interior space of the smaller vehicle. During the crash event, time and distance are convertible, and the time/distance value is inversely related to the magnitude of the load. As the time/distance value decreases, the load experienced during the event becomes greater.

Stephen Goch of TRW Vehicle Safety Systems, Inc. described the following sequence of events for a large-car inflatable restraint system during a typical 30-mph (48-km/h) Barrier Crash Test:[19]

T_0:   Front face of bumper makes contact with the barrier and begins to decelerate the vehicle. The occupants continue to travel at 30 mph (48 km/h).

T_{10}:   An arming sensor located in the occupant compartment detects rapid deceleration and closes contacts. Electrical signal is then transmitted to the discriminating sensors wired in series and located in the engine compartment.

T_{15}:   The discriminating sensors detect rapid deceleration and close contacts. Electrical signal is then transmitted to the diagnostic module, where it is ultimately transmitted to the air bag modules.

T_{20}:   Initiators in air bag modules receive electrical impulse and ignite, initiating bag deployment.

T_{25}:   Air bag covers open and bags begin to fill.

T_{30}:   Retractors lock, and occupants begin to load the lap and shoulder portions of the three-point systems.

T_{50}:   The air bags fill and pressurize, and begin contributing to the restraining force already being provided by the belt systems to the occupants.

T_{60}:   Knee bolsters begin to contribute restraining force. On the driver's side, the steering column begins to absorb the additional energy of the occupant.

T_{90}:   The occupants reach maximum forward excursion and begin rebound.

T_{120}:   The vehicle is at maximum excursion into barrier and begins to rebound.

T_{150}:   The vehicle and occupants come to rest.

Intelligent Safety Systems

Despite their effect on reducing traffic injuries and fatalities, safety systems also have drawbacks. For example, the NHTSA has identified 97 crashes wherein the deployment of the passenger air bag resulted in fatal head or neck injuries to a child. In addition, two children have been killed by driver-side air bags. Eighteen of these deaths were to infants in rearward-facing child safety seats. Most of the other 81 children were determined to have been completely unbuckled, "out of position," or wearing only the lap portion of the seat belt (improperly restrained) at the time of the crash. Since 1990, 149 deaths have been attributed to air bags that deployed in low-speed crashes. Close proximity to the air bag module at the moment of deployment is the greatest source of risk. This happens with drivers who sit too close to the steering wheel (either out of necessity or habit), children in rearward-facing infant seats, and children who are permitted to travel unrestrained. The necessarily high force of deployment is responsible for the hazard.

One of the greatest challenges of air bag design comes from the limited time within which impact detection and deployment must occur. Because of the physics involved, air bags must be designed for a particular mass located a given distance from the air bag module. For safe deployment, occupants should be no closer than 250 mm (10 in.) from the air bag at the moment of deployment. First-generation systems cannot detect occupants who may be too close, nor can they vary response intensity according to the mass of the occupant. As a result, the force of deployment can actually cause injury and even death to out-of-position occupants and small children.

Side-impact systems are often more challenging to design because of the limited time/distance available in which to detect and deploy. Consider that in a side-impact collision there may be as little as 250 mm (10 in.) between the occupant and the impacting vehicle at the time of actual impact. If the impacting vehicle were detected at two or three meters distance, valuable time would be gained during which systems could be deployed.

Air bag performance can be significantly improved through pre-event sensing and variable deployment schemes, which are designed to monitor variables such as occupant weight and position and then adjust air bag deployment accordingly. Lighter occupants require less inflation force, and out-of-position occupants may make it inadvisable to deploy air bags at all. Sensors can also detect an impending collision and begin deployment before the actual event occurs. By sensing an event in advance, deployment can begin early and with less force, which translates into more effective protection and a reduced potential for injury. Intelligent air bag systems can significantly increase effectiveness and reduce the potential for injury from the system.

CRASH MANAGEMENT STRATEGY FOR LOW-MASS VEHICLES

Low-mass vehicles can be designed to fulfill ECE and U.S. Barrier Crash Test requirements. Differences between low-mass and high-mass vehicle crash management strategies have to do with the condensed time budget and different methods of providing occupant ride-down space in the low-mass car. A typical barrier crash of a 450-kg (990-lb.) low-mass car will be a condensed event resulting in relatively high loads. Consequently, efficiency throughout the crash protection system must be optimized in order to prevent injury. Load spikes caused by a nonuniform deformation rate must be minimized so that occupant deceleration can be carried out at maximum loads without crossing into the injury zone.

Studies at the Swiss Federal Institute of Technology suggest that ride-down space in low-mass vehicles is best provided by increased free space inside the passenger compartment, rather than by relying on an exterior deformation zone.[20] Conventional cars are designed with an exterior deformation zone in combination with a rigid passenger compartment. Stroke, or ride-down space, comes partially from vehicle deformation and partially from free space inside the passenger compartment. Studies suggest that low-mass vehicles might be designed with an exterior that is largely identical to the rigid passenger compartment of a conventional car. Occupants would be decelerated by air bags and elastic restraints within the vehicle. The more rigid exterior of the low-mass vehicle would cause the less rigid deformation zone of the larger vehicle to yield during a collision. Inside, the low-mass car would be relatively open, smoothly contoured, and padded. Advanced vehicle control systems could eliminate the conventional steering column and pedal controls.[21]

Horlacher AG, Mohlin, Switzerland, has developed several promising low-mass vehicle designs that incorporate an impact belt to improve vehicle crashworthiness. The company has also provided vehicles for crash testing and evaluation by the Zurich Institute of Technology in cooperation with the Swiss Institute of Forensic Medicine and the Department of Accident Research at Winterthur Insurance Company in Switzerland. Tests indicate that Horlacher's "hard shell" concept, in which the vehicle is built for minimal deformation during a collision, provides maximum protection and best utilizes the crush zone of the high-mass vehicle. Barrier test results for Horlacher's design and two other low-mass vehicles—one unmodified and the

TABLE 7.6 ECE Barrier Crash Test of Low-mass Vehicles with Eurosid1-dummy

	Solec 90 (unmodified)	Solec 90 (reinforced)	Horlacher "Impact Belt"
Impact velocity	11 m/sec	11.2 m/sec	9.3 m/sec
Vehicle mass	621 kg	684 kg	552 kg
Head acceleration (3 ms)	128 g	51 g	45 g
Head injury criterion	2230	421	292
Max. floor acceleration	351 g	86 g	84 g
Max. impact deformation	340 mm	298 mm	138 mm
Remaining deformation	310 mm	230 mm	20 mm
Mean force during high energy absorption	160 kN	260 kN	250 kN

Source: Horlacher AG, Mohlin, Switzerland, www.horlacher.com

other modified structurally by the addition of a steering wheel air bag—are shown in Table 7.6.[22]

Lightweight construction and occupant safety are not necessarily mutually exclusive concepts. Low-mass vehicles utilizing improved materials and restraint systems, in conjunction with a more rigid exterior, can provide a comparatively safe environment for occupants during a crash. Crash tests were also conducted with a hard-shell Horlacher vehicle sent head-on against an Audi 100 of approximately twice the mass. Photographs of the crash dramatically illustrate that a small-car/large-car involvement need not result in a demolished small car. With the hard-shell concept, ride-down space comes from the vehicle's interior. Occupant deceleration is controlled by the restraint system.

A basically hard-shell, low-mass vehicle can also be equipped with a relatively small exterior deformation zone, perhaps on the order of 300 mm (12 in.). Interior ride-down space and/or deceleration loads might thereby be reduced. Passenger-side stroke is available from the typically large free space between the occupant and the dash. Driver-side stroke might include the controlled forward displacement of a more conventional steering assembly.* There are a variety of ways in which ride-down space can be provided. The

---

* Side-by-side seating is assumed. A tandem arrangement introduces interesting possibilities for protecting the rear passenger with the back of the driver's seat.

FIGURE 7.18 Crash Test of Horlacher Hard-Shell Vehicle. A hard-shell Horlacher impacting an Audi 100 head-on. Horlacher speed is 52 km/h (32 mph). Audi speed is 26 km/h (16 mph).

Courtesy: Horlacher AG, www.horlacher.com

FIGURE 7.19 Horlacher Hard-Shell Vehicle after Crash Test. Horlacher after the crash. Passenger compartment is intact. Windshield damage was caused by a crash recorder that broke loose.

Courtesy: Horlacher AG, www.horlacher.com

FIGURE 7.20 Audi 100 after Crash Test. The Audi 100 after the crash.

Courtesy: Horlacher AG, www.horlacher.com

percentage that comes from vehicle excursion into the barrier, in relation to the percentage that comes from the occupant's forward excursion inside the passenger compartment, is unimportant. Taken as a sum, the total stroke represents the total working space (time/distance) available in which to bring the occupant to rest uninjured. The primary consideration would be the degree to which the engineer deems it necessary to rely on the larger vehicle's crush zone to decelerate the smaller vehicle. A 30-mph Barrier Crash Test of a hypothetical low-mass vehicle is shown in Figures 7.21 and 7.22.

The hypothetical crash is essentially the same as a large-car barrier crash except that events take place within a more condensed time budget. Such a condensed schedule is technically feasible. Rapid-deployment air bags have been demonstrated, and application techniques are outlined in available literature on air bag technologies. The hazards of rapid-deployment systems primarily involve the potential for injury to out-of-position occupants during rapid inflation.[23] Different vehicle standards for low-mass cars could resolve such difficulties by mandating designs that keep occupants in place.

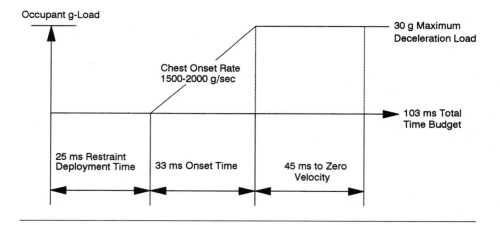

FIGURE 7.21 *Hypothetical Low-Mass Vehicle Deceleration Time Budget*

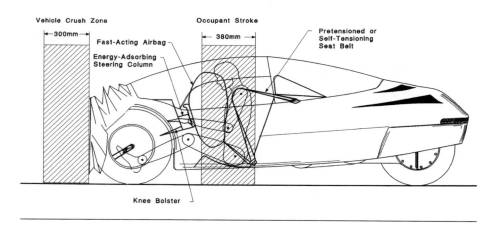

FIGURE 7.22 *Hypothetical Low-Mass Vehicle 30-mph Barrier Crash*

ADVANCED SYSTEMS FOR IMPROVING AUTOMOBILE SAFETY

Several years ago at Quincy-Lynn Enterprises, an envelope arrived containing a drawing that illustrated a rather humorous idea for surviving automobile crashes. Occupants were strapped into a spherical compartment, which rested in a receptacle in the vehicle on top of a large shoehorn-like device. The shoehorn handle extended forward so it protruded at the front of the vehicle, where it would be the first item to make contact in a frontal collision. Impact would cause the shoehorn to pivot and catapult the occupant sphere bodily away from the crash event. An elastic umbilical cord then arrested the sphere's ballistic flight to keep the occupants from being catapulted out of the neighborhood! Even now, the idea brings a chuckle whenever it comes to mind. But the image of modern automobiles careening out of control, sacrificing pieces of their structure as occupants are slammed against restraints at bone-crushing loads, seems all too primitive and perhaps not much different from the bizarre idea proposed in the drawing. Crash survival technology is but an interim step toward the much more civilized goal of avoiding automobile crashes in the first place. The potential to avoid vehicle crashes exists within the technologies and subsystems that are being developed under the Intelligent Transportation Systems (ITS, formerly Intelligent Vehicle and Highway System [IVHS]) program.

Several programs worldwide are underway to develop and deploy ITS. Intelligent Transport Systems is a multilayered assemblage of subsystems designed to provide a variety of traffic management and advisory functions, along with trip navigation, vehicle location, and ultimately vehicle guidance and control on automated highways. Deployment of select subsystems is already underway and will be expanded as technology matures. The Automated Highway System (AHS) is a central component. The AHS allows cars to travel along highways in platoons of 10 to 20 vehicles. Vehicles are under the control of computers, and cars are guided within special lanes at a set speed in tightly packed groups with approximately three meters headway. Reduced headway and improved traffic management are expected to produce a 300 percent increase in highway capacity (see Chapter Eight).

Collision avoidance is a central benefit of ITS. Several stand-alone subsystems and integral technologies have the potential to significantly reduce automobile collisions. In combination with Advanced Vehicle Control Systems (AVCS), new technologies could virtually eliminate automobile collisions altogether. Neither ITS nor AVCS are particular technologies, but rather several technologies that can be applied in various ways to improve road conditions and reduce driving hazards. Technologies for improving vehicle safety include the following:

- Obstacle detection
- Blind zone detection
- Forward-looking radar
- Driver-condition monitoring
- Headway control
- Collision detection (warning and active)
- Head-up display
- Voice activation

- Automatic lane change
- Automatic braking

Electronic safety features are already here. Technologies such as the antilock braking system (ABS) and active suspension systems are commonplace on production vehicles. Versions of obstacle detection and warning systems are already being offered in the form of backup warning devices. Such systems continuously monitor near-vehicle obstacles in side and rear blind spots. Early versions can provide head-up icon display and/or audible warning upon initiation of lane change or when shifting into reverse. Headway control is already showing up on production cars as part of adaptive cruise control. Adaptive cruise control detects a vehicle ahead and adjusts speed as needed to maintain a safe headway. Figure 7.23 illustrates near-vehicle surveillance using radar.

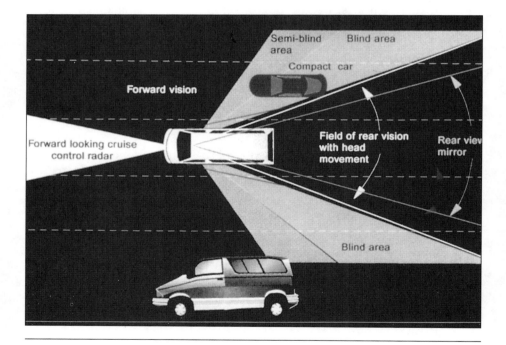

FIGURE 7.23 Near-Vehicle Surveillance Systems. TRW's radar-based near-vehicle surveillance systems can provide driver warning, or when integrated with AVCS, take control of the vehicle and steer around an obstacle or brake to avoid it.

Courtesy: TRW Safety Systems, www.trw.com

Advanced systems will be able to detect an impending collision and steer around an obstacle or brake to avoid it. Systems that can see through fog and past headlight illumination range and project icons or virtual images that appear to be located ahead of the vehicle already exist. Ultimately, systems will be integrated into AVCS to automatically brake or steer the vehicle away from an impending impact. Figure 7.24 shows an icon projection system

FIGURE 7.24 Ford's Night Vision Icon Projection System. A system by Ford can see through rain and fog and project icons ahead of the vehicle in the driver's field of vision to locate obscured objects.

Courtesy: Ford Motor Company, www.ford.com

developed by Ford for seeing through fog and other conditions of degraded visibility.

Sensors for obstacle and blind-zone detection are relatively inexpensive and compact. A 180-degree field of vision for intersection surveillance is also possible but requires more sophisticated technology. In order to see past traffic, the sensor must be located high on the vehicle or even above it. Looking around corners requires stationary sensors located at intersections. Again, early versions may be limited to head-up icon display and audible warning.

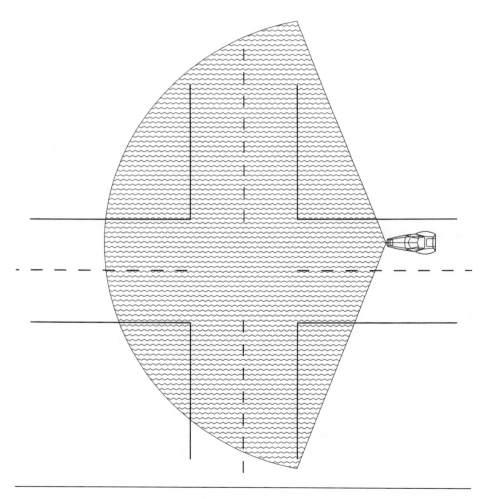

FIGURE 7.25 Collision Detection and Avoidance at Intersections

Later versions will brake or steer the vehicle to avoid an imminent collision. The lifesaving potential of collision-avoidance systems becomes apparent when one considers that approximately 90 percent of automobile accidents are caused by human error. Systems that warn or intervene to avoid collisions could significantly reduce driver-error accidents.

Driver monitoring is another technology that has far-reaching implications. A system developed by Nissan can detect driver fatigue by monitoring steering inputs.[24] Distinctive steering patterns are produced by the fatigued driver, and a continuous state of zero input is characteristic of drivers who have dozed off at the wheel. An audible warning could alert the fatigued driver, and automatic braking could stop the vehicle if the driver has dozed off. It is also possible to remotely monitor electroencephalograms (EEGs), heart rate, and body temperature with sensors in the steering wheel. In the event of a medical emergency (e.g., a heart attack), the system could take control of the vehicle and bring it safely to a stop. It could also automatically send a signal to notify an emergency medical center. Drivers who have been drinking alcohol might also be sensed and warned, or the vehicle could be deactivated if the operator is intoxicated. This feature alone has the potential to eliminate roughly half of the fatalities caused by automobile accidents.

Much of the technology for active collision avoidance has already been developed, but liability issues remain to be solved before this type of technology can be put to work saving lives. Liability is much less with warning devices, but when a system intervenes and takes over control of the automobile, manufacturers will face liability in any instance where the system fails to perform as intended, regardless of the number of lives that may have otherwise been saved. Today, legal issues are standing in the way of preventing most highway injuries and deaths.

NEW STANDARDS FOR NEW VEHICLE TYPES

In the past, there have been discussions about the advisability of changing the Code to include a blanket exemption for all vehicles under a specified weight (450 kg or 1,000 lbs., for example); however, the NHTSA has consistently resisted the idea of creating a categorical exemption based on vehicle weight, and instead has issued limited exemptions to low-volume specialty vehicle manufacturers in certain instances. But new vehicle types are likely to introduce new problems that must be offset by new solutions. Neither a blanket exemption nor an individual exemption addresses the question of how unconventional, low-mass vehicles might be safely integrated into traffic.

Ultimately, a new category may have to be created to account for the special considerations of unconventional vehicle types.

Crash management systems for sub-cars and extremely low-mass urban cars and commuter cars may require different approaches, both in relation to conventional passenger cars and in relation to each other. Open-bodied vehicles in the sub-car category may be so varied in design as to make consistent safety standards difficult to develop. Consequently, manufacturers might be required to demonstrate "reasonable precautions" in vehicle safety engineering. The NHTSA might then individually evaluate each design and provide certification on a case-by-case basis. Standards for three- and four-wheel open-bodied designs could specify occupant restraints, roll bars, protective side members, and high-density padding at points of potential contact with the vehicle structure; however, a certain level of risk is probably unavoidable and might therefore be accepted under the "obvious and voluntary" rationale that the NHTSA cites in safety issues involving motorcycles.

As for urban cars, the present barrier test may not be appropriate for a vehicle that operates primarily in the urban environment.* Within all fatal or disabling accidents (including both urban and rural involvements), 97 percent involve an angular impact, and a full 90 percent occur at speeds below a 48-km/h (30-mph) barrier equivalent velocity.[25] The (30 mph) barrier crash is a murderous event that equates to a real-world head-on crash between two cars that are each traveling at 48 km/h (30 mph), which translates into a 96-km/hr (60-mph) closing rate. Cars designed to be less dangerous in such a crash may actually be more dangerous in the far more common low-speed, angular-impact urban event. Average speeds are substantially lower in the urban environment. Consequently, current standards may deserve reevaluation as they apply to urban cars.

Electric cars, most likely urban cars, present unique problems. High-mass batteries in an otherwise low-mass vehicle require additional structural considerations. Batteries should be retained and protected within the vehicle during a crash. If batteries break away, vehicle mass will suddenly decrease, and the kinematics will no longer be the same. Also, occupants must be protected from spilled electrolyte. Runaway electrical fires, hydrogen explosions, and electrical shocks are additional hazards. Although the technology for containing electrolyte and preventing electrical shocks, fires, and explosions is reasonably straightforward, new standards for electric urban cars may be necessary.

---

* Urban cars are envisioned as restricted to surface streets. A commuter car is assumed to be a vehicle of unrestricted use. Therefore, it will likely operate in other than an urban environment.

Urban cars might be restricted to surface streets and 80 km/h (50 mph) in a manner similar to the K-car in Japan. Commuter cars can have no such restrictions because of the freeway operating environment in the United States. Additionally, electronic crash warning and avoidance technology could be mandated according to existing technology, with systems emphasis given according to conditions of the most likely operating environment. Rollover accidents tend to increase in relation to the decrease in vehicle wheelbase, but are less affected by reductions in vehicle mass. Rollover accidents are also significantly more common in rural areas than in cities, primarily because of greater speeds on rural roads. The NHTSA estimates that the vehicle is out of control in 50 to 80 percent of rollover accidents and that the risk of ejection is 3.5 times greater in fatal single-vehicle rollover accidents.

Countermeasures might therefore concentrate on preventing ejection and loss of control, as well as improving the dynamic margin of safety against rollover. Vehicles with ABS have a lower incidence of loss of control. A mandated margin of safety against rollover may reduce the potential incidence of such accidents involving commuter cars and three-wheel cars. In the 1990s, the NHTSA reviewed several crash-avoidance and crashworthiness options, including the idea of a mandated margin of safety against rollover. Tilt-table tests and side-pull tests with minimum stability requirements were considered. Similar tests were also reviewed in Europe.

Because of design limitations with high-profile vehicles, the NHTSA decided against rollover standards, and instead opted to require labeling on vehicles that are at higher risk of rollover. Bright-colored labels must now be placed in vans and SUVs and must read: "WARNING: Higher Rollover Risk," and "Avoid Abrupt Maneuvers and Excessive Speed." The NHTSA also provides consumers with information on vehicle rollover resistance. Rollover labeling might be appropriate for ULM cars of significantly reduced wheelbase.

Other options for commuter cars might include improved restraints, better headrests, and mandated electronic crash-avoidance systems. A five-point harness with a torso web would improve occupant survival. Headrests are typically too far from the head and do not provide support for the neck. Low-mass vehicles tend to be followed more closely, struck from the rear more often, and are subjected to higher deceleration forces when struck. Requirements for headrest designs might therefore be more stringent. In addition, close-following drivers might be warned by flashing taillights. Information from a rearward-looking sensor could determine when a closing-rate/speed/headway threshold is breached. Better engineering and a different orientation in standards can compensate for many of the disadvantages that result from reduced vehicle mass. Emphasizing improved technology in low-mass vehicles has the potential to largely offset the hazards associated with reduced vehicle mass.

Three-Wheel Cars

Three-wheelers are now classified as motorcycles in the United States, which essentially places them outside the domain of meaningful safety standards; however, three-wheel cars may be more deserving, rather than less, of carefully developed engineering guidelines. A dynamic margin of safety against rollover is not addressed in existing standards, and it may be necessary if three-wheel cars are to be safely integrated into traffic. In addition, the potential for longitudinal center-of-gravity displacement in three-wheel cars must be considered in developing appropriate standards. The idea of placing three-wheel vehicles in the passenger car classification may not be appropriate either. Three-wheelers require special engineering considerations. A brief review of the regulatory history and the prevailing attitudes toward motorcycle/three-wheeler safety will help bring three-wheeler classification into perspective and perhaps clarify the more broad issue of special standards versus blanket exemptions for unconventional, low-mass vehicles.

In the late 1960s, the U.S. Code of Federal Regulations contained an exemption for all four-wheel cars of less than 1,000-lb. (454-kg) curb weight. The exemption was removed in 1973 because on closer examination it appeared that smaller cars actually have a greater need for safety features than do larger cars. At the same time, the NHTSA also proposed to redefine the motorcycle classification to exclude three-wheel vehicles that have essentially the same end use as automobiles (i.e., three-wheel vehicles that are built with some protective structure resembling a body). Through a series of legislative false starts, intense debate, and industry resistance, three-wheel vehicles have not been excluded from the motorcycle definition as was originally intended. Today a motorcycle is still any vehicle "having a seat or saddle for the use of the rider, and designed to travel with not more than three wheels in contact with the ground."

Over the years, the issue of three-wheel vehicle classification has been closely associated with arguments in favor of a blanket exemption for low-mass, four-wheel cars. According to the argument, if motorcycles are allowed, and three-wheelers can operate under the motorcycle classification, it does not make sense to ban ultralight four-wheelers, just because they may not be able to meet passenger car safety standards. Ultralight four-wheel cars would surely be safer than motorcycles, and perhaps even safer than certain three-wheel designs that are presently allowed under the motorcycle classification. The NHTSA affirms that the idea of a blanket exemption is not valid because of the concept of "obvious and voluntary" risk.

The difference in risk types is the pivotal distinction behind the decision to allow motorcycles on the roadways, relatively unregulated, while regulating inherently safer automobiles. According to this reasoning, people are entitled to take voluntary risks such as climbing a mountain, sky diving, or riding a motorcycle. These activities carry a certain level of "obvious" risk, and the person who participates in them is presumed to "voluntarily" accept that risk. Involuntary risks have to do with using ordinary household appliances or driving automobiles. The risks involved in these activities may not be as obvious and, therefore, their safety must be more carefully regulated. Automobiles are equipped with a substantial body that might lead one to believe that they are safer than they really are. Motorcycles, on the other hand, are obviously hazardous and, therefore, people are not likely to be misled into believing that they are protected from injury when, in fact, they are not.

Arguments based on obvious and voluntary risk do not support the idea of placing full-bodied, three-wheel vehicles in the motorcycle category. Quite the contrary, such arguments seem to imply that three-wheel vehicles should be subject to appropriate safety standards. At the same time, it seems clear that three-wheel cars do not belong in the passenger car category. As discussed earlier, three-wheel vehicles require unique engineering and safety considerations that are not addressed by existing passenger car safety standards. Moreover, new standards that address the unique safety implications of unconventional vehicle types are probably necessary if extremely low-mass vehicles, of both three and four wheels, are to match the safety record of larger cars. In the past, debates have often focused on exemptions; however, exemptions do not really address the issue.

The final issue concerns product liability. Without appropriate regulations, the most qualified manufacturers may actually be discouraged from marketing three-wheelers, or other unconventional vehicle designs, in order to avoid potential liability difficulties. On the one hand, manufacturers are reluctant to say that marketing decisions are subject to legal intimidation. On the other hand, the realities of product liability are certainly relevant. Potential product liability problems were an important consideration in the decision by Harley-Davidson to shelve the three-wheel Trihawk project, even though the vehicle had excellent handling characteristics and the market appeared strong. Regulations that adequately address the safety implications of ULM and unconventional vehicle types could open the way for manufacturers to develop new, environmentally friendly, and energy-efficient vehicle types. In this regard, appropriate safety regulations for low-mass cars deserve a thoughtful and thorough review.

REFERENCES

1. "Shopping for a Safer Car," Insurance Institute for Highway Safety.

2. A. C. Malliaris, "Discerning the State of Crashworthiness in the Accident Experience," Office of Vehicle Research, NHTSA, Paper presented at the 10th International Conference on Experimental Safety Vehicles, DOT.

3. L. Evans, "Accident Involvement Rate and Car Size," General Motors Research Laboratories, Publication GMR-4453, August 1983.

4. Allan F. Williams et al., "Cars Owned and Driven by Teenagers," *Transportation Quarterly*, April 1987, pp. 177–188.

5. Paul Wasielewski, "Do Drivers of Small Cars Take Less Risk in Everyday Driving?" General Motors Research Laboratories, July 13, 1983, GMR-4425.

6. F. T. Sparrow, *The Coming Mini/Micro-Car Crisis: Do We Need a New Definition?* (New York: Pergamon Press, 1984).

7. F. T. Sparrow, "The Coming Mini/Micro-Car Crisis: Do We Need a New Definition?" Automotive Transportation Center, Purdue University, 1984.

8. National Safety Council, "Accident Facts," 1990 edition.

9. *See* note 5.

10. E. Chelimisky, "Automobile Weight and Safety," GAO/T-PEMD-91-2 (Washington, DC: U.S. General Accounting Office). (Weight categories are reported in pounds and have been converted by the author.)

11. U.S. Congress, Office of Technology Assessment, "Improving Automobile Fuel Economy: New Standards, New Approaches," OTA-E-504 (Washington, DC: U.S. Government Printing Office, October 1991).

12. Harold J. Mertz and James F. Marquart, "Small Car Air Cushion Performance Considerations," SAE Paper No. 851199 presented at the Government/Industry Meeting and Exposition, Washington, D.C., May 1985.

13. National Safety Council, "Accident Facts," 1991 edition.

14. This was predicted as early as 1973 based on technology known at that time. See final statements in: Donald Friedman, "Development of Advanced Deployable Restraints and Interiors," Paper presented at the Vehicle Safety Research Integration Symposium, Washington, D.C., May 30–31, 1973.

15. Donald Friedman, "Development of Advanced Deployable Restraints and Interiors," Paper presented at the Vehicle Safety Research Integration Symposium, Washington, D.C., May 30–31, 1973. DOT HS-820 306.

16. Test criteria of FMVSS 208 and NCAP procedures include a Head Injury Criterion formula, which includes load/time limitations, a maximum chest acceleration of 60 g, a maximum chest compression of 3 inches, and a maximum femur load of 2,250 pounds (1,021 kg). These limitations are approximated by assuming a maximum full-body limitation of 60 g. Refer to Code of Federal Regulations, Title 49, Part 571-FMVSS and Part 572-ATD's for complete description of tests and criterion.

17. *See* note 12.

18. *See* note 15.

19. Stephen Goch et al., "Inflatable Restraint System Design Considerations," SAE Paper No. 901122.

20. Robert Kaeser and Felix H. Walz, "New Safety Concepts for Low-Mass Electric/Hybrid Cars," Paper presented at the 25th ISATA meeting, Florence, Italy, June 1992; and Robert Kaeser, "Safety Potential of Urban Electric Vehicles in Collisions," Paper presented at The Urban Electric Vehicle Conference, Stockholm, Sweden, May 25–27, 1992.

21. Felix H. Walz et al., "Occupant and Exterior Safety of Low-Mass Cars (LMC)," Paper presented at the IRCOBI Conference, Berlin, Germany, September 1991.

22. Robert Kaeser et al., "Collision Safety of a Hard-Shell Low-Mass Vehicle," Paper presented at the IRCOBI Conference, September 9–11, 1992, Verona, Italy.

23. *See* note 12.

24. Shigeo Aono, Nissan Motor Co. Ltd., "Electronic Applications for Enhancing Automotive Safety," SAE Paper No. 90137.

25. Donald Friedman, Minicars Inc., "Development of Advanced Deployable Restraints and Interiors," Paper presented at the Vehicle Safety Research Integration Symposium, Washington, D.C., May 30–31, 1973, DOT HS-820-306.

·

CHAPTER EIGHT

INTELLIGENT TRANSPORTATION SYSTEMS

by Dr. Wonshik Chee

1. WHAT IS ITS?

According to *Merriam-Webster's Collegiate Dictionary*, one of the meanings of *intelligent* includes being guided or controlled by a computer; "especially: using a built-in microprocessor for automatic operation, for processing of data, or for achieving greater versatility." The intelligent transportation systems (ITS) refer to transportation systems whose performance is enhanced by computers. Transportation systems include vehicles, road signals, and other road infrastructures. By using computers, these individual elements of transportation systems can be made "intelligent" to achieve higher performance or new functionality.

ITS can be divided into three categories: (1) intelligent vehicles, (2) intelligent infrastructure, and (3) intelligent communication systems. Intelligent vehicles are those that can autonomously drive on roadways. Key elements of intelligent vehicles include sensors, actuators, and computers. Sensors are used to monitor the condition of the vehicle (e.g., vehicle speed, acceleration, wheel speed, vehicle spacing). Actuators are electromechanical devices that can change the mechanical condition of the vehicle, such as steer-by-wire (SBW) and drive-by-wire (DBW). Computers are supposed to make a decision for the actuators based on the information provided by the sensors. This structure of intelligent vehicles is known as *mechatronic systems*. A mechatronic system is referred to as a mechanical system that can be created when electronics take on the decision-making function formerly performed by mechanical components. The typical structure of a mechatronic system is shown in Figure 8.1. Here, the mechanical system in Figure 8.1 can be an engine, a wheel, and the vehicle itself.

Typical tasks performed by intelligent vehicles include lane-following maneuvers, lane-changing maneuvers, and vehicle-following maneuvers. For vehicle-following maneuvers, the computer installed on the vehicle receives information about distance to the preceding vehicle, wheel speed, and acceleration from various sensors like radar, accelerometers, and a wheel-speed sensor. Then the computer generates commands to various actuators such as DBW. The decision made by the computer is stored as software. The software is developed by controller engineers to perform desired tasks, such as

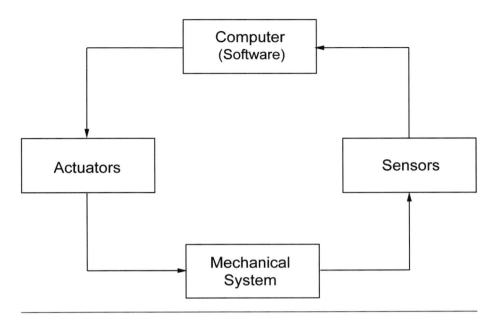

FIGURE 8.1 *Structure of Mechatronic Systems*

following the preceding vehicle or cruising a slower-moving vehicle. For lane-following maneuvers, the computer receives information about the location of the vehicle on the road using magnets embedded into the roadways or a video camera, and generates a steering-angle command to the SBW.

The intelligent infrastructure is referred to various electronic devices to provide traffic information and/or information about individual vehicles. Loop sensors are popularly used to measure the traffic volume and the mean speed of the traffic. The measured information is delivered to the traffic management center by hardwires. The traffic management center controls the traffic signals to maintain the speed of the traffic on the roadway to a specific level or to detect possible traffic accidents. One of the new systems of intelligent infrastructure is the magnetic road-reference system,[1] developed by California Partners for Advanced Transit and Highways (PATH)—a joint venture of Caltrans, the University of California, other public and private academic institutions, and private industry. The magnetic road-reference system consists of magnets embedded into the roadways and magnetometers installed on the vehicles to detect the magnetic fields generated by the magnets in order to find the position of the vehicle with respect to the centerline of the roadway. A typical scenario for the magnetic road-reference system and the loop sensor is shown in Figure 8.2.

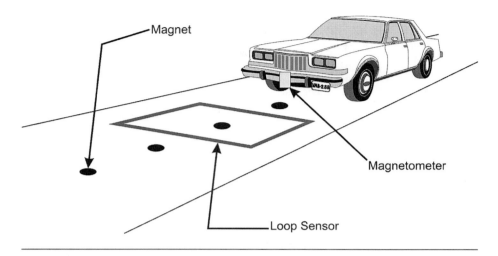

FIGURE 8.2 Magnetic Road Reference System and Loop Sensors

The intelligent communication systems are referred to the system that communicates between the intelligent infrastructure and/or the intelligent vehicles. One of the communication systems between the infrastructure and intelligent vehicles is the transmitter, which broadcasts road-curvature information and traffic accidents information.[2] Current popular navigation systems and telematics are other examples of intelligent communication systems. Between the vehicles, acceleration and speed are transmitted to be used for so-called platoon maneuvers.

The main objectives of ITS include safety enhancement, improvement of highway efficiency, and reduction of fuel consumption and air pollution. Because more than 70 percent of traffic accidents are caused by human error, intelligent vehicles can reduce traffic accidents by compensating. For example, intelligent vehicles are installed with radar to scan the incoming traffic situation. When the distance to the preceding vehicle is less than the safe level, the control system of intelligent vehicles can reduce the speed. Furthermore, by exchanging the speed and the acceleration through the intelligent communication system, intelligent vehicles can react to the changes in incoming traffic more quickly than any human could. As a result, spacing between vehicles can be significantly reduced if it is compared with the one any human can maintain. California PATH showed the technical feasibility of intelligent vehicles by maintaining the spacing at 3 meters (10 ft.) while running at 90 km/h (55 mph).[3] If spacing between vehicles is reduced, more vehicles can run on the same highways. Therefore, the efficiency of highways is improved.

Because the vehicle status information, such as acceleration and speed, can be exchanged with other vehicles, intelligent vehicles can be operated more smoothly by avoiding hard acceleration and hard braking. By operating engines smoothly, toxic emission material, which is usually generated during hard acceleration, as well as the consumption of fuel, can be significantly reduced. Furthermore, improved efficiency of freeways can save the individual's driving time. Therefore, time for operating vehicles can be reduced, which results in saving fuel and reducing air pollution.

Among the various components of ITS, this chapter concentrates on the advanced vehicle control system (AVCS) for intelligent vehicles. In Section 2, a systemwise discussion of AVCS is presented. Then, longitudinal maneuvers for intelligent vehicles (controlling intelligent vehicles to follow the preceding vehicles) are discussed in Section 3. Lateral maneuvers (controlling intelligent vehicles to follow the lanes and change the lanes) are discussed in Section 4. Finally, a brief discussion of the future of ITS is presented in Section 5.

2. What Is AVCS?

Various tasks of intelligent vehicles can be achieved by various control systems. Such control systems are called advanced vehicle control systems (AVCS). For each task, a feedback control system is developed. The overall structure of the AVCS becomes a mechatronic system. Thus AVCS consists of hardware and software, as discussed previously. The hardware of AVCS includes sensors, actuators, and computers. The software is the computer codes of the control algorithm for each task.

Typical maneuvers performed by AVCS are categorized as longitudinal maneuvers and lateral maneuvers. Longitudinal maneuvers are referred to tasks to control intelligent vehicles in order to maintain spacing between vehicles and the speed of each vehicle. A typical scenario of longitudinal maneuvers is shown in Figure 8.3. In this scenario a group of vehicles are driving like a train. The group of vehicles is called a *platoon*. Inside of the platoon, the spacing between the vehicles is tightly maintained (e.g., 3 meters [10 ft.] when they are driving at 90 km/h [55 mph]). In addition to the spacing measurement, information about each vehicle, such as acceleration and velocity, is transmitted to other vehicles in the same platoon to improve performance.

Figure 8.4 shows the typical scenario for lateral maneuvers. Lateral maneuvers include lane-following maneuvers and lane-changing maneuvers. For lane-following maneuvers, the vehicle position needs to be measured with

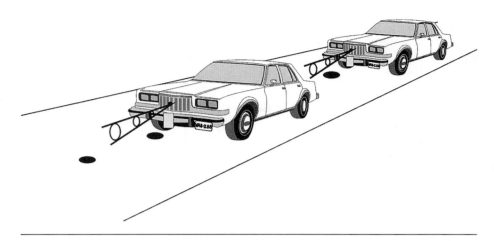

FIGURE 8.3 Typical Scenario of Longitudinal Maneuvers

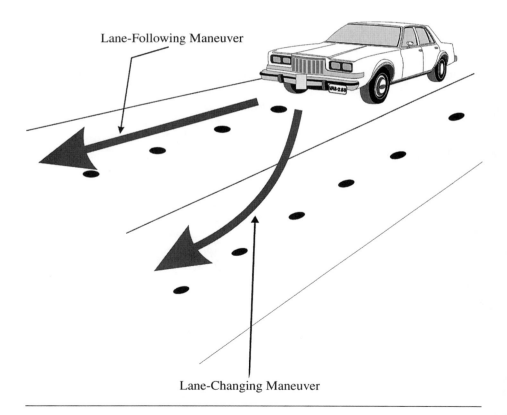

FIGURE 8.4 Typical Lateral Maneuvers

respect to the center of the lane. Two methods are used to measure the position: magnets and video cameras. There are pros and cons for each method. For the magnet method, the vehicle position can be obtained in any weather condition; however, the measurement range for magnets is limited because the magnetic field generated by the magnets cannot be strong enough to cover the entire width of a lane. The construction cost of installing the magnets is tremendously high because the magnets need to be embedded into the surface of a roadway at every 1 meter.

For the video camera method, there is no limit on the position-sensing range. In addition, because the video cameras are looking ahead of the vehicles, incoming curve information can be obtained; however, video cameras cannot be used for certain weather conditions (e.g., rain and snow) and lighting conditions (i.e., silhouette). Computational load is significantly high for the video camera method because each frame of the video images is analyzed by the computer in real time, which means that all of the necessary computations should be finished before the next frame of data is obtained (typically within 30 milliseconds). One of the main advantages of using video cameras is that the road infrastructure does not need to be modified.

Because the range of the position measurement is limited for magnets, lane-changing maneuvers need to be performed using only onboard sensors, such as a yaw-rate sensor, and the accelerometers must be installed laterally. Therefore, the controller for lane-changing maneuvers needs to be different from the one for lane-following maneuvers. In contrast, for the video camera method, the controller for lane-following maneuvers can be used for lane-changing maneuvers as well. A more detailed discussion of this topic is presented in Section 4.

A typical configuration of the hardware for intelligent vehicles is shown in Figure 8.5. Sensors of an intelligent vehicle include onboard sensors (e.g., yaw-rate sensor, longitudinal accelerometer, and lateral accelerometer), radar, video cameras, and magnetometers.

Onboard sensors include a yaw-rate sensor and accelerometers. They are used to measure inertial information of the vehicle. In most cases, two accelerometers are installed: one in the longitudinal direction and the other in the lateral direction. Both accelerometers are installed at the center-of-gravity (cg) location. More recently, three-axis accelerometer chips are becoming popular because the cost is significantly cheaper than that of mechanical accelerometers. The yaw-rate sensor measures the rate of the yaw angle of the vehicle. The yaw-rate sensor is usually installed in the trunk because the yaw-rate measurement is the same on the centerline of the vehicle.

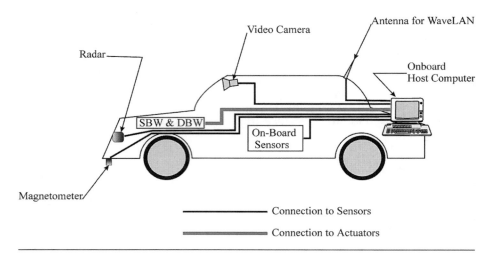

FIGURE 8.5 *Configuration of Hardware of Intelligent Vehicles*

Radar (i,e., laser radar, radio frequency (RF) radar, or infrared radar) is installed behind the engine grille and measures the distance to the preceding vehicles. Some sophisticated laser radar can map and track several vehicles simultaneously to provide a bird's-eye view of the traffic status. It is worthwhile noting that the radar cannot work properly if it snows or rains heavily because precipitation can reflect and/or diffract the emitted waves.

Magnetometers are used to measure magnetic fields generated by the magnets embedded into the roadways. Because these sensors are specific to the magnetic road-reference systems,[4] they are not installed on all intelligent vehicles. A detailed explanation of the magnetic road-reference systems is discussed in Section 4.

Video cameras were not a popular choice for intelligent vehicles until the mid-1990s. The main barrier to adopting video cameras was the computational load to process video images in real time. A typical NTSC video camera captures 30 frames per second. In order to control the vehicle using video images, the images of each frame should be processed to obtain vehicle position and road curvature before the next frame image is captured. Therefore, until the mid-1990s when the typical central processing unit (CPU) clock speed was less than 100 MHz, very expensive digital signal processor (DSP) boards were used to process the images; however, controlling intelligent vehicles using video cameras, called *visual guidance,* is advantageous because no special road infrastructure, like the magnets in the magnetic road-reference system, is necessary. As the computer cost goes down, adoption of visual guidance to intelligent vehicles is becoming more popular.

The wave local area network (LAN) antenna is used to form a localized network among the intelligent vehicles in a platoon. Each vehicle in a platoon communicates the key vehicle information, such as velocity and acceleration, with other intelligent vehicles. By exchanging such information, performance of longitudinal maneuvers of intelligent vehicles can be improved. The details of this topic are discussed in Section 3.

The actuators of an intelligent vehicle include steer-by-wire (SBW) and drive-by-wire (DBW). SBW and DBW are the systems that replace mechanical linkages between the human operator and the steering or braking/traction system. In fact, SBW is another mechatronic system. When the driver turns the steering wheel, the rotational angle of the steering wheel is measured by encoders. The measured rotational angle is sent to the microprocessor for SBW to generate the desired command to the hydraulic actuator or the motor, which turns the pinion gear of the front axle. Therefore, the mechanical connection between the pinion gear and the steering wheel is replaced by electric wire.

DBW works in the same way. For example, if the driver presses the gas pedal, the amount of the gas pedal movement is measured by encoders. Then the measurement is sent to the microprocessor for DBW to generate the desired command to the motor that turns the throttle valve. The main objective of using SBW and DBW is to reduce the component weight. In addition, SBW and DBW respond faster and more accurately than conventional mechanical parts. Furthermore, when DBW is incorporated with an engine control system, more efficient operation of the engine is possible.

In order to integrate the various types of hardware explained previously, all of the sensors and actuators are connected to a computer through various interface devices. The onboard computer shown in Figure 8.5 is installed in the trunk and used for this purpose. In addition, the computer stores the software for all of the control algorithms.

Figure 8.6 shows a typical software structure for controlling intelligent vehicles. The software has two levels: The low-level software is in charge of communication with various devices and service of requests from the high-level software. The device driver layer is one of the layers of the low-level software, as well as a collection of computer programs to communicate with other devices through various interface hardware such as CAN, A/D converter, RS

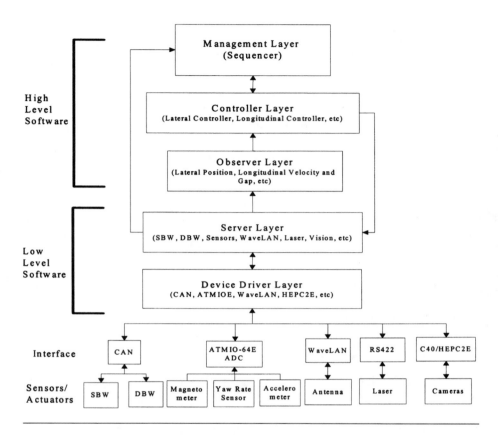

FIGURE 8.6 Structure of Software for AVCS

422, Wave LAN, and DSP boards (C40/HEPC2E board). The device driver programs connect to the corresponding hardware at a specified sampling time and write the data from the hardware at the specific location of the computer memory. The server layer programs serve the requests from processes of the high-level software by transmitting the data from the device driver to the requesting process and control commands to the corresponding actuator device drivers.

The high-level software consists of three layers: observer layer, controller layer, and management layer. One of the key roles of the observer layer is signal processing. Because the sensor signals are contaminated by noise, the noise needs to be attenuated by various signal-processing algorithms. One of the popular algorithms is the Kalman filter, which is an algorithm to estimate the states of a dynamic system using the noise-contaminated measurement. The estimate of the states is possible by using the mathematical model of the dynamic system whose states are estimated.

Another key role of the observer layer is to convert various sampling times of the sensors and actuators to the sampling time of the control algorithm. The various sampling times for the sensors and actuators are needed because of the integration of the different hardware using different communication methods. For video cameras, image data are updated every 30 milliseconds (msec); however, the SBW and DBW are updated at 20 msec. The sensors connected to the A/D converter are sampled at 2 msec. Therefore, some buffer zone is necessary to accommodate various sampling times. The role of the buffer zone is provided by the observer layer.

The role of the controller layer is to generate the control commands to the actuators. The control algorithms are designed to satisfy various objectives, such as tracking performance, ride comfort performance, and robust performance. Tracking performance refers to the performance of the control algorithm to make the vehicle follow the desired speed profile or the lane using various sensor data. By the action of the control algorithm, the tracking error becomes zero asymptotically. In order to make the control system track the desired value faster, the controller should generate larger control action; however, one cannot increase the control action arbitrarily large because the larger control action harms the passengers' ride comfort as well as stability of the system. Thus not compromising tracking performance and ride comfort performance is one of the important design considerations for controllers.

Robust performance is inevitable for the control of intelligent vehicles. For the vehicles, parameters of the vehicle used for controller designs are usually changing in time because of aging of the vehicle and various operation conditions. For example, vehicles run over dry roads, wet roads, icy roads, and snow-covered roads. Thus the friction force generated between tires and the road surface varies significantly based on the condition of the road surfaces. Therefore, controllers for intelligent vehicles should overcome the effects of inaccurate parameters of the vehicle. When the controller shows similar performance despite inaccuracies of the system information, the controller shows robust performance. The details of control algorithms are discussed in Sections 3 and 4.

At the top of the high-level software, there is the management layer. The management layer interacts with the human driver and manages the operation of various controllers to perform the scenario for various maneuvers. For example, if the radar finds an obstacle ahead, it is reported to the management layer. Then the management layer generates a command to the lateral-motion controller to change to an adjacent lane if an adjacent lane is empty or gives a command to the longitudinal-motion controller to stop the vehicle before it collides. In other words, the management layer performs the role of managing various maneuvers based on the surrounding driving condition of the intelligent vehicle.

Another important role of the management layer is to detect the faults of sensors and actuators and to reconfigure the systems to overcome the effects of the faults. Even though no sensors are installed redundantly, some of the vehicle status can be redundantly detected by different sensors at the same time. For example, suppose that an intelligent vehicle is following a curve turning counterclockwise. Before entering the curve, the video camera will detect the existence of the curve and compute the radius of the curvature. While the vehicle drives on the curve, the yaw-rate sensor will detect the counterclockwise rotation, and the lateral accelerometer will detect the centrifugal force. If a fault occurs in the yaw-rate sensor, its output will not cohere to other sensors' outputs. Therefore, the fault in the yaw-rate sensor can be detected. In the same way, faults in video cameras and lateral accelerometers can be detected.

In the following sections, details of control algorithms for longitudinal maneuvers and lateral maneuvers are discussed.

3. LONGITUDINAL MANEUVERS

In order to perform longitudinal maneuvers, the engine and the brakes need to be controlled. In fact, the control system for longitudinal maneuvers can be considered as an advanced cruise control system. The first task of the longitudinal maneuver is to control the vehicle speed. In contrast to current cruise control systems, the vehicle speed varies based on the traffic condition. For example, if the distance from the preceding vehicle is not large enough to avoid collision, the control system for longitudinal maneuvers will control the vehicle to reduce the speed in order to increase the distance. By exploiting this functionality, intelligent vehicles can form a platoon, as shown in Figure 8.7.

Actually, the intelligent vehicles in the platoon can be considered as a train connected together electronically. The vehicles shown in Figure 8.7 were driving at 55 mph. As seen in Figure 8.7, the distance between the vehicles is significantly smaller than the one human drivers can maintain. In other words, the space of road wasted in order to maintain safety distance can be

***FIGURE 8.7** Demonstration of Platoon Control (Adopted from "Intellimotion"[5])*

reduced. Therefore, the efficiency of the freeway, which is measured as the number of vehicles passing one lane per hour, can be significantly improved.

Figure 8.8 shows the simplified model for longitudinal motion of the vehicle. The definitions of the symbols are listed in Table 8.1. By applying Newton's second law of motion to the free body diagrams shown in Figure 8.8, an equation of motion can be obtained. The resulting equation becomes a second-order nonlinear differential equation. The nonlinearity of the equation is because of the model of the engine and the tire. (Among the various models of tire forces, Pacejka's magic formula is often used. Detailed information about the magic formula can be found in "Race Car Vehicle Dynamics."[6]) A model of the engine is developed in order to relate the throttle angle with the engine speed and the torque. Details of modeling the engine and controlling the engine can be found in "Design of a Robust Controller for Automotive Engines: Theory and Experiment"[7] and "An Observer-Based Controller Design Method for Improving Air/Fuel Characteristics of Spark Ignition Engines."[8]

Using the radar, intelligent vehicles can measure the distance to the preceding vehicle. Then a controller can be designed to maintain the distance as a certain value. For example, suppose that a driver sets the cruise speed at 55 mph. The intelligent vehicle will maintain the speed of the vehicle until it

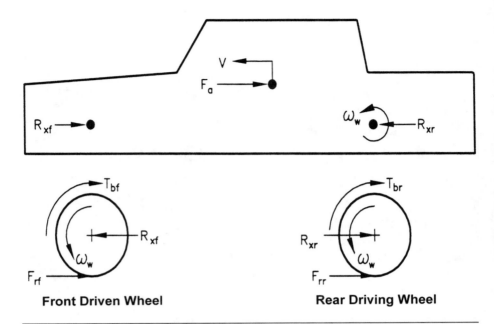

Front Driven Wheel   **Rear Driving Wheel**

FIGURE 8.8 Simplified Model for Longitudinal Vehicle Motion Control (Braking, adopted from "Longitudinal Control Development for IVHS Full Automated and Semi-Automated Systems: Phase I"[9])

detects a preceding vehicle. If the preceding vehicle runs slower, the distance to the preceding vehicle will be reduced. Because the distance can be measured by the radar, the controller can decide an appropriate speed to maintain the safe distance to the preceding vehicle in order to avoid collision. Thus the intelligent vehicle will reduce its speed. Now, suppose that the preceding vehicle accelerates. Then the distance measured by the radar will be

TABLE 8.1 Definitions of Symbols of Longitudinal Vehicle Motion Model

Symbol	Definition	Symbol	Definition
V	Speed of the vehicle	T_{bf}, T_{br}	Brake torque of front wheel and rear wheel
F_a	Inertial Reaction Force	F_{rf}, F_{rr}	Friction force of front wheel and real wheel
ω_W	Wheel speed	R_{xf}, R_{xr}	Reaction force

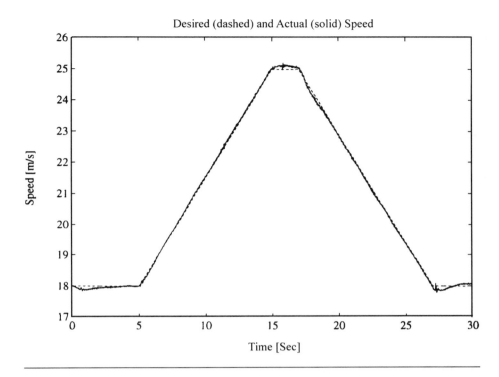

FIGURE 8.9 *Experimental Result of Autonomous Cruise Control (Speed of Vehicle, adopted from "Longitudinal Control Development for IVHS Fully Automated and Semi-Automated Systems: Phase I"[10])*

increased. Because the desired cruise speed is 55 mph, the intelligent vehicle will accelerate until it reaches this speed. This type of control is sometimes called autonomous (or advanced) cruise control (ACC).

Figures 8.9 and 8.10 show the experimental results of ACC performed on a prototype vehicle.[11] In this experiment, the preceding vehicle accelerates from 18 m/s to 25 m/s. The distance to the preceding vehicle was measured, and the controller was designed to maintain the distance as 10 ft. The speed of the vehicle under control is shown as a solid line in Figure 8.9. The spacing error is shown in Figure 8.10. Note that the maximum spacing error is less than 0.2 m, and the maximum error occurs when the preceding vehicle starts to accelerate or decelerate.

One of the innovative characteristics of intelligent vehicles is the platoon maneuver. As mentioned previously, the objective of the platoon is to enhance

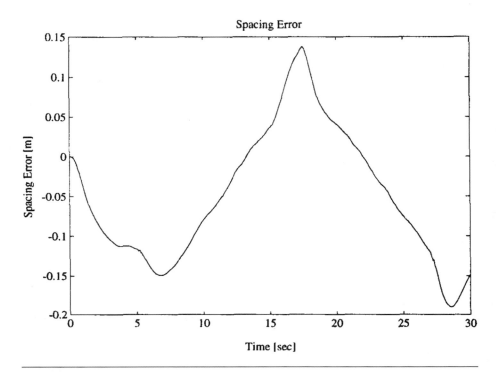

FIGURE 8.10 Experimental Result of Autonomous Cruise Control (Spacing Error, adopted from "Longitudinal Control Development for IVHS Fully Automated and Semi-Automated System: Phase I"[12]).

the efficiency of freeways by reducing the spacing between vehicles in the platoon. Figure 8.11 shows the definition of the distances used in the platoon control. The *intra-platoon distance* is the spacing between the vehicles in a platoon, and the *inter-platoon distance* is the spacing between platoons. The goal of platoon control is to minimize the intra-platoon distance as much as possible. Each vehicle is installed with radar to measure the intra-platoon distance. The controller for longitudinal motion is designed to maintain the measured intra-platoon distance at a specified value, for example 10 ft. Based on the intra-platoon distance, the controller generates an appropriate throttle angle to control the engine or brake master cylinder pressure to control the brake.

If the platoon consists of four or five vehicles, satisfactory performance can be achieved by controlling the vehicles using only the intra-platoon distance; however, if the platoon is formed with more than five vehicles, maintaining the intra-platoon distance during acceleration or deceleration becomes difficult. The difficulty originates in the "string stability" problem.[13] Suppose that

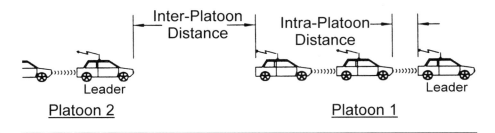

FIGURE 8.11 Definition of Platoon Distance (Adopted from "Longitudinal Control Development for IVHS Fully Automated and Semi-Automated System: Phase II"[14])

the lead vehicle of a platoon suddenly applies its brakes. Then the following vehicles will apply their brakes only if the intra-platoon distance is reduced over a certain threshold value. In other words, the longitudinal controller of the following vehicles cannot apply the brakes at the same instant when the lead vehicle applies its brakes. Therefore, the following vehicles apply their brakes after some time has elapsed. Because the time delay in the application of brakes increases for the vehicles in the tail of the platoon, the intra-platoon distance becomes smaller for the vehicles in the tail of the platoon. Then if the number of vehicles in a platoon is more than a certain number, the trailing vehicle will collide with the preceding vehicle.

This phenomenon exists in a platoon because each vehicle can only react to the change of the intra-platoon distance with some delay. Figure 8.12 shows the intra-platoon distance errors (spacing error in the figure) of the following vehicles in a five-vehicle platoon controlled by the longitudinal controller without using lead vehicle information. Here, lines 1 to 4 represent the spacing error of the following vehicles. Vehicle 1 is the one following the lead vehicle of the platoon, vehicle 2 is following vehicle 1, and so on. As can be seen, the spacing error becomes larger for the vehicles in the tail of the platoon, especially the spacing error of vehicle 4 at 10 seconds, which is almost twice that of vehicle 1 at the same time. Therefore, string stability is not achieved for this situation.

In order to avoid this type of collision, string stability should be considered when the controller for longitudinal maneuvers is designed. Furthermore, by adopting a communication device like a wave LAN and exchanging vehicle information, such as vehicle speed and acceleration, the controller can achieve better string stability. With better string stability, the intra-platoon distance can even be lowered. By exchanging vehicle information, the controller can react to the speed change before the change of intra-platoon distance is

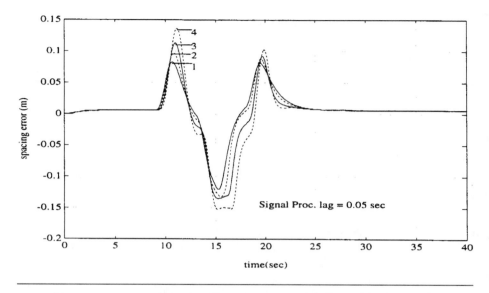

FIGURE 8.12 Control of a Five-Vehicle Platoon without Lead Vehicle Information (Adopted from "Longitudinal Control Development for IVHS Fully Automated and Semi-Automated Systems: Phase II"[15])

detected. Therefore, the controller can react to the braking of the preceding vehicle faster with the information.

Figure 8.13 shows the spacing error when the same platoon discussed for Figure 8.12 is controlled by exchanging preceding vehicle information. In this case, speed and the acceleration of the lead vehicle of the platoon are broadcasted to other vehicles in the platoon. In contrast to the results shown in Figure 8.12, the spacing error becomes smaller for the vehicles in the tail of the platoon. Therefore, the string stability of the platoon can be assured with exchange of the speed and the acceleration information.

4. LATERAL MANEUVERS

Lateral maneuvers include lane-changing maneuvers and lane-following maneuvers. In contrast to longitudinal maneuvers, lateral maneuvers invoke more critical issues. First, unlike longitudinal dynamics, lateral dynamics are complicated. Even though the order of the equation of motion is the same, the equations of motion for lateral dynamics are coupled equations. Furthermore, one of the states of the model, lateral body slip angle, cannot be measured.

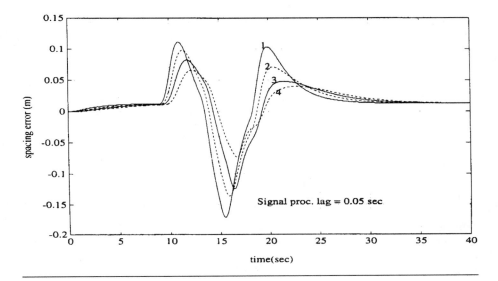

FIGURE 8.13 *Control of a Five-Vehicle Platoon with Lead Vehicle Velocity and Acceleration (Adopted from "Longitudinal Control Development for IVHS Fully Automated and Semi-Automated Systems: Phase II"[16])*

The equation of motion for lateral dynamics can be derived from the free body diagram shown in Figure 8.14. The velocity is denoted as V. Yaw angle measured counterclockwise with respect to the X-axis of the global coordinate is defined as ε. The steering angle of the vehicle is denoted as δ. Details of the equation of motion can be found in "Race Car Vehicle Dynamics."[17]

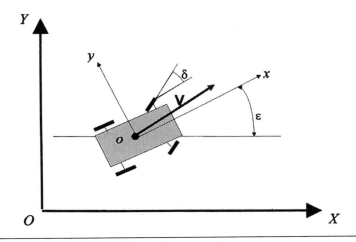

FIGURE 8.14 *Simplified Model for Control Algorithm Development*

For lateral motion control, obtaining the vehicle position with respect to the center of a lane is one of the critical tasks. For intelligent vehicle applications, video cameras and the magnetic road-reference systems are two methods that can be used to obtain the vehicle position reliably.

The position-sensing system that uses magnets was proposed in the California PATH program and has been successfully deployed with prototype vehicles.[18] The magnetic road-reference system of the California PATH program consists of magnets embedded into the roadways and magnetometers installed on the vehicles to measure the magnetic field generated by the magnets. Typically, three magnetometers are installed on each vehicle. One of the magnetometers is installed on the centerline of the front bumper of the vehicle. The remaining two magnetometers are installed on the front bumper at the same distance from the centerline to the left and right, respectively. The measured magnetic fields are sent to the computer. Then the position of the vehicle is obtained by searching the table that presents a functional relation between the lateral position and the magnetic field strength.[19] One of the disadvantages of this position computation method is that the instant when the center magnetometer passes on top of a magnet needs to be found, but it requires taking measurements from the magnetometer using expensive hardware. Also, this computation method provides only one position measurement from a magnet. Therefore, should this method fail to find the instant, the vehicle's position cannot be found.

Figure 8.15 shows the location of the magnetometers installed on the front bumper of the vehicle. The magnetic fields generated by the embedded magnets are measured by three magnetometers. If a magnet is close to one of the three magnetometers, the magnetometer closest to the magnet detects a stronger magnetic field than the other two magnetometers. For the situation shown in Figure 8.15, the left magnetometer will detect the strongest magnetic field. In order to obtain the distance, a calibration should be performed to make a table to search using the measured magnetic fields. Because searching a value for a table requires large computational loads, an analytical method was developed in Chee's "Unified Lateral Motion Control of Vehicles for Lane Change Maneuvers in Automated Highways and Visual Guidance."[20]

In the analytical method, two magnetometers are used.[21] Figure 8.16 shows the experimental setup to calibrate the analytical method for the vehicle position. Here, two magnetometers are installed on the front bumper of the vehicle. In order to calibrate the system, a magnet is placed at the known positions labeled with numbers. The positions of the magnets are measured before calibration is performed. Calibration results are shown in Figure 8.17.

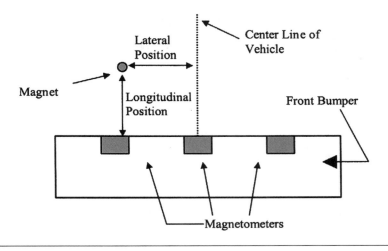

FIGURE 8.15 *Installation Location of Magnetometers*

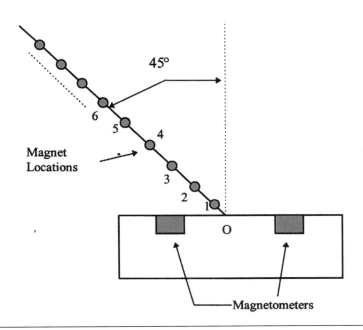

FIGURE 8.16 *Experimental Setup for Calibration of Magnetic Position-Sensing System (Adopted from "Unified Lateral Motion Control for Lane Change Manuevers in Automated Highways and Visual Guidance"[22])*

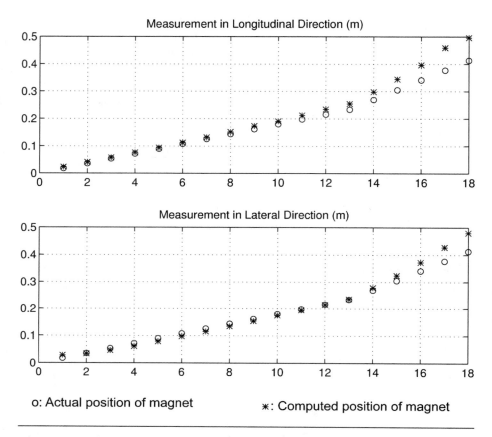

o: Actual position of magnet   ∗: Computed position of magnet

FIGURE 8.17 Calibration of Magnetic Road-Reference System (Adopted from "Unified Lateral Motion Control of Vehicles for Lane Change Maneuvers in Automated Highways and Visual Guidance"[23])

In the two plots of Figure 8.17, the abscissa is the location number of the magnets shown in Figure 8.16. As expected, as the magnet moves away from the magnetometers, the position error becomes larger because the measured magnetic field is small and is dominated by the sensor noise; however, the analytical method provided a reliable position measurement for ±0.3 m, which is enough for lane-following maneuvers.

Compared with the video camera, the magnetic road-reference system obtains the position information by looking down to the road. Thus road-curvature information cannot be found from the road magnetic road-reference system. When the vehicle is controlled to follow the lanes using the

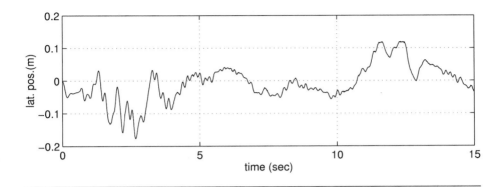

FIGURE 8.18 *Lateral Tracking Error During a Lane-Following Maneuver (Adopted from "Unified Lateral Motion Control of Vehicles for Lane Change Manuevers in Automated Highways and Visual Guidance"[24])*

magnetic road-reference system, the controller should react fast to the tracking error in order to maintain the tracking error within the measurement range. Otherwise, the vehicle may run out of the lane when the curve is incoming because the curve increases the tracking error as a result of geometry. The fast reaction inevitably results in frequent change in acceleration (i.e., generates large jerk). Therefore, a fast-reacting controller deteriorates passengers' ride comfort.

In order to improve the ride comfort during lane-following maneuvers using the magnetic road-reference system, additional magnetometers are installed on the rear bumper,[25] and the poles of magnets are alternated to form a code to provide intelligent vehicles with the upcoming curvature information.[26]

Figure 8.18 shows the tracking error when the intelligent vehicle is controlled to follow the test track using the magnetic road-reference system. The curvature information of the track is shown in Figure 8.19. The vehicle speed was 35 mph, and only the front bumper magnetometers were used. Considering the measurement range (0.3 m), the vehicle was controlled well within the range.

The alternative to the magnetic road-reference system is to use video cameras. As mentioned previously, video camera information can provide curvature. In addition, because the video cameras can see adjacent lanes at the same time, the measurement range is not limited. Therefore, larger tracking error is allowable. As a result, the controller can react to the tracking error slowly. If the control reacts slowly, the acceleration and the jerk generated by

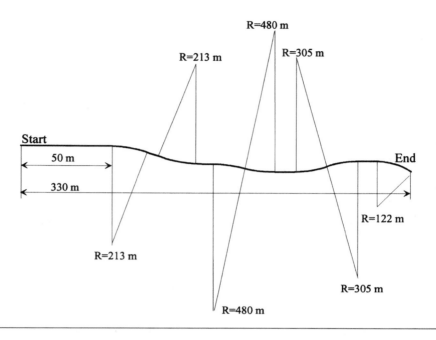

FIGURE 8.19 Curvature Information of Test Track (Adopted from "Unified Lateral Motion Control of Vehicles for Lane Change Maneuvers in Automated Highways and Visual Guidance"[27])

the control action can be reduced. Thus the controller action can be smoother and the ride comfort can be improved.

Figure 8.20 shows the typical driver's view of the road scene of freeways. In order to obtain the vehicle position with respect to the center of the lane and the incoming curve information, the image needs to be processed. Instead of processing the whole image, very small areas—denoted by the far-field image-processing window and the near-field image-processing window—are processed to find the location and the direction of the portion of the lane marker. Note that the windows are floating horizontally because images of the lane markers are moving in the video image when the vehicle drives; however, the vertical positions of the windows are fixed such that the windows are located at the constant distance from the video camera. This distance from the camera to the position corresponding to the image-processing window is called the *look-ahead distance*. Typically, 10 m is used for the look-ahead distance of the near-field image-processing window, and 20 m is used for the one of the far-field image processing window.

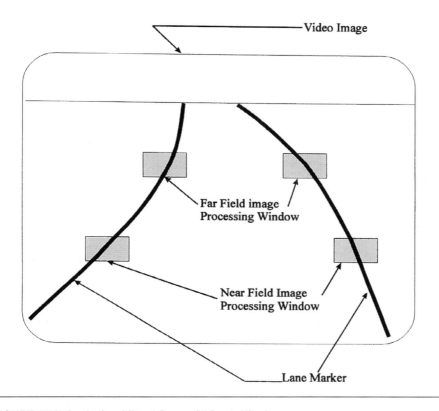

FIGURE 8.20 Analysis of Road Scene (Driver's View)

Figure 8.21 shows a bird's-eye view of the same road scene shown in Figure 8.20. Definitions of the symbols in Figure 8.21 are listed in Table 8.2. From the image of the windows, the horizontal position and the direction of the lane markers are computed by the image-processing program. Then the position and the direction obtained from the image-world coordinate, which is in pixels from the upper-left corner, are transformed into the ones in the real-world coordinate, which will be y_{LCi} and y_{RCi}. Using y_{LCi} and y_{RCi}, y_{Ci} can be found. Then $\Psi_{i/C}$ can be found using the measured direction of the lane markers. If these values are obtained for the two different positions with different look-ahead distances, y_{CG}, tracking error of the vehicle, and $\Psi_{R/C}$, directional error of the vehicle, can be found. A detailed derivation can be found in "Unified Lateral Motion Control of Vehicles for Lane Change Maneuvers in Automated Highways and Visual Guidance,"[28] "Autonomous High Speed Road Vehicle Guidance by Computer Vision,"[29] and "Recursive 3-D Road and Relative Ego-State Recognition."[30]

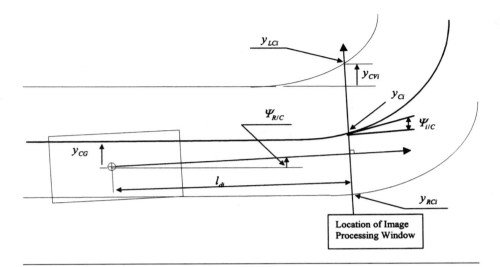

FIGURE 8.21 *Analysis of Road Scene (Bird's-Eye View)*

TABLE 8.2 *Definition of Symbols of Figure 8.21*

Symbol	Definition	Symbol	Definition
l_{di}	Look-ahead distance of ith window	y_{Ci}	Location of the center of the lane at l_{di} with respect to cg of the vehicle
y_{CG}	Location of the center of the lane at cg with respect to cg of the vehicle	y_{LCi}	Location of the left lane marker of the lane at l_{di} with respect to cg of the vehicle
$\Psi_{R/C}$	Direction of road at cg with respect to the direction of the vehicle	y_{RCi}	Location of the right lane marker of the lane at l_{di} with respect to cg of the vehicle
$\Psi_{i/C}$	Direction of road at l_{di} with respect to the direction of the vehicle	y_{CVi}	Shift of lane marker due to the curve

The controller for the visual guidance is designed to make y_{CG} and $\Psi_{R/C}$ equal zero. In other words, the vehicle is controlled to achieve zero tracking error as well as zero directional error. Because the curvature information can be obtained from the video image data, a smoother maneuver can be achieved.

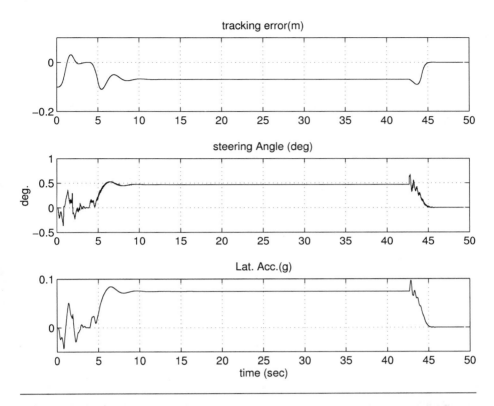

FIGURE 8.22 *Performance of Visual Guidance (Adopted from "Unified Lateral Motion Control of Vehicles for Lane Change Maneuvers in Automated Highways and Visual Guidance"[31])*

Figure 8.22 shows the simulation results of visual guidance. The speed of the vehicle is 20 m/s, and the vehicle is trying to follow the curve shown in Figure 8.23. The radius of the curvature is 500 m. As shown in Figure 8.22, tracking error is maintained less than 0.15 m. Note that the maximum lateral acceleration is less than 0.1 g throughout the maneuver.

Lane-changing maneuvers are the other key maneuvers of the lateral motion control of intelligent vehicles. With video cameras, lane-changing maneuvers can be performed using the controller designed for lane-following maneuvers. Note that the controller for lane-following maneuvers makes y_{CG} and $\Psi_{R/C}$ equal zero. If we replace y_{CG} with (y_{CG} − location of the desired trajectory) and $\Psi_{R/C}$ with ($\Psi_{R/C}$ − direction of the desired trajectory), the controller can make the vehicle follow the desired trajectory for lane-changing

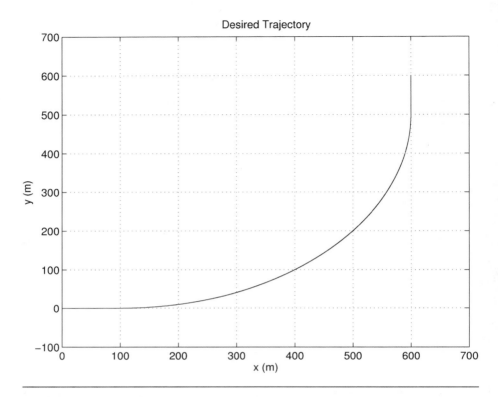

FIGURE 8.23 Desired Trajectory for Visual Guidance Simulation (Adopted from "Unified Lateral Motion Control of Vehicles for Lane Change Manuevers in Automated Highways and Visual Guidance"[32])

maneuvers. Therefore, lane-changing maneuvers can be achieved without designing a new controller for lane-changing maneuvers.

The desired trajectory for lane-changing maneuvers is designed considering the passengers' ride comfort. Figure 8.24 shows the acceleration profile of the trapezoidal acceleration trajectory, which is one of the popular choices for lane-changing maneuvers. Because the width of the lane of freeways is specified, the maneuver time (T) can be found if the maximum allowable jerk (J_{max}) and the maximum allowable acceleration (a_{max}) are selected. Once the trapezoidal acceleration profile is selected, the trajectory can be obtained by integrating the acceleration profile twice with respect to time. Note that the choice of larger values of J_{max} and a_{max} results in shorted T. Therefore, more comfortable lane-changing maneuvers (i.e.,

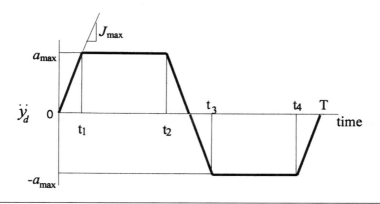

FIGURE 8.24 *Trapezoidal Acceleration Trajectory (Adopted from "Lane Change Maneuver of Automobiles for the Intelligent Vehicle and Highway System [IVHS]"[33])*

smaller J_{max} and a_{max}) take longer time. Because the trajectory for lane-changing maneuvers does not physically exist or is not installed on the freeways, the trajectory is called virtual desired trajectory (VDT).

With the magnetic road-reference system, a new controller for lane-changing maneuvers needs to be designed. Because the measurement range of the magnetic road-reference system is limited to 0.3 m, the vehicle needs to pass through the region where the magnetic road-reference system cannot provide the position of the vehicle. Therefore, the controller for lane-changing maneuvers is designed using only the onboard sensors, such as yaw-rate sensor and the lateral accelerometer. Using the onboard sensors, the lateral position of the vehicle is estimated during the maneuver, and the estimated position is compared with the desired trajectory. Thus the lane-changing controller tries to follow the desired trajectory, such as the trapezoidal acceleration trajectory, using the estimated position.

Another issue of lane-changing maneuvers with the magnetic road-reference system is the transition between lane-changing maneuvers and lane-following maneuvers. Ideally, the transition should be performed without any jerky motion. In order to achieve the ideal transition, the smooth control command (steering angle) should be found during the transition, especially the slope of the steering-angle variation at the end of the lane-changing maneuver should be matched with the one in the beginning of the lane-following maneuver, and vice versa; however, matching the change rates of the steering angles generated by different controllers is very difficult. If any discrepancy exists between them, it will result in jerky motion.

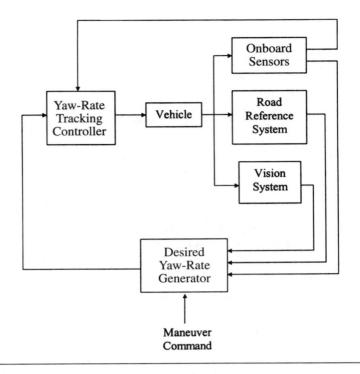

FIGURE 8.25 Schematics of Unified Lateral Control System

In order to minimize the potential jerky motion during the transition, the unified lateral control system idea was developed.[34] Figure 8.25 shows the schematics of the unified lateral control system. The key ideas of the system are design of the controller for yaw rate, called the *yaw-rate-tracking controller,* and introduction of the desired-yaw-rate generator. The desired-yaw-rate generator computes the value of the yaw rate of the vehicle based on the maneuver command. If the maneuver command requests a lane-changing maneuver, the desired yaw rate will be computed using the VDT. If the maneuver command requests a lane-following maneuver, the desired yaw rate will be computed to make the tracking error, measured either by the vision system or by the magnetic road-reference system, zero. The yaw-rate-tracking controller generates steering angle to make the measured yaw rate of the vehicle follow the desired yaw rate computed by the desired-yaw-rate generator. One of the advantages of this system is that there is no need to switch the controllers during the transition between lane-changing maneuvers and lane-following maneuvers. Therefore, the jerky motion that can be caused by the mismatched rate of the steering angle during the transitions can be avoided.

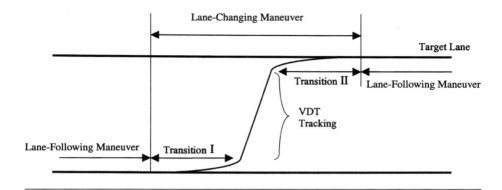

FIGURE 8.26 Strategy for Lane-Changing Maneuvers Using Magnetic Road-Reference System (Adopted from "Unified Lateral Control System for Intelligent Vehicles"[35])

Figure 8.26 shows the strategy for lane-changing maneuvers using the magnetic road-reference systems. During the transitions, the desired yaw rate is obtained as the weighted sum of the desired yaw rate for lane-following maneuvers and the one for lane-changing maneuvers. Here, the weight changes in time to avoid the jerky motion. For example, during the transition from a lane-following maneuver to a lane-changing maneuver, the weight for the lane-following maneuver is initially 1 and gradually decreases to 0. In contrast, the weight for the lane-changing maneuver is initially 0 and gradually increases to 1. Therefore, in the beginning of the transition, the desired yaw rate of the transition is the same as that of the lane-following maneuver, and, in the end of the transition, the desired yaw rate of the transition will be the same as that of the lane-changing maneuver. During the VDT tracking shown in Figure 8.26, the desired yaw rate is computed using the VDT.

Figure 8.27 shows the experimental results of a lane-changing maneuver performed by the unified lateral control system using the magnetic road-reference system. The lane-changing maneuver starts at 0.5 second. During 0.5 second and 1.5 seconds, transition from the lane-following maneuver to the lane-changing maneuver is performed. During 1.5 seconds and 5 seconds, the VDT tracking is performed. Then, during 5 seconds and 6 seconds, the transition from the lane-changing maneuver to the lane-following maneuver is performed. The top plot shows the yaw-rate response during the maneuver, and the second plot from the top presents the steering-angle variation. The bottom plot shows the velocity during the maneuver. As shown in the steering-angle variation plot, the steering-angle variation is very smooth during

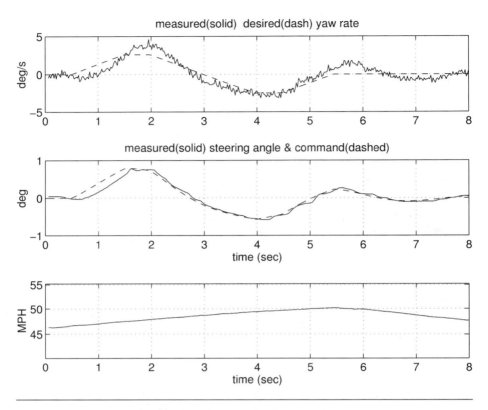

FIGURE 8.27 Performance of Lane-Changing Maneuver Using Magnetic Road-Reference System (Adopted from "Unified Lateral Control System for Intelligent Vehicles"[36])

the maneuver. Note that the wiggles in the yaw-rate response are caused by sensor noise.

5. Future of ITS

In this chapter, ITS was discussed with an emphasis on AVCS. The research activities of AVCS intrinsically have more restrictions than other areas of ITS. One of the restrictions is the reliability of AVCS. Compared with other systems of ITS, failure or malfunction of AVCS results in fatal accidents because AVCS takes over the control of the vehicle. Thus any failure of AVCS should be avoided, which is extremely difficult.

A technical feasibility demo was hosted by the National Automated Highway System Consortium (NAHSC) in 1997.[37] In the demo, fully automated intelligent vehicles were built and tested on highways. Even though the vehicles performed successfully, the NAHSC was disintegrated after the demo. Since then, researchers of AVCS have aimed at development of safety-enhanced systems using the technology achieved during the development of fully automated driving vehicles.

In the near future, it is unlikely that we will see autonomous driving vehicles running on freeways. Instead, many new systems will be introduced to compensate for the human errors of driving in order to improve safety. Recently, the ACC discussed in Section 3 has been successfully commercialized, and the systems are sold for high-end luxury vehicles. Other technologies discussed in this chapter will eventually be implemented on commercial vehicles. Then we will save the driving time for other activities as we do by using washing machines for laundry.

SPECIAL ACKNOWLEDGMENT

The author would like to extend a special note of gratitude to Dr. Wonshik Chee for writing this chapter. Dr. Chee is an assistant professor in the Department of Mechanical Engineering at the University of Texas, San Antonio. He has worked on intelligent vehicles since 1993. He received his Ph.D. in 1997 from the University of California, Berkeley. He also worked on Demo '97, which was held in San Diego, California in 1997. His main research expertise is in mechatronic system development with various applications, including intelligent vehicles.

REFERENCES

1. W. Zhang, R. Parsons, and T. West, "An Intelligent Roadway Reference System for Vehicle Lateral Guidance/Control," *Proceedings of the American Control Conference*, San Diego, California, pp. 281–286, 1990.

2. S. Tsugawa, M. Aoki, A. Hosaka, and K. Seki, "A Survey of Present IVHS Activities in Japan," Paper presented at the 13th IFAC World Congress, Vol. Q, pp. 147–152, 1996.

3. *Intellimotion*, Vol. 6, No. 3, PATH publications, California PATH, 1997.

4. *See* note 1.

5. *See* note 3.

6. W. F. Milliken and D. L. Milliken, *Race Car Vehicle Dynamics* (Warrendale, PA: SAE International Publications Group, 1995).

7. S. B. Choi, "Design of a Robust Controller for Automotive Engines: Theory and Experiment," Ph.D. Dissertation, University of California, Berkeley, 1993.

8. S. B. Choi, "An Observer-Based Controller Design Method for Improving Air/Fuel Characteristics of Spark Ignition Engines," *IEEE Transactions on Control Systems Technology*, Vol. 6, No. 3, pp. 325–334, May 1998.

9. J. K. Hedrick et al., "Longitudinal Control Development for IVHS Fully Automated and Semi-Automated Systems: Phase 1," California PATH Research Report, UCB-ITS-PRR-95-4, 1995.

10. *Ibid.*

11. *Ibid.*

12. *Ibid.*

13. D.V.A.H.G. Swaroop, "String Stability of Interconnected Systems: An Application to Platooning in Automated Highway Systems," California PATH Research Report, UCB-ITS-PRR-97-14, 1997.

14. J. K. Hedrick et al., "Longitudinal Control Development for IVHS Fully Automated and Semi-Automated Systems: Phase 2," California PATH Research Report, UCB-ITS-PRR-96-1, 1996.

15. *See* note 13.

16. *Ibid.*

17. *See* note 6.

18. *See* note 1.

19. *Ibid.*

20. W. Chee, "Unified Lateral Motion Control of Vehicles for Lane Change Maneuvers in Automated Highways and Visual Guidance," Ph.D. Dissertation, University of California, Berkeley, 1997.

21. *Ibid.*

22. *Ibid.*

23. *Ibid.*

24. *See* note 20.

25. J. Guldner, H. Tan, and S. Patwardhan, "Analysis of Automatic Steering Control for Highway Vehicles with Look-Down Lateral Reference Systems," California PATH Technical Note, UCB-ITS-PTN-96-03, 1996.

26. J. Guldner, S. Patwardhan, H. Tan, and W. Zhang, "Coding of Road Information for Automated Highways," California PATH Working Paper, UCB-ITS-PWP-97-7, 1997.

27. *Ibid.*

28. *Ibid.*

29. E. D. Dickmanns and A. Zapp, "Autonomous High-Speed Road Vehicle Guidance by Computer Vision," *Proceedings of the IFAC 10th Triennial World Congress*, Munich, FRG, pp. 221–226, 1987.

30. E. D. Dickmanns and B. D. Mysliwetz, "Recursive 3-D Road and Relative Ego-State Recognition," *IEEE Translation of Pattern Analysis and Machine Intelligence*, Vol. 14, No. 2, pp. 199–213, 1992.

31. *See* note 20.

32. *Ibid.*

33. W. Chee and M. Tomizuka, "Lane Change Maneuver of Automobiles for the Intelligent Vehicle and Highway System (IVHS)," *Proceedings of the 1994 American Control Conference*, Baltimore, MD, pp. 3586–3587, 1994.

34. W. Chee, "Unified Lateral Control System for Intelligent Vehicles," SAE Technical Paper Series, 2000-01-3074 (Warrendale, PA: SAE International Publications Group, 2000).

35. *Ibid.*

36. *Ibid.*

37. *See* note 3.

CHAPTER NINE

ALTERNATIVE CARS IN EUROPE

Courtesy: DaimlerChrysler AG, Smart Car, www.smart.com

> *The mass transit system of choice world-wide is the automobile. To the dismay of idealists and urban saviours, it has no serious competitors, even though its continued use depends on the precarious supply of elaborately refined petroleum products.*

—Syd Mead

After World War II, a type of ultralight microcar, called the "bubble car" in the United Kingdom and the "scooter car" in Germany, was sold throughout much of Western Europe. Designs for these unique three- and four-wheel alternative cars grew out of the shortage of motor fuel, the poor economy, and the high automobile taxation that characterized the region after World War II. Curb weight on the order of 225–375 kg (500–825 lb.) and fuel economy of 20–25 km/L (50–60 mpg) were typical. First costs, as well as operating and maintenance costs, were substantially below the costs of owning and operating a conventional car of the period. But regardless of their cost-saving benefits, these miniature cars became extinct or evolved into larger vehicles as conditions improved. The fate of bubble cars in Europe seems to ask a fundamental question about the idea of low-mass, mission-specific cars today: Is there a significant market for ULM cars in a modern, affluent society?

The European experience with midget cars can provide important insights for manufacturers of alternative vehicle types today. Undoubtedly, the issue is more complex than simply counting the successes and failures of bubble cars and projecting the results onto modern transportation products of similar curb weight. A variety of attributes having to do with the regional economies, taxation policies, consumer attitudes, and vehicle theme played significant and complex roles. Small size and limited utility were part of the equation. Equally important is the unmistakable vehicle theme of basic utility, which these cars offered to consumers who had their sights set on better economic times. An improving economy in Europe produced consumers who first wanted improved performance and greater utility from their single vehicle, and ultimately larger, more luxurious automobiles. Bubble cars occupied a position at the bottom of the economic ladder. In addition, they existed long before phrases such as "ecological breakdown," "gridlock," "greenhouse effect," and "ozone holes" were invented. Today, new consumer attitudes, new societal concerns, and new economic conditions point to a different product and a different marketing approach.

The potential market for ULM cars in Europe will grow in relation to the growth in multicar households, which is a result of the region's greater affluence. A successful "alternative car" must therefore appeal to affluent consumers. Neither European nor American consumers are likely to be attracted by a vehicle theme that implies a compromise in lifestyles. Today's consumers want more than basic utility. Complex personal factors having to do with lifestyle and self-image are behind most automobile purchasing decisions. In this respect, bubble cars can serve more as an antithetical model.

EARLY EUROPEAN MICROCARS AS A METAPHOR OF THEIR ENVIRONMENT

If there is one characterizing attribute of Europe in the 20th century, it might be that of volatility. In America, both the political history and the course of development have been comparatively stable and consistent. Although Americans may still remember Vietnam as a shock to their national self-confidence, Europeans experienced a series of Vietnams in their own backyards throughout much of the 20th century. Although World War II may have been a boon to the U.S. economy, it was a disaster for economies in Europe. After the War, countries that were not physically ruined were in economic shambles. Despite poor economies, new cars were nevertheless taxed at 40 to 60 percent of the purchase price, and already punitive license fees increased in proportion to engine displacement and vehicle weight. Gasoline was rationed, and what little was available (6 gallons per month in England) was extremely high priced. Rationing also occurred in the 1950s, long before Americans received a similar dose of reality with the OPEC oil embargo in the early 1970s.

In addition, Europe has never enjoyed the enormous consumer base that exists in the United States. Although the population of Western Europe as a whole was (and still is) slightly larger than that of the United States, this population of consumers was divided into 14 main national units, each with its own frontiers and stringent trade barriers. Within this varied and difficult-to-reach market, too many manufacturers were competing for too few sales. Ninety percent of Europe's cars were built by 14 principal factories, and the remaining 10 percent of cars emerged from an additional 36 manufacturers. Yet the combined sales of automobiles in Western Europe in 1953 barely equaled the sales of Chevrolets in the United States during the same year.[1] Benefits that flow from the economy of scale were therefore unavailable to European manufacturers, at least in regards to the domestic market. Naturally, the automobile market was ripe for entrepreneurial ventures that may not have resulted in the best products, nor been good for established manufacturers that were in for the long haul.

Against this backdrop, several differences have traditionally existed between North American and European populations and transportation needs. The densely packed cities of Europe are much more adaptable to public transportation. As a result, transit systems have been better financed and used by more commuters. The proliferation of the automobile after World War II can be at least partially blamed on the United States. As soon as the economy began to recover, Europeans launched their own version of America's highway mania, partly because of the American idea of unbridled mobility, but also because it is much less expensive to build roads than public transportation systems. Today, Europe has an extensive network of modern roads and the cars to occupy them as well as public transportation systems that rate among the world's finest.

As for automobile designs, European designers have traditionally been more innovative than their American counterparts. In 1945, one in four European cars already incorporated the front-wheel-drive powertrain, approximately a quarter-century before Americans discovered the inherent benefits of the layout. The live rear axle had been virtually abandoned in Europe even before the war. After the war, new ideas in vehicle configuration were rampant. Small cars developed by German aircraft manufacturers were marvels of mechanical simplicity. The Heinkel and the Messerschmitt were vehicles that carried the name of their airborne predecessors. Many came from existing automobile manufacturers, and several designs came from small companies with no history in automobile manufacturing at all. Far too many were prime examples of what can happen when an inexperienced and under-financed company rushes to market with an unrefined product, and companies with little or no service or sales organizations often left products on their own once they were in the field.

Isetta and Heinkel

The Isetta and Heinkel are two examples of well-engineered bubble cars. Designed in Italy and ultimately manufactured by BMW in Germany, the 343-kg (755-lb.) diminutive Isetta (Figure 9.1) was destined to become an inexpensive transportation workhorse for thousands of Europeans during the 1950s. The less popular Heinkel was designed and built in Germany by the manufacturers of the famous Heinkel aircraft. The Isetta and Heinkel were strikingly similar vehicles, and both utilized a single front door; however, the more powerful and deluxe four-wheel Isetta featured a pivoting steering column that attached to the door and would thereby swing out of the way when the door was opened. The Heinkel's steering column was fixed, and as a result partially blocked entrance into the cabin.

FIGURE 9.1 *Isetta*. The Isetta, one of the most popular bubble cars, weighed only 343 kg (755 lb.) and sipped fuel at a thrifty 21.25 km/L (50 mpg).

The Isetta began life with a 236-cc two-stroke engine. Later, a more powerful 300-cc engine was offered. The engine was as unorthodox as the rest of the car. Its two parallel cylinders shared a common combustion chamber. The inlet port was in one cylinder and the exhaust port was in the other—a design that reportedly gave the engine excellent scavenging characteristics. Power from the rear-mounted engine was delivered through a four-speed syncromesh gearbox and then transferred by a duplex chain drive to the two rear wheels. The drive wheels were set on a common axle, and a single drum brake provided stopping power to both rear wheels.

Many people mistakenly believed that the Isetta was a three-wheel vehicle. Close inspection, however, revealed four wheels. The rear wheels were placed only 216 mm (8-1/2 in.) apart in order to eliminate the need for a differential. As a result, the four-wheel Isetta ended up with the worst of both worlds: three-wheel rollover characteristics at the price of four wheels. The Heinkel, on the other hand, was a true three-wheeler (Figure 9.2).

FIGURE 9.2 Heinkel. The look-alike Heinkel was manufactured by the German aircraft manufacturer of the same name. Both the Heinkel and Isetta were designed with a single front door.

In a road test of the Iso Isetta manufactured in Milan, Italy, *The Motor* magazine, in September 1954, reported a top speed of nearly 50 mph (80.5 km/h) and fuel economy slightly greater than 50 mpg (21.25 km/L). In those days there were no official driving schedules, so fuel economy tests consisted of simply driving the vehicle over a random course that was deemed representative of regional motoring habits. "Driven hard" for 219 miles (352.37 km), *The Motor* reported 50.4 mpg (21.43 km/L). In a later test published in the magazine's April 11, 1956, issue, top speed had been increased to 60 mph (96.5 km/h) and fuel economy had dropped to 44.2 mpg (18.8 km/L) over a similar varied course. Total drag at 10 mph measured 29 lbs. At 60 mph, drag was extrapolated at 83 lb., which equates to a road load of about 9.9 kW (3.28 hp). Of all the era's so-called bubble cars, the Isetta had one of the best reputations for economical and trouble-free operation.

Messerschmitt

Having been put out of the aircraft business after the war, Messerschmitt began manufacturing three-wheel cars under the same name (Figure 9.3). An unconfirmed rumor claimed that early designs used the same canopy installed on the famous German fighters of World War II. The car's close-fitting

FIGURE 9.3 Messerschmitt. The three-wheel Messerschmitt was one of the most unorthodox designs of the period. Note the blown canopy and curved windshield. The engine was located in the rear.

FIGURE 9.4 *Messerschmitt Interior.* The canopy swings to one side to expose the entire interior. Direct-link steering was via downturned handlebars.

canopy was noted for quickly fogging over in cold weather. Pivoting the canopy to one side exposed a gaping interior that was blocked only by the relatively low sides of the vehicle (Figure 9.4). Power came from a 174-cc single-cylinder, two-stroke Fitchel & Sachs engine. Later, displacement was increased to 191 cc for more power. Power was delivered to the single rear wheel through a four-speed transmission, which had no reverse gear. In order to propel the car in reverse, the engine was stopped and then restarted in the reverse direction. Top speed was 50 mph (80.45 km/h) with the original 174-cc engine. *The Motor* magazine, in an April 1956 article entitled "Air Cooled Outings," reported fuel consumption over a varied course of 72.5 mpg (30.8 km/L) with the smaller engine.

The Messerschmitt carried two passengers in tandem; however, the rear seat was wide enough for one adult and a small child. Curb weight was reported as 462 lb. (210 kg). The Messerschmitt was equipped with a fully independent suspension, and all three wheels were sprung with rubber elements mounted in torsion, rather than conventional springs. The advantages of rubber springs include low cost and inherent self-damping properties. Poor life is an inherent disadvantage.

MARKET WAS HEADED IN THE OPPOSITE DIRECTION

Designs of the period were numerous and diverse. Although many were well engineered, bubble cars were positioned exactly 180 degrees from where the market was headed. In addition, consumers never really accepted the bubble car as a legitimate alternative to a "real car." In 1956, miniature car sales in Germany and France had risen to nearly 15 percent of total new-car sales. In Europe as a whole, 386,742 of the 1,888,000 total cars sold in 1956 were bubble cars, which means that one in five cars were miniatures.[2] But most so-called miniatures were cars like the Citroen 2CV and the Fiat 500 and 600 series, which were actually smaller, underpowered versions of more conventional cars. The possible exception would be the Isetta, which sold about 30,000 units in 1956, despite its unorthodox design.

Reliant was another successful manufacturer of three-wheel miniature cars. Reliant's success, however, came primarily from the fact that its cars grew up with the market, while still offering consumers the benefits of reduced taxation (Figure 9.5). The tax advantages of three-wheelers were substantial. For example, the purchase tax in the United Kingdom on a three-wheel car was 30 percent of the nominal wholesale value, compared to a 60 percent purchase tax on a four-wheel car. The road fund tax on a three-wheeler, which is similar to the yearly state-level license fee in the United States, was about half the rate of a four-wheel car. By 1978, the Reliant had become essentially a conventional car, which happened to have only three wheels.

If an appeal can be attributed to alternative cars that emerged in post-war Europe, it is undoubtedly that of basic utility as a way to cope with the prevailing economic conditions. But the economy was improving, and consumers were looking forward to better times. As early as 1952, *The Motor* magazine wrote:

> There seems to be sufficient evidence to show that the general public prefers a roomy car of presentable appearance to a "baby car," regardless of the higher initial costs and running expense. . . . Questions of prestige and "keeping up appearances" also play an important part, and the customer's aim of displaying (or assuming) by his car, especially when used mainly for business purposes, a certain degree of prosperity, frequently seems to override economic considerations.

FIGURE 9.5 Reliant Robin. This 1978 three-wheel Reliant Robin Saloon actually began life as a miniature car similar in concept to the Heinkel. It evolved along with the market while still retaining three wheels. Production was finally discontinued early in 2001.

Although consumer rationales may be based on utility and costs, appeals to bare utility are weak motivators, even in a poor economy. Moreover, modern consumers are far more affluent, and products must satisfy complex psychological needs and reflect modern lifestyles. Alternative cars must therefore attract consumers on their own merits while suggesting through design the social goal of conserving energy and preserving the environment. Such cars must satisfy personal needs and provide utility in a way that supports personal values. The messages contained in the demise of bubble cars can be applied to modern products. An alternative personal mobility product that is positioned as a low-cost substitute for a "real car" is likely to produce disappointing sales. When personal values are packaged in an upscale and intrinsically appealing product, sales will likely follow. Correct positioning and vehicle theme are crucial to the product's success. In addition, new products must also compete with a variety of transportation alternatives and be compatible with the existing social, business, and transportation infrastructure.

TRANSPORTATION IN A REVITALIZED EUROPE

High population densities, shorter driving distances, growing pollution, and mounting traffic congestion all suggest a good environment for small, energy- and space-efficient runabouts. Automobile tax structures also favor smaller cars. Conversely, fewer multivehicle households, a consumer preference for larger, more luxurious cars, and a greater reliance on public transportation for primary and secondary transportation needs point to just the opposite result. The prevalence of leasing and the widespread use of business cars are other important factors that could either encourage or discourage mission-specific cars.

Transportation systems in Europe are universally oriented toward the private automobile. Car ownership in Europe grew rapidly throughout the last half of the 20th century. But Europeans still own fewer cars per capita (or per household) and drive only 60 to 80 percent as far as Americans on a trip-average basis. The shorter distances typical of Europe's high-density cities, along with higher automobile operating costs and greater access to public transportation, all work together to reduce total vehicle-kilometers traveled by car. When reduced automobile ownership is combined with lower driving distances, Europeans actually drive only 40 percent of the distance driven by Americans, on a kilometer per capita basis.[3] Still, automobiles in Europe account for approximately 80 percent of the total passenger-kilometers traveled. Table 9.1 provides a comparison of roadway, automobile, and population densities in Europe, Japan, and the United States.

TABLE 9.1 *Automobile and Population Density of Select Countries*

Country	Area (km²)	Population	Population Density (pop/km²)	Cars	People/ Car	Cars/ km²
UK	244,046	75,701,000	236	20,069,437	2.9	82
Germany	365,755	79,753,000	218	34,051,299	2.3	93
France	547,026	56,614,000	103	23,010,000	2.4	42
Italy	301,225	57,663,000	191	24,300,000	2.4	81
Japan	377,708	123,537,000	327	32,621,046	3.8	86
U.S.	9,155,579	249,633,000	27	143,081,443	1.7	16

Widespread use of private cars, along with more densely packed cities and shorter driving distances, are all conditions that favor electric cars and space-efficient urban cars; however, purchasing decisions are based on individual preferences or business needs, rather than on potential benefits to society. The most significant constraint affecting the potential market for ULM cars is undoubtedly that of multicar households. In Europe, far fewer households have two vehicles. In the United Kingdom in year 1990, 33 percent of households had no car, 44 percent had one car, and 23 percent had two or more cars. In France approximately 28 percent of households have two or more cars. Statistics are similar throughout much of Western Europe. This compares with approximately 60 percent of U.S. households that have two or more cars, and only 9 percent of households that have no car.

When only one vehicle is available, the single vehicle must serve the broadest possible utility. Vehicle ownership and travel patterns in Europe have therefore tended to discourage limited-utility vehicles. In the United Kingdom, even the two-passenger sports car has essentially become extinct. Because prestige and self-image are important motivators, and these needs must often be satisfied by a single vehicle, vehicles have tended to become larger, more powerful, and more luxurious.

Another important attribute of European car purchasing habits has to do with the prevalence of company cars. In West Germany and Sweden, more than one-third of new cars are purchased as company cars. In the United Kingdom, more than 50 percent of new cars purchased in recent years have been company cars. Holland also has significant numbers of company-purchased cars. Business cars accounted for 27.7 percent of the more than 13 million new cars sold throughout Europe in 1990 and roughly 35 percent of the market in year 2000. Most companies allow employees to use their company cars for private purposes, and only 17 percent charge the employee for private use. Employees typically pay a flat fee in their income tax for the benefit of these cars; however, many employees who drive company cars do not pay for gasoline and are not taxed according to how much gas they use.

A survey of company car policies conducted by the British Institute of Management revealed that 90 percent of the organizations surveyed allocated cars to managing directors, 86 percent to senior managers, and 60 percent to middle managers.[4] For junior managers, the need for an automobile in business was the primary criterion for car allocation in 32 percent of the companies. The type of car provided is usually determined by job status and internal hierarchy, salary, and the need to project a company image. During a business downturn, companies seek ways to reduce their fleet costs, which generally include withdrawing company cars or switching to vehicles that are less expensive to purchase and operate. While the economics of ULM cars may be appealing to businesses, job requirements and hierarchy disputes may tend to work in the opposite direction, unless the alternative product also has prestige. By providing the option to expand car allocations or forego withdrawal during tough times, smaller ULM vehicles could penetrate the business market. The second-car market that remains when the primary car comes from the employer also offers a niche for alternative mobility products.

The leasing market is another important factor. Leased cars in general are much more prevalent in Europe. In the Netherlands, more than 35 percent of new cars are leased. In the United Kingdom, leased cars account for 25 percent of new-car sales, and in Belgium 15 percent are leased. The lease market is rapidly growing in Germany, France, and elsewhere. The most popular lease arrangement is "contract for hire," in which total vehicle operating expenses, minus fuel, are included in the monthly payments. This trend has significant implications for manufacturers of electric cars in general, as well as for unconventional IC vehicles. Independent lessors may overestimate the maintenance costs of EVs in contract-for-hire agreements, or companies may be reluctant to extend contract-for-hire leasing on nontraditional vehicle types. Residual values are another important aspect. Residual values may not be as high with unconventional vehicle types until lenders become more familiar with depreciation values.

Automobile purchase tax policies support a move to smaller, less powerful cars. Existing tax policies could have a mixed effect on electric cars because of the higher taxes that might result from potentially higher initial costs, and the lower taxes resulting from having circumvented IC-engine-based taxation. In Europe, cars in general are taxed much more heavily than in the United States, and the rate and type of tax varies between countries. Value-added tax (VAT), which is based on purchase price, favors less expensive cars. Governments can influence the EV market through additional preferential taxation. Table 9.2 shows automobile taxation in selected European countries.

Available parking space and living arrangements also affect the automobile market. In the United Kingdom, most of the population lives in single-family homes, but most homes are older and many do not have garages. In many areas, streets are narrow and street parking is limited or prohibited. People are therefore more restricted in the number of cars by restricted or inadequate parking space. In Germany and Italy, 72 percent of households live in apartments. As a result, the recharging infrastructure for electric cars cannot rely as heavily on home recharging. Shorter trip distances make electric cars more viable in Europe. Reduced availability of home recharging in regions of

TABLE 9.2 Car Tax Rates in Selected Countries

Country	VAT	Car Tax	Vehicle Excise Duty
Belgium	25%/33%	None	£26.53–£675.44
France	22%	None	Regional tax
Germany	15%	None	Engine size tax
Ireland	21%	21.7% < 2 litre 24.5% > 2 litre	Engine tax size
Italy	19% < 2 litre 38% > 2 litre	None	Horsepower tax
Netherlands	18.5%	16%–24% based on list price	Av. £171.47
UK	17.5%	5% on 5/6 of price	£110

Source: European Vehicle Leasing 1992/93

greater apartment dwellers and fewer multicar households tend to make them less viable.

EUROPEAN TRANSIT SYSTEMS

Although transportation in Europe is strongly oriented toward the private car, the region's more densely packed populations are much more adaptable to transit systems, as well as to cycling and walking. Consequently, Europeans travel by bus or rail far more than do commuters in the United States. In Europe, transit systems (all modes) typically carry 15 percent of the total passenger-kilometers traveled, compared to about 2.5 percent in the United States. Europeans also walk or cycle to work, services, and leisure activities far more often than their American counterparts. In comparison to Americans, Europeans generally do not view cars as the same necessity to living and working.

In addition, the proliferation of private automobiles may soon begin to peak in many areas of Western Europe. Experiments by several European communities with prohibiting or restricting automobiles in inner cities could profoundly change attitudes toward cars and public transportation over the coming decades. In the United States, wide-open spaces and low land values have encouraged the ever-increasing spread of the suburbs, which in turn has made American transportation systems more committed to the private automobile.

The future of transit systems in Europe may be with the railways. In the United States, railways have been much maligned since the ascendancy of the automobile. After World War II, rail services were generally neglected in Europe as well. By concentrating on new road networks, policies favored trucks for short-haul freight transport and the private automobile for passenger transport; however, the oil embargo in the early 1970s sent transportation planners back to the railroads nearly everywhere except in the United States. Most developed countries have a greater density of railways than exists in the United States, both in terms of kilometers of rail per square kilometer of land area and in relation to the density of roads.

Table 9.3 provides a comparison of roadway and railway densities in several European countries, as well as those in the United States and Japan. Table 9.4 compares passenger transport by roadway with transport by railway. Although Europeans travel far less distance than their counterparts in the United States, they travel approximately 7 to 15 times more distance by railway primarily because cities and nations are much more condensed in Europe than is typical of the United States.

Although railways are much more costly to build than roadways, they are much more efficient carriers, both in terms of space and in terms of energy per passenger-kilometer. A passenger traveling by rail uses only one-fifth of the fuel used to traverse the same distance by air and less than half the fuel

TABLE 9.3 1990 Road and Rail Infrastructure of Selected Countries

Country	Roads		Rail Network	
	Kilometers ($\times$1000)	Density (km/1000 km^2)	Kilometers ($\times$1000)	Density (km/1000 km^2)
UK	382	1561	16.9	69
Germany*	549[†,‡]	1538[†]	44.0[‡]	123
France	806	1481	34.1	63
Italy	304[§]	1009[§]	19.6	65
Japan	1115	3015	—	—
U.S.	6244	667	192.7	21

Source: Transportation Statistics Great Britain 1992, The Department of Transport

*Includes former Federal Republic of Germany (East Germany).

[†]Excludes urban roads in former Federal Democratic Republic (West Germany).

[‡]Estimated from previous year.

[§]Excludes urban roads.

TABLE 9.4 *Passenger Transport Per Head of Population in Kilometers*

Country	Travel by Road		Rail (Excluding Metro)	All Modes
	Passenger Cars and Taxis	Buses and Coaches		
Great Britain	10,586	817	595	11,997
West Germany	9,388	885	709	10,982
France	10,413	732	1,130	12,274
Italy	8,555	1,472	788	10,815
Japan	4,491	867	2,915	8,272
U.S.	17,002	156	85	17,243

Source: Transportation Statistics Great Britain 1992, The Department of Transport

of an equivalent trip by car. Trains are also one of the safest modes of travel. In the United Kingdom in 1991, there were 1.4 fatalities per billion passenger-kilometers resulting from railway accidents. Motor vehicle accidents during the same year resulted in 57 fatalities per 100 million vehicle-kilometers, or more than 300 fatalities per billion passenger-kilometers, which makes travel by car about 225 times more hazardous than the same rail trip.

FUTURE OF MICROCARS IN EUROPE

A purely utilitarian rationale would point to transit systems as the most space- and energy-efficient option for Europeans. But as Renault's spokesman, George Cagnard, said: "Now that people have the freedom to pick up and go where and when they please, do you really think they will give it up?"[5] Regardless of the practical and economic reasons to use transit systems, even in transit-rich Europe, people are unwilling to abandon the freedom of private cars. True, the measure of freedom available in gridlocked traffic, or while confined to a vehicle in search of a parking place, may be more illusory than real. But if one's city were walled off and inhabitants were prohibited from leaving, the entire population would suddenly be overtaken by a compulsion to escape. Choice and a sense of freedom are basic to human nature, and in general, automobiles provide a sense of freedom, as well as real and discretionary mobility.

Today, when one speaks of alternative cars in Europe, the subject is just as likely to be electric cars as it is microcars. For many years, electric vehicles have been more extensively used in Europe than in the United States, but more on a commercial and fleet level than for private transportation. Only within the last decade has the idea of consumer BEVs become more seriously discussed as the ultimate answer to national energy sufficiency and inner-city air pollution. According to the European Electric Road Vehicle Association (AVERE), Europe's counterpart to the Electric Vehicle Association of the Americas (EVAA) in North America, BEVs are expected to significantly penetrate European markets over the coming decades.

AVERE (www.avere.org) is active in several areas to ensure that the BEV infrastructure develops along sound lines and is coherent throughout all of Europe. Items on the agenda encompass a broad range of technical, legal, and political factors that are both necessary for market development and best considered before many BEVs are in service. On a technical level, the goal is to promote safe product design and standardization among suppliers, manufacturers, and public recharging stations. Ideas for special credit cards for recharging stations are also in the early stages. On a legal and political level, AVERE works to encourage tax incentives, special roadway and parking accommodations, and government participation in EV development programs. Efforts have been largely successful.

Utilities companies have been firmly behind the switch to electric cars (both in North America and Europe). Enthusiasm on a government level is also high. The combination of shorter trip distances and higher population densities makes BEVs even better suited for European cities than for those in the United States; however, a consumer market for electric cars does not naturally exist. Just as with urban- and commuting-specific IC vehicles, there is little or no pent-up consumer desire waiting to be filled. Consumers are not hungry for transportation products of less capacity and performance, and in the case of BEVs, higher first costs. European attitudes, economic conditions, and transportation preferences and infrastructures differ among regions and differ from those in the United States. On a region-by-region basis, transportation patterns and local infrastructures provide niches and marketing opportunities for manufacturers of alternative cars, but the fundamental challenge in Europe is essentially identical to the challenge in other developed regions. If there is to be a large market for BEVs, it may have to be created through artificial means such as taxation policies and preferential roadway accommodations, and by innovative product design and new marketing appeals.

The best environment for creating a market for urban/commuter cars exists where the choice of individual vehicle utility is highly discretionary and trip distances are relatively short. These conditions are characteristic of affluent

consumers residing in areas of high population density. The economies of
high-density Western European populations have been steadily improving,
and multicar ownership is growing. A successful European Union may ultimately produce the world's strongest economic block, and over the long
term, improve living conditions throughout the region. The conditions necessary for cultivating a market for urban- and commuting-specific cars in
Europe already exist to a significant degree, and trends point to more favorable conditions in the future.

The practical advantages of significantly smaller vehicles have already been
explored. The hypothetical ULM car mentioned in Chapter Two (see Figure
2.2) could achieve triple the fuel economy of the average new car in Europe,
and thereby equal or surpass the energy efficiency of transit systems. Transportation expenses would be lower, and the transportation system would
become more efficient and less polluting. Arguments in favor of electric cars
are perhaps even more powerful. But in terms of first costs to consumers,
BEVs will undoubtedly be more costly alternatives, at least until the technology is more mature and volumes are much greater. In terms of vehicle performance, electric cars may also entail at least some compromise in vehicle
range capabilities. A shift in emphasis away from multipurpose conventional
vehicles and toward urban-specific vehicles may be the most essential factor
in persuading consumers to purchase battery-electric cars. Consumers must
see BEVs, not in comparison to conventional cars, but in relation to actual
trip requirements.

Today, the so-called green market, in combination with a real and growing
crisis in inner-city traffic congestion and air pollution, provides a rich environment for pioneering alternative ULM vehicle types. In addition, much of
Europe is well suited to BEVs. Although urban cars and commuter cars powered by clean-running IC engines may provide similar benefits with greater
vehicle capabilities, Europe is suited to BEVs. The electric car's technical distinction creates a natural shift in consumer perceptions, which could be
encouraged through advertising appeals that center on product differentiation and global benefits. Vehicle themes must suggest an upscale orientation
that is compatible with modern lifestyles.

A product's emotional value becomes increasingly more important as consumers become more affluent. Ultimately, psychological values tend to direct
the choice of facts on which purchasing rationales are based. Alternatives to
high-mass private cars become marketable first to the degree that multicar
households provide a universe of consumers with discretionary vehicle
needs, and second to the degree that consumers' psychological needs are satisfied by the new product. Favorable economic conditions, as well as product

appeal, were both in short supply when bubble cars were marketed after World War II. Bubble cars appealed primarily to consumers' utilitarian needs and relied on a poor economy to create a universe of potential consumers.

The battle between utilitarian and psychological values is exemplified by the historic switch that took place between Henry Ford and GM's Alfred Sloan in the early days of the automobile industry. The automobile began essentially as a plaything for the wealthy, rather than the consumer product that it is today. Ford envisioned a much larger market if the average consumer could purchase a car. He therefore produced a single, low-priced car and appealed to consumers on the basis of price and utility. By the early twenties, Ford Motor Company dominated the market with 60 percent market share compared to GM's 12 percent. Sloan was an admirer of Ford's pioneering efforts, but he also understood that Ford's preoccupation with the idea of a basic, low-cost car was his greatest weakness in the emerging consumer market. Sloan began his onslaught of Ford by revamping GM's line to offer consumers a choice of cars, starting with a slightly higher-priced but more stylish version of Ford's low-priced car and progressing through a line of products, each offering increased luxury and style. The yearly model change was introduced, and cars could be purchased in a choice of colors. Although Ford's product still carried the lowest price, within five years GM was the dominant and profitable car maker, and Ford was losing money.

Ultimately, Ford had to adopt Sloan's methods or perish. In the face of improving economic conditions and greater consumer expectations, appeals to consumers' psychological needs had created a combination that proved fatal to the idea of the purely utilitarian automobile. Bubble cars were the victims of similar events. A modern design based on a similar appeal to basic utility is likely to meet with a similar fate. Today's alternative car must incorporate modern values and appeal to a more sophisticated consumer (Figure 9.6).

REFERENCES

1. Laurence Pomeroy, "The Size, Structure and Shape of European Automobiles," *The Motor*, March 17, 1954.

2. Laurence Pomeroy, "Miniature Motorcars," *The Motor*, February 5, 1958.

3. Lee Schipper et al., "Fuel Prices, Automobile Fuel Economy, and Fuel Use for Land Travel: Preliminary Findings from an International Comparison," Lawrence Berkeley Laboratory.

FIGURE 9.6 The Future of Microcars

Rendering by Barbara Monge

4. Michael Woodmansey, "Business Cars Survey 1985," British Institute of Management.

5. Quote cited by Stephen Kindel, "What Price Freedom," *Financial World*, August 22, 1989.

INDEX

ABOUT THE AUTHOR

Robert Q. Riley is an industrial designer and a mechanical engineer with design and engineering successes in a wide range of product categories. His automotive experience includes vehicle styling, layout, packaging, and powertrain design, focusing on low-energy-demand passenger cars. He is widely recognized for designing high-performance three-wheel road vehicles, electric and hybrid cars, and fuel-efficient internal combustion engine (ICE) automobiles of up to 128-mpg fuel economy. Nonautomotive designs range from high-performance watercraft and hovercraft to fitness equipment and medical products.

He is also known as the "Father of the Modern Bent" for his pioneering work in modern recumbent bicycle design. He recently set new standards of excellence in recumbent design with the release of his Ground Hugger XR2 carbon fiber machine. A solar-assist version of the Ground Hugger XR2 took first place in Category B (unfared recumbents) in the 2001 Australian World Solar Cycle Challenge.

Mr. Riley consults on new product design and product strategies. He promotes environmentally friendly technologies and writes and speaks on the subject of alternative automobile design. He has led conference workshops and speaks at industry, scientific, and academic events. He consulted on the *Different Roads* automobile exhibit at New York's Museum of Modern Art, and he was the lead speaker at the museum's daylong symposium on the future of the automobile. He was one of two U.S. technical consultants selected by Delcan Corporation to contribute to Transport Canada's *Sustainable Transportation Technology Forecast*.

Mr. Riley is a member of the Society of Automotive Engineers, the Industrial Designers Society of America, the Intelligent Transportation Society—Arizona, and the Marine Technology Society. His website is located at www.rqriley.com.

513